Deterministic and Stochastic Scheduling

NATO ADVANCED STUDY INSTITUTES SERIES

Proceedings of the Advanced Study Institute Programme, which aims at the dissemination of advanced knowledge and the formation of contacts among scientists from different countries

The series is published by an international board of publishers in conjunction with NATO Scientific Affairs Division

A	Life Sciences	Plenum Publishing Corporation
B	Physics	London and New York
C	Mathematical and Physical Sciences	D. Reidel Publishing Company Dordrecht, Boston and London
D	Behavioural and Social Sciences	Sijthoff & Noordhoff International Publishers
E	Applied Sciences	Alphen aan den Rijn and Germantown U.S.A.

Series C – Mathematical and Physical Sciences

Volume 84 - Deterministic and Stochastic Scheduling

Deterministic and Stochastic Scheduling

Proceedings of the NATO Advanced Study and Research Institute on Theoretical Approaches to Scheduling Problems held in Durham, England, July 6-17, 1981

edited by

M. A. H. DEMPSTER
Dalhousie University, Halifax, Canada and Balliol College, Oxford, England

J. K. LENSTRA
Mathematisch Centrum, Amsterdam, The Netherlands

and

A. H. G. RINNOOY KAN
Erasmus University, Rotterdam, The Netherlands

D. Reidel Publishing Company

Dordrecht : Holland / Boston : U.S.A. / London : England

Published in cooperation with NATO Scientific Affairs Division

Library of Congress Cataloging in Publication Data

CIP

NATO Advanced Study and Research Institute on Theoretical Approaches to Scheduling Problems (1981 : Durham)
Deterministic and stochastic scheduling.

(NATO advanced study institutes series. Series C, Mathematical and physical sciences ; v. 84)
"Published in cooperation with NATO Scientific Affairs Division".
Includes index.
1. Scheduling (Management)–Congresses. I. Dempster, M. A. H. (Michael Alan Howarth), 1938- . II. Lenstra, J. K. III. Rinnooy Kan, A. H. G., 1949- . IV. North Atlantic Treaty Organization. Division of Scientific Affairs. V. Title. VI. Series.
TS157.5.N26 1981 658.5'3 82-407
ISBN 90-277-1397-9 AACR2

Published by D. Reidel Publishing Company
P.O. Box 17, 3300 AA Dordrecht, Holland

Sold and distributed in the U.S.A. and Canada
by Kluwer Boston Inc.,
190 Old Derby Street, Hingham, MA 02043, U.S.A.

In all other countries, sold and distributed
by Kluwer Academic Publishers Group,
P.O. Box 322, 3300 AH Dordrecht, Holland

D. Reidel Publishing Company is a member of the Kluwer Group

Printed in The Netherlands

CONTENTS

1 Luk van Wassenhove
2 Ismail Turksen
3 S.D. Flåm
4 Chris Potts
5 Wolfgang Zetsche
6 Alexander Rinnooy Kan
7 Rolf Möhring
8 Josep Casanovas
9 Barbara Simons
10 Franz Jos. Radermacher
11 Manfred Kunde
12 Dinis Pestana
13 Jan Karel Lenstra
14 Lahrichi Abdelkader
15 Camille Price
16 Paolo Camerini
17 Hans Ziegler
18 Dave Worthington
19 N. Abu el Ata
20 Brigitte Moraux
21 Roger Cliffe
22 Michael Pinedo
23 Mike Carter
24 Jacques Carlier
25 Ivette Gomes
26 Chris Wright
27 Philippe Chretienne
28 Sema Alptekin
29 Aat Schornagel
30 Jim Templeton
31 Rick Burns
32 Denise Wagstaff
33 Kaj Holmberg
34 Gene Lawler
35 Chip Martel
36 Ton de Kok
37 Ali Kiran
38 Barbara Manners
39 Eric Ole Barber
40 Richard de Wit
41 Gideon Weiss
42 Hamilton Emmons
43 Leen Stougie
44 Odile Marcotte
45 Rafael Solis
46 Dorotea De Luca
47 Rolf Hommes
48 Fernando Ruis Diaz
49 Grazia Speranza
50 Michael Dempster
51 Emilio Berrocal
52 Colette Merce
53 Michael Cain
54 Ilias Tatsiopoulos
55 Gilbert Harrus
56 Werner Helm
57 Rommert Dekker
58 Geoff McKeown
59 Esther Jakimovski
60 Adrie Beulens
61 Johannes Siedersleben
62 Rui Oliveira
63 Thomas Rubach
64 Rolf Wiebking
65 Kevin Glazebrook
66 Geneviève Jomier
67 Ken Sevcik
68 Peter King
69 José Viegas
70 Annie Hayes
71 Marshall Fisher
72 Vic Rayward-Smith
73 Greg Frederickson
74 Bill Thomson
75 Ben Lageweg
76 Eric Muller
77 Sheldon Ross
78 Roger England
79 Linus Schrage
80 Wayne Smith

Photo-Mayo, Gateshead, England, 1981.

PREFACE

This volume contains the proceedings of an Advanced Study and Research Institute on Theoretical Approaches to Scheduling Problems. The Institute was held in Durham, England, from July 6 to July 17, 1981. It was attended by 91 participants from fifteen different countries.

The format of the Institute was somewhat unusual. The first eight of the ten available days were devoted to an Advanced Study Institute, with lectures on the state of the art with respect to deterministic and stochastic scheduling models and on the interface between these two approaches. The last two days were occupied by an Advanced Research Institute, where recent results and promising directions for future research, especially in the interface area, were discussed.

Altogether, 37 lectures were delivered by 24 lecturers. They have all contributed to these proceedings, the first part of which deals with the Advanced Study Institute and the second part of which covers the Advanced Research Institute. Each part is preceded by an introduction, written by the editors.

While confessing to a natural bias as organizers, we believe that the Institute has been a rewarding and enjoyable event for everyone concerned. We are very grateful to all those who have contributed to its realization.

Our first words of thanks must be to our sponsors. The generous support from the NATO Advanced Study Institutes Programme and from the NATO Systems Science Panel provided a firm financial foundation for the Institute. We are particularly indebted to Dr. M. di Lullo and Dr. A. Bayraktar for their encouragement and support. The Institute of Mathematics and Its Applications (IMA), in the person

M. A. H. Dempster et al. (eds.), Deterministic and Stochastic Scheduling, xi–xii.

of its Deputy Secretary, Catherine M. Richards, was in charge of the local organization and did a marvellous job. The Mathematisch Centrum in Amsterdam maintained its usual high standard in providing leaflets, posters and other printed material for distribution among the participants.

Collingwood College in Durham proved to be the perfect setting for the Institute, thanks to the kind hospitality of the Bursar, Mr. C.D. Coombe, and his friendly and helpful staff.

The administrative burden before, during and after the Institute was to a large extent carried by Marjolein Roquas of the Mathematisch Centrum, Nelly Kauffman-Dumoulin and Roland Kahmann of the Erasmus University, and Barbara Manners and Denise Wagstaff of the IMA.

A central role at the Institute was played by Dr. Wayne E. Smith of the University of California in Los Angeles. His scheduling rule, developed 25 years ago, formed the common basis of many lectures. We are happy that, although he did not attend, he was willing to appear on the group picture.

Finally, our thanks must go to lecturers and participants for contributing to the excellent atmosphere that was maintained throughout the Institute. Their enthusiasm is largely responsible for our recollection of twelve pleasant days in Durham.

Michael Dempster
Jan Karel Lenstra
Alexander Rinnooy Kan

December 1981

Part I. Advanced Study Institute Proceedings

INTRODUCTION

The fifteen contributions to the first part of these proceedings provide an extensive survey of the area of deterministic and stochastic scheduling, and also review a few prominent interfaces between these two approaches.

DETERMINISTIC SEQUENCING AND SCHEDULING

Scheduling may be loosely defined as the art of assigning resources to tasks in order to insure the termination of these tasks in a reasonable amount of time. When all the data of a scheduling problem are known in advance, the model is termed *deterministic* and its investigation belongs to the area of *combinatorial optimization.*

Combinatorial optimization involves the study of problems in which an optimal ordering, selection or assignment of a finite set of objects has to be determined. Aside from scheduling problems, it also encompasses many routing, distribution, location and allocation problems. While combinatorial optimization, as a subarea of operations research, is rooted in the theory of *mathematical programming*, the last years have witnessed an increasing application of tools from *computer science*. Due to this development, particular attention has been paid to the *computational complexity* of combinatorial problems and to the resulting implications for the *design and analysis of algorithms* for their solution.

It is now commonly accepted to refer to such a problem as being *well solved* or *easy* if it can be solved by an algorithm whose running time is bounded by a *polynomial* function of problem size [Lawler 1976]. Such an algorithm need not exist, and if the problem in question can be proved to be *NP-hard*, its solvability in polynomial time is very unlikely indeed [Karp 1972; Garey & Johnson

M. A. H. Dempster et al. (eds.), Deterministic and Stochastic Scheduling, 3–14.

1979]. These complementary concepts have been very useful in the analysis of scheduling problems, the large majority of which have now been classified as belonging to either one or the other category.

If a problem is NP-hard (as most practical problems are), two different approaches to its solution suggest themselves. Any method to solve the problem to *optimality*, such as *dynamic programming* or *branch-and-bound*, will require an often time consuming search through the set of feasible solutions, and its worst-case running time will invariably be *exponential*. The alternative approach is to use a fast *heuristic* method to find an *approximate* solution, and then to undertake an empirical or mathematical analysis of the quality of the solutions obtained.

In the first paper of this volume, Fisher discusses the latter approach, and in particular the *worst-case* analysis of the behavior of a heuristic. While this may be overly pessimistic from a practical point of view, such an analysis establishes a rigorous *performance guarantee* to the user of the method, that may be complemented by a probabilistic analysis of its *expected* performance for a given probability distribution of the problem instances.

Fisher describes the fundamentals of worst-case analysis in the context of the *knapsack* problem and reviews the existing results for *parallel machine scheduling* and *bin packing*. The latter model is studied in greater detail in Coffman's contribution, which will be discussed below.

Lawler, Lenstra and Rinnooy Kan present a comprehensive survey of *deterministic scheduling theory*, which is a revised and updated version of [Graham *et al.* 1979]. The presentation is based on a detailed *problem classification*.

In the generic *single machine* scheduling problem, a number of jobs, each with a given processing time, has to be executed on a machine that can handle at most one job at a time, subject to a variety of constraints that may include release dates, deadlines, and precedence constraints; it is also specified whether preemption (job splitting) is allowed or not. Each job incurs at its completion time a certain cost, where the cost function is nondecreasing over time. The problem is to find a schedule that minimizes a given optimality criterion, which is usually the maximum or the sum of the job completion costs. This problem class contains many well-solved and NP-hard problems, to which all the tools of combinatorial optimization have been applied.

The single machine model can be generalized in various directions. In the first generalization, each job has to be processed on any one of a number of *parallel* machines. These machines can be *identical*, *uniform*, or *unrelated*; *i.e.*, if the speed at which machine i processes job j is denoted by s_{ij}, then either $s_{ij} = 1$ for all i and j, or $s_{ij} = q_i$ for all j, or the s_{ij} are arbitrary. *Nonpreemptive* problems in this class tend to be NP-hard, and many approximation algorithms for their solution have been designed and analyzed. There exist some polynomial-time algorithms for the minimization of total completion time (flowtime) and, in particular, for the minimization of maximum completion time (makespan) of unit-time jobs subject to certain types of precedence constraints. *Preemptive* problems turn out to be easier; the most notable recent advances in scheduling theory, concerning polynomial-time algorithms as well as NP-hardness proofs, have been made in this area.

In another generalization of the single machine model, each job consists of a set of operations, each of which has to be executed on a specific machine. In an *open shop* the operations of a job may be processed in any order, in a *flow shop* the same machine order is prescribed for each job, and in a *job shop* the machine orders are specified but job-dependent. Except for a couple of well-solved special cases, these problems are very difficult, and research has focused on the systematic development of enumerative methods for their solution.

The long list of references accompanying this paper bears witness to the impressive research effort that has been directed to deterministic scheduling problems in recent years. Although some vexing open problems remain, it is fair to say that this particular class of models is well understood from a theoretical point of view. This lends additional interest to its *stochastic* extensions, which are discussed throughout this volume, and to extensions involving more complicated *resource constraints* than those exemplified by simple machines of capacity one [Blazewicz *et al.* 1980].

The next three contributions deal with recent algorithmic results for deterministic *parallel machine* problems. Simons treats the problem of determining the existence of a *feasible nonpreemptive* schedule on m *identical* machines when each of n jobs has to be executed during a given processing time between a given integer release date and a given integer deadline. It is well known that the single machine version of this model is NP-complete, but solvable in O(n log n) time by *Jackson's rule* if all release dates are equal. This rule has been extended to the case of arbitrary m, arbitrary release dates and unit processing times. Simons' contribution is a polynomial-time algorithm for the case of equal processing times, which solved a long-standing open question.

Her method can be viewed as repeated application of the extension of Jackson's rule, where at each iteration constraints on job starting times are added. Thus, the method cannot be *on line*, but it is easy to see that this reflects an inevitable feature of the problem itself.

Martel considers another feasibility problem: does there exist a feasible *preemptive* schedule on m *uniform* machines for n jobs with given processing times, release dates and deadlines? The case of identical machines can be solved in $O(n^3)$ time by standard network flow techniques.

Martel's polynomial-time algorithm is based on a *polymatroidal* network flow model, in which capacity constraints are defined by submodular functions on sets of arcs rather than by limits on single arcs. Although the degree of the running time is unusually high, the approach derives additional interest from its applicability to a wide range of combinatorial optimization problems [Lawler & Martel 1980].

In the final paper on deterministic scheduling, Lawler studies several problems involving the *preemptive* scheduling of *precedence-constrained* jobs on parallel machines. These problems are preemptive counterparts of problems involving the nonpreemptive scheduling of unit-time jobs, previously solved by Brucker, Garey and Johnson, and there is a striking correspondence between polynomial-time algorithms for the latter and the former types of problems.

This contribution, as well as the two preceding ones, illustrate the technical sophistication of many of the algorithms that have recently been developed in this area, both in terms of design and implementation.

STOCHASTIC SEQUENCING AND SCHEDULING

The technical complications which arise in scheduling problems when the resources are scarce or there are constraints on the execution of the tasks become even more difficult when there is some uncertainty in availability of resources, task durations or in the constraints associated with allowable schedules. This is the situation in which a *probabilistic* model is usually necessary. Such scheduling problems occur in the context of technological manufacturing problems, in processes involving transport and storage, and in the behavior of computer systems - not to mention numerous applications whose origin is more bellicose.

When deterministic scheduling models are extended to more realistic stochastic models by assuming various data are random variables, several options in *problem formulation* arise immediately. Five important alternative assumptions present themselves:

1. Is *expectation* of random costs or rewards an appropriate optimality criterion or must some *more complicated* stochastic objective be used?
2. Are optimizing decisions to be taken *before* or *after* the random variables are realized?
3. Are all data, such as processing time distributions, available *at the outset* or is a *stochastic process generating arrivals* to the system involved?
4. Is the problem posed over a *finite* or *infinite* horizon?
5. Are probability distributions of random variables or stochastic processes *known completely* in advance or are they known only up to certain parameters which must be *estimated* as the data is realized?

All these questions are familiar in stochastic system theory regarding the optimization of systems involving continuous decision variables, *i.e. stochastic programming* and *stochastic optimal control* models [Dempster 1980; Fleming & Rishel 1975]. Their importance however may not as yet be fully appreciated by researchers in related fields. Here we are principally interested in the analysis of stochastic scheduling models and algorithms, and in other questions in theoretical computer science and the mathematics of operations research such as the control of queueing systems and networks.

Question 1 concerns the stochastic nature of the optimality criterion involved in the model. For most stochastic scheduling models this will be the *expected value* of the (now random) criterion, such as makespan or flowtime, used in the corresponding deterministic model. Such a criterion is entirely appropriate for these models in that they generally apply to *repetitive* situations in which relatively *small* costs or gains are involved per unit time. In the contrary situation, when a *once for all* decision must be taken in the face of uncertainties involving relatively *large* gains or losses, total preference ordering of reward distributions using von Neumann-Morgenstern *expected utility theory* is a more general tool [Luce & Raiffa 1957]. For certain scheduling models it is possible to establish optimality in *distribution* for the random criterion: the probability of achieving a given criterion level is everywhere at least as great under the optimal policy as for any other. This is a very stringent optimality criterion, which guarantees optimality for any expected utility criterion involving a monotone utility function.

Question 2, concerning the timing of the realization of the random variables in the problem, is of crucial importance for the nature of the analysis. In the context of combinatorial optimization, if the random data is realized *before* optimization or approximation is performed, the investigation of the resulting distribution problem - find the distribution of the criterion value obtained, or its expectation, or other moments - is termed *probabilistic analysis* of an algorithm. That is, a multivariate distribution is assumed for the problem data and the question of interest is often the *a priori* expected performance of the algorithm. For certain simple scheduling problems and parameter distributions such results may be obtained for an *optimization* algorithm. More often, in order to evaluate the performance of an *approximation* algorithm, an upper (say) bound on the expected heuristic criterion value is compared with a lower (say) bound on the expected optimal criterion value. Once it is assumed that some of the problem data is realized *sequentially* only *after* some decisions have been taken, the resulting decision problems generally become more difficult to analyze. In some simple cases - *e.g.* for list scheduling heuristics applied to certain parallel machine problems - the analysis remains the same independent of the timing of the data realization. However, this situation seldom applies to the case of a scheduling policy optimal in expectation for a problem with any complexity of structure.

Question 3 - whether or not a stochastic process of arrivals to the system is assumed - delimits the boundary between *static* models which form the bulk of problems analyzed in operations research and computer science and the dynamic *queueing* models analyzed in both disciplines. In computer science, the study of queueing systems has recently become of fundamental importance for computer system and network *performance modelling* [Kleinrock 1976]. Most of the models used so far differ from the static stochastic scheduling models in that no *active* scheduling policy - other than arrival order, possibly by priority class - is normally assumed. A current research topic involves the extension of recent results for stochastic scheduling problems to models with arrival processes. Such models are natural stochastic extensions of deterministic scheduling models incorporating job *release dates*. The natural setting for their analysis is in continuous time, although certain optimality results for scheduling problems are so far only established in discrete time. Ultimately - for example, to improve our understanding of real job shops - it would be useful to analyze a *network* of multimachine scheduling problems. Each node of the network would be not simply a single server but rather a scheduled multimachine system, so that node input and output processes would be more complicated than have so far been analyzed. Nevertheless, the recent general theory of the optimal control of stochastic systems driven by point processes should be relevant [Walrand & Varaiya 1978].

Question 4, concerning the length of the planning horizon, is closely connected to the assumption of an arrival process in that often the only tractable analysis - for example, for queueing models - refers to the *asymptotic* state distribution of the system as the underlying stochastic processes tend to their long run stationary *equilibrium* distributions at infinity. In the probabilistic analysis of algorithms for finite horizon scheduling problems, asymptotic analysis as the number of jobs tends to infinity is also useful and commonly involves (implicitly or explicitly) a time horizon tending to infinity.

Question 5 concerning estimation of distributional parameters, usually by recursive *Bayesian* methods simultaneous with decision making, has so far received scant attention in the scheduling and queueing literature despite the fact that it is an area of current research effort in stochastic programming and stochastic control theory.

With this background the content and methods of the remaining papers may be briefly described.

The first three contributions survey the state-of-the-art in *stochastic scheduling* theory. In the first of these, Gittins treats a general approach to a number of problems in discrete time. A basic example is the *preemptive single machine* scheduling problem with random processing times and an optimality criterion involving expected total discounted rewards associated with individual job completion times. In addition, optimal staging of search for a hidden stationary object, of synthesis of a certain chemical compound, of multiattribute testing of a chemical formulation, and for Bellman's gold mining problem are discussed.

For all these problems the theory of semi-Markov decision processes may be used to show the optimality of a *forwards induction* policy which maximizes the expected discounted reward per unit time up to a random time (depending on job or task completions) when the state of the system changes, and then continues recursively. Such policies usually reduce to giving *priority* at each stage to the task for which the value of an *index* - which typically varies as work on the task progresses - is largest.

As well as outlining the general theory, Gittins gives the forms of the appropriate priority indices for the problems mentioned above. For the single machine problem with increasing job completion rates, the policy is a stochastic generalization of Smith's *ratio rule* for the deterministic case; certain extensions to precedence-constrained jobs are possible as well.

Weiss treats a class of stochastic scheduling problems in continuous time. These are *preemptive parallel machine* problems with negative exponential processing times and three different optimality criteria: minimization of expected flowtime and makespan and maximization of the expected time to first machine idleness.

The optimal policies are priority scheduling rules which assign at job completions uncompleted jobs to machines in ascending (in case of the first criterion) or descending (in case of the other criteria) order of expected processing time. These policies are for obvious reasons termed *shortest expected processing time* (SEPT) and *longest expected processing time* (LEPT). For the case of two identical machines direct proofs are given, while for the case of m uniform machines optimality proofs depend on results from the theory of Markov decision processes.

Weiss also considers the case of m identical machines in discrete time, when processing times are drawn from a common distribution with monotone completion rate, but different jobs are available for scheduling after having been previously processed for different amounts of time. SEPT and LEPT policies are again directly shown to be optimal. (For analogous results in continuous time, based on the Bellman-Hamilton-Jacobi theory of optimal control, see [Weber 1981].)

Pinedo and Schrage survey recent results for a number of *nonpreemptive shop* scheduling problems. They are principally concerned with models involving two machines, nonidentical jobs with negative exponential processing time distributions, and expected makespan as well as expected flowtime criteria.

Optimal policies are presented for open shops, flow shops (both with and without intermediate storage) and job shops. Some results are also presented for flow shops with more than two machines.

The next three contributions concern recent results in the theory of *queueing systems and networks*. In abstract terms, such dynamical systems may be represented at any time t by equations of the form $Y_{t+} = F(Y_{t-}, s, w)$, which may be read as "the state of the system after some time t is determined by its state before and up to t, by the schedule s (known for all time) and by an exogenous factor w (known for all time) which represents the realization of randomness.

If the set S of values taken by s is a singleton, then the system operates under a predetermined *fixed schedule*, which - although it can give rise to some complex considerations - does not permit what one usually calls optimization. If the set W of values taken by w is a singleton, then the system is deterministic. If W is larger, then the system is *nondeterministic*, which does not make it stochastic. For instance, Petri nets and Karp-Miller vector addition systems are nondeterministic but not stochastic. To make the system *stochastic* or probabilistic we must assign a probability measure to (a sigma field of subsets of) W. In particular, this will generate a probability distribution over the values taken by Y_{t+} for each t.

For each w we have a *realization* of the dynamical system's behavior: thus w is the representation of the (possibly infinite sequence of) random quantities during one particular realization of the underlying stochastic process. For instance, w could contain all of the arrival instants and service or task execution times for a particular realization.

Ross discusses properties of the long run *steady state* or *equilibrium* distributions of various *multiserver* queueing systems. These are models in which S is rather simple but where W already has a relatively rich content, even though standard restrictive assumptions are made (in particular with respect to independence of interarrival times and of service or task execution times). Equilibrium conditions for such (semi-)Markov processes are generated by the *stationary* process which is the stochastic analogue of a (unique) equilibrium point of a globally stable deterministic dynamical system in canonical form.

The proof techniques utilized by Ross illustrate the variety of approaches currently available for the study of queueing systems. *Reverse process* arguments, recently popularized by Kelly [Kelly 1979] for the analysis of queueing networks, are employed for the *Erlang loss* model - in which arrivals finding all servers busy are lost to the system - and the *shared processor* model - in which service is shared amongst all customers - with Poisson arrival processes and general independent service distributions. Simple *expected cost* arguments and appropriate independence assumptions are used to analyze the Erlang model when a given length queue must occur before customers are lost. Further, the *embedded Markov chain* approach is used to study the multiserver model with an arbitrary renewal process generating arrivals and negative exponential service times.

Sevcik surveys analytical results on *queueing networks* and their application to *computer performance measurement*. In the last twenty years a considerable body of literature has developed on the theory of networks of queues. A set of tasks, possibly infinite due to arrivals to and departures from the system, circulates among a finite or infinite set of resources according to some probabilistic routing scheme. Under certain assumptions the stationary state distribution of these systems exhibits the *product form*: it is the product over the resources of the corresponding marginal distributions of the number of tasks awaiting or obtaining service at the resources. From the probability distribution of system states, performance measures such as *resource utilizations*, *queue length distributions* and *mean response times* can be derived. Several types of *schedules* may be allowed at the resources; these remain currently limited but useful. For example, priority scheduling disciplines or even first-come-first-served service violate the assumptions that lead to product-form solutions. This is unfortunate in that this special solution structure yields efficient algorithms for computing the stationary distribution. When all the required restrictions for such a solution to exist are not satisfied, it is nevertheless sometimes possible to obtain solutions via approximate methods, in particular using *decomposition*. This has led to the solution of optimal scheduling problems for virtual memory systems in which throughput is a function of multiprogramming activity.

The product-form solution gives rise to intriguing questions concerning the reasons for the apparent resource independence in the stationary distribution of system states. An alternative approach, termed *operational analysis* in the literature, is to consider a set of measurements of the flow of a finite number of tasks through the system. Although this sampling approach has not yet been fully elucidated, given suitable sample independence assumptions analogous to those of the complete probabilistic model, the sampling distribution of system states exhibits the product form under much less severe restrictions on the underlying service distributions.

Gelenbe's paper is motivated by the same *measurement* orientation to the study of queueing systems as operational analysis. It introduces the notion of *timed vector addition systems* - a dynamical version of a concept of Karp and Miller developed for the investigation of parallel program schemata - and shows their equivalence with a dynamic description of *Petri nets* and flow (event) descriptions of stochastic queueing networks under general probabilistic assumptions. Relative frequency measures for flows through the system are defined and some basic propositions on time averages are set out.

One of the most interesting probabilistic issues that these measurement-oriented approaches to queueing systems raise is the question of which system properties hold *almost surely*. Similarly, one sees a number of properties of scheduled queueing networks that remain true for all schedules. There is currently room for interesting investigations in this area.

INTERFACES BETWEEN DETERMINISTIC AND STOCHASTIC SCHEDULING

In the remaining four contributions to the first part of the proceedings attempts are made to relate the deterministic and stochastic approaches to specific planning and scheduling problems.

Schrage discusses the deterministic *multiproduct lot scheduling* problem. Various products have to be scheduled on a single machine of limited capacity to meet given demand schedules over a finite horizon. Associated with each product are setup, production and storage costs.

Listings are given of single product lotsizing problems that are solvable in polynomial time as well as of those that are NP-hard. Multiproduct problems, which are of more practical interest, appear to become hard very quickly. Schrage next reviews dynamic and linear programming approaches to the problem and analyzes the quality of the LP approximation relative to the integer optimum. Although the demands are realistically random, so far little analysis of this stochastic extension has been undertaken.

Coffman is concerned with performance analysis of heuristics for various *bin packing* problems. In the first part of the paper he treats approximation algorithms for three combinatorial optimization problems: minimizing the common capacity of a set of bins required to pack a given list of pieces (in another guise, minimizing makespan on identical parallel machines for a given list of jobs), minimizing the number of unit-capacity bins needed to pack a given list of pieces, and minimizing the height of a unit-width rectangle needed to pack a collection of squares whose sides must be parallel to the bottom or sides of the enclosing rectangle. The emphasis is on the proof techniques - problem *reduction*, *area* arguments and *weighting function* techniques - used to establish *worst-case performance* bounds for various heuristics that pack pieces as they are drawn from a specific list.

In the second part of the paper, Coffman turns to the probabilistic analysis of the *expected performance* of some of these heuristics, when the piece sizes are independent identically distributed random variables. He considers the *lowest-fit* algorithm for the first problem mentioned above in terms of the Markov process of *variations* - increments to the capacity required by the current pieces - defined on piece number. He analyzes the asymptotic state of this process as the number of pieces tends to infinity and shows that its convergence is geometric, so that its equilibrium distribution allows an accurate approximation of expected capacity even for a rather small number of pieces. Similar techniques are applied to the *next-fit* algorithm for the second problem. By specializing to uniform distributions, more precise analyses for more complex heuristics can be performed.

Dempster surveys recent results in stochastic discrete programming models for *hierarchical planning problems*. Practical problems of this nature typically involve a sequence of decisions over time at an increasing level of detail and with increasingly accurate information. These may be modelled by *multistage stochastic programs* whose lower levels (later stages) are stochastic versions of familiar NP-hard combinatorial optimization problems. Hence, they require the use of heuristics for obtaining approximate solutions.

After a brief survey of distributional assumptions on processing times under which SEPT and LEPT policies remain optimal for parallel machine scheduling problems, results are presented for various *two-level scheduling* problems in which the first stage corresponds to the *acquisition* of machines and the second one to the *allocation* of machines to jobs. Heuristics which are asymptotically optimal both in expectation and almost surely are analyzed for problems with identical or uniform machines. A *three-level location, distribution and routing* model in the plane is also discussed.

In the final paper, Gaul treats a *project network* model in which activity durations are random variables. In the deterministic case, the problem of finding the minimum value of the overall *project duration* consistent with the given activity precedence constraints is equivalent to the longest path problem on the corresponding acyclic digraph and requires linear time. In the stochastic case, the problem of finding the *distribution function* of the minimum project duration, while analytically completely solved, is computationally intractable - even approximations are difficult to obtain efficiently.

The author gives an efficient algorithm for the (stochastic programming) *decision* problem which sets *a priori* completion time estimates for individual activities so as to minimize piecewise linear realized completion time discrepancy costs. The method consists of approximating the random activity durations by discrete random variables and applying a modified *out-of-kilter* algorithm to a deterministic network (whose size is independent of the accuracy of the discrete approximation) a finite number of times (which depends on the approximation accuracy required). The method is illustrated with a numerical example.

REFERENCES

J. BLAZEWICZ, J.K. LENSTRA, A.H.G. RINNOOY KAN (1980) Scheduling subject to resource constraints: classification and complexity. Report BW 127, Mathematisch Centrum Amsterdam; *Discrete Appl. Math.*, in process.

M.A.H. DEMPSTER (ed.) (1980) *Stochastic Programming*, Academic Press, London.
W.H. FLEMING, R.W. RISHEL (1975) *Deterministic and Stochastic Control*. Springer, New York.
M.R. GAREY, D.S. JOHNSON (1979) *Computers and Intractability: a Guide to the Theory of NP-Completeness*, Freeman, San Francisco
R.L. GRAHAM, E.L. LAWLER, J.K. LENSTRA, A.H.G. RINNOOY KAN (1979) Optimization and approximation in deterministic sequencing and scheduling: a survey. *Ann. Discrete Math.* 5,287-326.
R.M. KARP (1972) Reducibility among combinatorial problems. In: R.E. MILLER, J.W. THATCHER (eds.) (1972) *Complexity of Computer Computations*, Plenum Press, New York, 85-103.
F.P. KELLY (1979) *Reversibility and Stochastic Networks*, Wiley, New York.
L. KLEINROCK (1976) *Queueing Systems, Vol. II: Computer Applications*, Wiley, New York.
E.L. LAWLER (1976) *Combinatorial Optimization: Networks and Matroids*, Holt, Rinehart and Winston, New York.
E.L. LAWLER, C.U. MARTEL (1980) Computing maximal "polymatroidal" network flows. Research Memorandum ERL M80/52, Electronics Research Laboratory, University of California, Berkeley; *Math. Oper. Res.*, to appear.
R.D. LUCE, H. RAIFFA (1957) *Games and Decisions*, Wiley, New York.
J. WALRAND, P. VARAIYA (1978) The outputs of Jacksonian networks are Poissonian. Memorandum ERL M78/60, Electronics Research Laboratory, University of California, Berkeley.
R.R. WEBER (1981) Scheduling jobs with stochastic processing requirements on parallel machines to minimize makespan or flowtime. *J. Appl. Probab.*, to appear.

WORST-CASE ANALYSIS OF HEURISTIC ALGORITHMS FOR SCHEDULING AND PACKING

Marshall L. Fisher

University of Pennsylvania, Philadelphia

The increased focus on heuristics for the approximate solution of integer programs has led to more sophisticated analysis methods for studying their performance. This paper is concerned with the worst-case approach to the analysis of heuristic performance. A worst-case study establishes the maximum deviation from optimality that can occur when a specified heuristic is applied within a given problem class. This is an important piece of information that can be combined with empirical testing and other analyses to provide a more complete evaluation of a heuristic. In this paper the basic ground rules of worst-case analysis of heuristics are reviewed, and a large variety of the existing types of worst-case results are described in terms of the knapsack problem. Then existing results are surveyed for the parallel machine scheduling and bin packing problems. The paper concludes with a discussion of possibilities for further research.

1. INTRODUCTION

Much has been written about the gap between research and practice in the field of Management Science. In integer programming and combinatorial optimization, this gap has resulted largely from the different emphasis researchers and practitioners have placed on optimization and heuristic methods. Researchers have focused on optimization methods while practitioners have been forced to resort to heuristics to "solve" the real integer programming problems that arise in areas such as scheduling, routing, and location analysis. This gap has been substantially reduced in recent years as a result of two important developments.

M. A. H. Dempster et al. (eds.), Deterministic and Stochastic Scheduling, 15–34.

1. Practitioners have begun to take optimization methods more seriously.

This is largely because these methods have improved to the point where they can successfully solve real problems. Evidence for this is provided by a growing list of effective applications of optimization methods including Cornuejols, Fisher and Nemhauser [5], Geoffrion and Graves [25], Mairs, Wakefield, Johnson and Spielberg [41], Marsten, Muller and Killion [42], and Westerberg, Bjorklund, and Hultman [52].

2. Researchers have begun to take heuristics more seriously.

The work of Cook [4] and Karp [34], [35] in the early 70s crystallized the growing impression that most combinatorial optimization problems are extremely difficult. It is now apparent that while optimization methods may succeed in some cases, they will probably fail in others, necessitating the use of a heuristic. Moreover, even if an optimization method is applicable, the first step in such a method is usually the application of a heuristic to obtain a good starting solution.

This paper is about the second of these two developments. The increased focus on heuristics has led to more sophisticated methods for studying their performance. Originally, the performance of a heuristic could only be evaluated by programming the heuristic and trying it on a "representative" sample of problems. Now, two analytic approaches are also available for evaluating heuristics: worst-case analysis and probabilistic analysis. A worst-case study establishes the maximum deviation from optimality that can occur when a specified heuristic is applied within a given problem class. In a probabilistic approach, one assumes a density function for the problem data and establishes probabilistic properties of a heuristic, such as a bound on the probability that the heuristic finds a solution within a prespecified percentage of optimality.

It should be clear that these three approaches, empirical testing, worst-case analysis, and probabilistic analysis, offer different advantages in achieving the basic objective of predicting the performance of a heuristic. Therefore, they should be viewed as complementary rather than competitive. The principal advantage of empirical testing is accuracy. At least for the problems that are run, the performance of the heuristic is known with certainty. On the other hand, empirical testing is expensive in its use of computer time and provides only statistical evidence about the performance of the heuristic for problems that were not run. Worst-case analysis has the advantage of providing a guarantee on the maximum amount that the heuristic will deviate from optimality for any problem instance within the class studied. A worst-case study also involves the construction of examples for which the heuristic's performance is as bad as the guarantee. These examples are useful in understanding when and why the heuristic will perform at its worst. The main disadvantage of worst-case analysis is that the worst-case performance is usually not predictive of average performance. Probabilistic analysis corrects this deficiency by showing how the heuristic will perform for a "typical" problem instance. The major limitation of a probabilistic analysis is that one must be willing to specify a density function for the problem data, usually of a fairly simple form. Additionally, most probabilistic results are asymptotic in problem size, and are not strictly applicable to problem instances of any finite size.

This paper provides an introduction to the worst-case analysis of heuristics. In §2 the basic ground rules of worst-case analysis are reviewed and a large variety of the existing types of worst-case results are described in terms of the knapsack problem. §3 gives a selected sample of results for two other problems: parallel machine scheduling and bin packing. Conclusions and a forecast of future developments are provided in §4.

Although the list of references for this paper is long, it is not intended to be a comprehensive survey. For a comprehensive listing of references through mid-1976, consult the bibliography by Garey and Johnson [17]. Other surveys include Garey and Johnson, Chapter 6 [21], Kannan and Korte [33], and (for scheduling problems) Graham, Lawler, Lenstra, and Rinnooy Kan [27]. An earlier survey by this author is given in [12]. A general description of the probabilistic approach is available in Karp [36], and specific applications are described in [35], [7], and [13].

2. FUNDAMENTALS OF WORST-CASE ANALYSIS

The standard knapsack problem is concerned with optimally filling a knapsack of capability b using n items. Item j has value p_j and weight a_j. The problem is formulated

$$Z = \max \sum_{j=1}^{n} p_j x_j, \tag{1}$$

$$\sum_{j=1}^{n} a_j x_j \le b, \tag{2}$$

$$x_j \geq 0 \text{ and integral, } j=1,\ldots,n. \quad (3)$$

We assume that n, b, p_j and a_j are positive integers satisfying $a_j \leq b$ for all j. The 0-1 knapsack problem is a variation in which (3) is replaced by the requirement that $x_j = 0$ or 1.

Despite its simplicity, a number of heuristics and worst-case results have been obtained for the knapsack problem. These heuristics and results are highly representative of the more general research on worst-case performance of heuristics, and they will be used to introduce the fundamental concept of worst-case analysis of heuristics and to illustrate many of the various types of results that have been obtained.

The ground rules of a worst-case analysis are simple. The analysis is always conducted with respect to a well-defined heuristic and set of problem instances. Let P denote the set of problem instances, $I \varepsilon P$ a particular problem instance, Z(I) the optimal value of problem I, and $Z_H(I)$ the value obtained when heuristic H is applied to problem I. We assume that heuristic H is suitably specified to uniquely define $Z_H(I)$. In terms of the standard knapsack problem, I is defined by a 2n + 1-tuple of positive integers $p_1,\ldots,p_n,a_1,\ldots,a_n$, b that satisfy $a_j \leq b$ for all j. P is the set of all such 2n + 1-tuples for all positive integers n.

Worst-case results establish a bound on how far $Z_H(I)$ can deviate from Z(I). For simplicity, assume $Z(I) \geq 0$. Then for maximization problems, such a result would prove for some real number $r \leq 1$ that

$$Z_H(I) \geq rZ(I) \text{ for all } I \varepsilon P. \quad (4)$$

For minimization problems we wish to prove for some $r \geq 1$ that

$$Z_H(I) \leq rZ(I) \text{ for all } I \varepsilon P. \quad (5)$$

In analyzing a heuristic we would like to find the largest r for which (4) holds or the smallest r for which (5) holds. This value is called the <u>worst-case performance ratio</u>. We know that a given r is the best possible if we can find a single problem instance I for which $Z_H(I) = rZ(I)$ or an infinite sequence of instances for which $Z_H(I)/Z(I)$ approaches r.

One can equivalently express heuristic performance in terms of <u>worst-case relative error</u>. The <u>relative error</u> ε is defined in terms of the performance ratio r by $\varepsilon = 1 - r$ for maximization problems and $\varepsilon = r - 1$ for minimization problems. Both r and ε are used frequently, and the choice between the two measures is largely a matter of taste. The worst-case ratio r will be used in this paper.

For notational simplicity we will drop the argument I on $Z_H(I)$ and $Z(I)$ throughout the remainder of the paper with the understanding that Z_H and Z are always defined for a particular problem instance. The first heuristic we will consider for the knapsack problem is called a greedy heuristic. For convenience, assume that items are indexed so that $p_1/a_1 \geq p_2/a_2 \geq \ldots \geq p_n/a_n$. A greedy heuristic can be given for either the standard or 0-1 knapsack problems. For the standard problem items are considered in index order (decreasing p_i/a_i). We place as many integral units of each new item into the knapsack as will fit in the remaining capacity. In the case of the 0-1 problem, a single unit of each new item is placed into the knapsack if it fits.

A formal description of the greedy heuristic that applies to both versions of the knapsack problem can be given by introducing an upper bound u_j on x_j. For the 0-1 knapsack problem $u_j = 1$, while for the standard problem $u_j = [b/a_j]$ where $[y]$ denotes the largest integer less than of equal to y.

Greedy Heuristic

1. Set $j = 1$ and $\bar{b} = b$.
2. Set $x_j = \min(u_j, [\bar{b}/a_j])$ and $\bar{b} = \bar{b} - a_j x_j$.
3. Stop if $j = n$ or by $\bar{b} = 0$. Otherwise set $j = j + 1$ and go to 2.

If the items have been presorted so that $p_1/a_1 \geq p_2/a_2 \geq \ldots \geq p_n/a_n$, greedy runs in $O(n)$ time. Sorting the items requires $O(n \log n)$ time.

It is easy to show that for the standard knapsack problem greedy has a worst-case performance ratio of $r = 1/2$. Let Z_G denote the value of a greedy knapsack packing and Z the value of an optimal packing. Clearly $Z_G \geq p_1[b/a_1]$. Solution of (1)-(3) with the integrality requirement on x_j relaxed provides the upper bound $p_1(b/a_1) \geq Z$. Together these results imply

$$Z_G/Z \geq \frac{[b/a_1]}{b/a_1} = \frac{[b/a_1]}{[b/a_1]+\{b/a_1-[b/a_1]\}} \geq 1/2.$$

The second inequality follows from $[b/a_1] \geq 1 \geq \{b/a_1-[b/a_1]\}$. To show that the bound of 1/2 is tight, consider the series of problems with $n = 2$, $p_1 = a_1 = k + 1$, $p_2 = a_2 = k$, and $b = 2k$ for $k = 1,2,\ldots$. For this series $Z = 2k$ and $Z_G = k + 1$, so Z_G/Z can be arbitrarily close to 1/2.

Obviously for any heuristic and maximization problem class the worst-case performance ratio must be at least 0. The greedy heuristic applied to the 0-1 knapsack problem is a case which shows that the worst-case performance can be as bad as 0. The series of examples $n = 2$, $p_1 = a_1 = 1$, $p_2 = a_2 = k$, $b = k$, $k = 1,2,\ldots$ has $Z_G = 1$ and $Z = k$ so Z_G/Z can be arbitrarily close to zero. Such a heuristic is said to be arbitrarily bad for the given maximization problem class. Analogously, heuristic H is arbitrarily bad for the minimization problem if there is a sequence of examples for which the ratio Z_H/Z goes to infinity.

Since greedy is arbitrarily bad for the 0-1 knapsack problem, one might ask whether there is any polynomial time heuristic for this problem with $r > 0$. The answer is yes; in fact there are heuristics with r arbitrarily close to 1. The first method of this type to be suggested involves partial enumeration of subsets of items. The heuristic is parameterized by an integer k, has a worst-case performance ratio of $r = k(k+1)$ and requires $O(kn^{k+1})$ computation. For $S \subseteq \{1,\ldots,n\}$ let $Z_G(S)$ denote the value of a greedy packing of items from $\{1,\ldots,n\} - S$ into a knapsack of capacity $b - \Sigma_{j\varepsilon S} a_j$. The kth level partial enumeration algorithm obtains a solution by solving

$$Z_{E(k)} = \max_S \{ \sum_{j\varepsilon S} p_j + Z_G(S) \},$$

$$\sum_{j\varepsilon S} a_j \leq b\ ,$$

$$|S| \leq k.$$

This problem is solved by enumeration of all sets S of cardinality k or less. There are $\Sigma_{j=1}^{k} \binom{n}{j} < kn^k$ such sets and computation of $Z_G(S)$ for each set requires at most $O(n)$. Thus the time to compute $Z_{E(k)}$ is bounded by $O(kn^{k+1})$.

Partial enumeration was first proposed and studied by Johnson [30] for the case $p_j = a_j$ for all j. Sahni [50] extended this work to the general case. It is shown in [50] that $Z_{E(k)} \geq (k/(k+1))Z$. The series of problem instances $n = k + 2$, $p_1 = 2$, $a_1 = 1$, $p_i = a_i = j$ for $i \geq 2$, $b = (k+1)j$ for $j=3,4,\ldots$ has $Z = (k+1)j$ and $Z_{E(k)} = kj + 2$, which establishes that this bound is tight for all k.

A heuristic of this type is called a polynomial approximation scheme. Any performance ratio can be obtained, and for a fixed performance ratio, the computational requirements are polynomial in problem size. As a function of the worst-case relative error $\varepsilon = 1 - r$ and the problem size, the computation required by the partial enumeration scheme is $O[(1/\varepsilon)n^{(1/\varepsilon)}]$.

Thus, although the scheme can achieve r arbitrarily close to 1, the computation grows exponentially in the inverse of 1 - r.

This disadvantage is overcome in the following scheme for the 0-1 knapsack problem in which r can be arbitrarily close to 1 and computational requirements grow polynomially in the inverse of 1 - r. A method with these properties is called a <u>fully polynomial approximation scheme</u>.

There is a well-known dynamic programming algorithm in which the 0-1 knapsack problem is solved by recursively solving a family of knapsack problems that use some of the items and a portion of the knapsack. A variation on this approach will be given that forms the basis of the fully polynomial approximation scheme. Given integers y and k with $1 \leq k \leq n$, let

$$f_k(y) = \min \sum_{j=1}^{k} a_j x_j ,$$

$$\sum_{j=1}^{k} p_j x_j \geq y ,$$

$$x_j \in \{0,1\},\ j = 1,\ldots,k.$$

The function $f_k(y)$ gives the minimum knapsack space required to achieve value y using the first k items. By convention set $f_k(y) = \infty$ if $y > \Sigma_{j=1}^{k} p_j$. We also know by definition that $f_k(y) = 0$ if $y \leq 0$. Given these initial values $f_k(y)$ can be computed recursively for $k = 1,\ldots,n$ and $y = 1,2,\ldots$ by

$$f_1(y) = \begin{cases} a_1, & 1 \leq y \leq p_1, \\ \infty , & y > p_1, \end{cases}$$

$$f_k(y) = \min \{f_{k-1}(y), f_{k-1}(y-p_k) + a_k\}, \quad k = 2,\ldots,n.$$

The 0-1 knapsack problem can be solved by computing $f_n(y)$ for $y = 1,2,\ldots$. Let K + 1 be the least integer for which $f_n(K+1) > b$. Then the optimal value is given by Z = K. The optimal solution can also be obtained if a 0-1 index $x_k(y)$ is maintained as $f_k(y)$ is computed. The index is given by

$$x_k(0) = 0, \qquad k = 1,\ldots,n,$$

$$x_1(y) = 1, \qquad 1 \leq y \leq p_1,$$

and

$$x_k(y) = \begin{cases} 1, & f_{k-1}(y) > f_{k-1}(y-p_k) + a_k, \\ 0, & \text{otherwise} \end{cases}$$

for $k = 2,\ldots,n$ and $y \geq 1$. It is easy to verify that the solution $\bar{x}_n,\ldots,\bar{x}_n$ computed recursively by $\bar{x}_n = x_n(K)$ and $\bar{x}_k = x_k(K - \Sigma_{j=k+1}^n p_j\bar{x}_j)$, $k = n-1,n-1,\ldots,1$, satisfies $\Sigma_{j=1}^n p_j\bar{x}_j = K$ and $\Sigma_{j=1}^n a_j\bar{x}_j = f_n(K) \leq b$.

Let $d = \Sigma_{j=1}^n p_j$. Since $f_n(d+1) > b$ this dynamic programming scheme for solving the 0-1 knapsack problem requires at most $O(nd)$ evaluations of $f_k(y)$. Since the computational requirements are proportional to $\Sigma_{j=1}^n p_j$, one might consider the possibility of economizing on computation by using a coarser scale for measuring the p_j. Specifically, for some $\lambda > 1$ we might replace each p_j by $<p_j/\lambda>$ where $<a>$ denotes the smallest integer not less than a. With this substitution, computational requirements become $O(nd/\lambda)$. For example, if p_j is measured in dollars, then using $\lambda = 100$ is equivalent to measuring values in hundreds of dollars. If it's really impossible to measure item values more accurately than to the nearest hundred dollars, introducing such a scale change makes perfect sense; we save two orders of magnitude in computational requirements with no meaningful loss in accuracy. However, even if item values can be determined precisely to the nearest dollar, using scaled values may be a desirable heuristic if the loss in optimality due to scaling is outweighed by the savings in computation.

To assess the loss in optimality due to scaling, let $\bar{x}$ denote an optimal knapsack solution for values $<p_j/\lambda>$, define $Z_{DP(\lambda)} = \Sigma_{j=1}^n p_j\bar{x}_j$, and consider the ratio $Z_{DP(\lambda)}/Z$. Clearly $Z \geq d/n$ since $\max_j p_j \geq d/n$ and $a_j \leq b$ for all j. Also, since $\lambda\ p_j/\lambda <\geq p_j>$ and x is optimal for values $\lambda<p_j/\lambda>$,

$$Z \leq \sum_{i=1}^{n} \lambda<p_j/\lambda>\bar{x}_j \leq \sum_{j=1}^{n} (p_j+\lambda)\bar{x}_j \leq Z_{DP(\lambda)} + n\lambda.$$

Together these two results give

$$Z_{DP(\lambda)} \geq \left(1 - \frac{n\lambda}{Z}\right)Z \geq \left(1 - \frac{n^2\lambda}{d}\right)Z.$$

Thus to guarantee a specific performance ratio r we would set $\lambda = (1-r)d/n^2$ which implies computational requirements $O(nd/\lambda) = O(n^3/(1-r))$. Since these requirements are polynomial in n and the inverse of (1-r), we have obtained a fully polynomial approximation scheme.

The version of the fully polynomial approximation scheme presented here is simple and easy to explain. There are more complicated schemes based on similar principles that have running time less than $O(n^3/(1-r))$. The first of these was given by Ibarra and Kim [28]. Their scheme is improved upon in a paper by Lawler [38].

Fully polynomial approximation schemes exist for other knapsack-like problems (e.g., see Sahni [51]). On the other hand, Garey and Johnson [18] have obtained a negative result that precludes the existence of a fully polynomial approximation scheme for many important problems, including the bin packing, uncapacitated location, and traveling salesman problems discussed in §3. In [18] they define a special subset of the NP-hard problems that they call strongly NP-hard. They show that a fully polynomial approximation scheme does not exist for any strongly NP-hard problem that satisfies a mild condition on the growth rate of the optimal value relative to the magnitude of the problem data.

Dynamic programming with data scaling is clearly more attractive than partial enumeration for the 0-1 knapsack problem, at least for $r > 2/3$. The general method of partial enumeration is still of interest, however, because it is applicable to other problems for which a <u>fully</u> polynomial approximation scheme is precluded by the Garey-Johnson result. For example, references [11] and [43] describe a polynomial approximation scheme for the uncapacitated location problem that is based on partial enumeration.

The foregoing analyses of knapsack heuristics illustrate a number of features that are typical of many worst-case analyses. Specifically:

1. For a given maximization heuristic H, a bound on Z_H/Z is actually established by bounding the ratio of Z_H to an upper bound on Z. This is not surprising since the reason for resorting to a heuristic is that it is hard to get information about Z.

2. There is no single problem instance for which Z_H/Z achieves its worst-case value. Rather, an infinite series of instances exist over which Z_H/Z approaches its worst-case value.

3. Several types of worst-case performance are possible. At one extreme, a perfectly plausible heuristic has arbitrarily bad performance ($r = 0$). Another heuristic has a fixed performance ratio r between 0 and 1. Other heuristics are parameterized and can have any performance ratio r satisfying $0 < r < 1$.

4. For fixed r, all heuristics have a running time that is polynomial in problem input. Most heuristics either run in polynomial time or are iterative improvement methods that can be made to run in polynomial time by early termination. For heuristics where r may be varied, the running time can vary either exponentially in $1/(1-r)$ (polynomial approximation scheme) or polynomially in $1/(1-r)$ (fully polynomial approximation scheme).

5. The examples that achieve the worst-case performance are quite specialized, and one might suspect that the worst-case performance is not predictive of performance for a 'random' problem. It is possible to compute more predictive performance bounds by making the bound a function of one or more summary parameters of the problem inputs. For example, as a function of $m = [b/\max_j a_j]$, the bound for greedy for the standard knapsack problem can be improved to

$$Z_G/Z \geq \frac{[b/a_1]}{b/a_1} \geq \frac{[b/a_1]}{[b/a_1] + 1} \geq \frac{m}{m+1} .$$

The performance ratio of 1/2 obtained earlier represents the worst case in which $m = 1$. The more general performance ratio in which m appears explicitly would provide useful information (i.e., a better performance ratio) to someone contemplating the use of the greedy heuristic for the repetitive solution of knapsack problems for which it is known that the weight of all items are small relative to the knapsack size. Despite their apparent usefulness, past research has not emphasized data dependent bounds.

6. After executing the greedy heuristic one can compute what might be called an 'a posteriori' bound. For the standard knapsack problem and the greedy heuristic this bound is equal to $Z_G/(b/a_1)p_1$ and is the tightest of the possible bounds we have discussed. It would be useful in deciding after the initial application of greedy whether additional computational effort using more elaborate heuristics or optimizing procedures is warranted.

3. SELECTIVE SURVEY OF RESULTS FOR OTHER PROBLEMS

This section will give a brief description of some results for two other problems: (1) parallel machine scheduling and (2) bin packing. These particular results have been selected to illustrate the diversity of worst-case research on heuristics.

3.1. Parallel Machine Scheduling

In pioneering papers published in 1966 and 1969, Graham [25], [26] describes what is apparently the first worst-case analysis of a heuristic. Graham was concerned with the following parallel machine scheduling problem. Given m identical machines and n jobs, where each job j is to be processed on one of the machines for an uninterrupted interval of length p_j, determine an assignment of jobs to machines that minimizes the earliest time at which all jobs are complete.

For simplicity, assume that jobs are indexed so that $p_1 \geq p_2 \geq \ldots \geq p_n$. The longest-processing time (LPT) rule is a heuristic for this problem that assigns jobs sequentially in increasing index order. Each job is assigned to the machine with the least processing already assigned. Assume ties in the choice of a machine are resolved in favor of the machine with smallest index. Figure 1 depicts the LPT and optimal assignments for the example with m = 3, n = 7, and the process times 5, 5, 4, 4, 3, 3 and 3.

In this example the completion time of the LPT assignment is 11/9 times the minimum completion time. Graham showed in [25] that this is the worst that the LPT rule can do when m = 3. More generally, if Z_{LPT} is the completion time of the LPT assignment and Z the minimum completion time, then

$$Z_{LPT} \leq (4/3 - 1/3m)Z.$$

Graham's work has spawned a variety of results giving improved heuristics (such as [3] and [20]) or dealing with variations of the basic model. This research is surveyed in [19]. In addition, Graham [25] has given a polynomial approximation scheme for this problem based on partial enumeration. Fully polynomial approximation schemes based on dynamic programming and data scaling have been developed for various scheduling problems by Sahni [51]. The comprehensive survey of scheduling theory by Graham, Lawler, Lenstra and Rinnooy Kan [27] also contains a thorough discussion of approximation results.

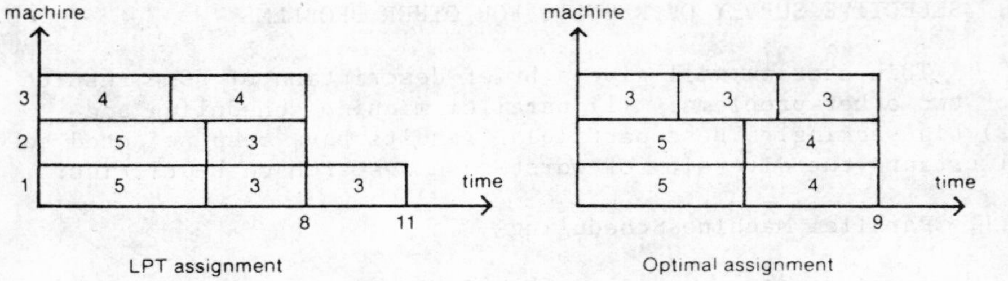

Figure 1. Assignments for an example.

3.2. Bin Packing

Although there has been a comparative flood of worst-case results on heuristics beginning in the mid-70s, the second worst-case analysis did not appear until seven years after Graham's 1966 scheduling work. This second study concerned the bin packing problem and was reported in a series of papers by Garey, Graham, Johnson and others. The principal references are [31] and [32].

In the bin packing problem we are given n items with weights $a_1,\ldots,a_n$ and an unlimited supply of bins, each with capacity b. The problem is to pack the items into the bins so as to minimize the number of bins used.

It is convenient to assume that items have been sorted so that $a_1 \geq a_2 \geq \ldots \geq a_n$. The first-fit-decreasing (FFD) rule is a heuristic for this problem that assigns items to bins in increasing index order. The bins are indexed and each new item is assigned to the lowest index bin into which it will fit. Figure 2 depicts the FFD and optimal packings for an example in which $n = 30$, $b = 40$ and we have 6 items of weight 21, 6 items of weight 12, 6 items of weight 11, and 12 items of weight 8.

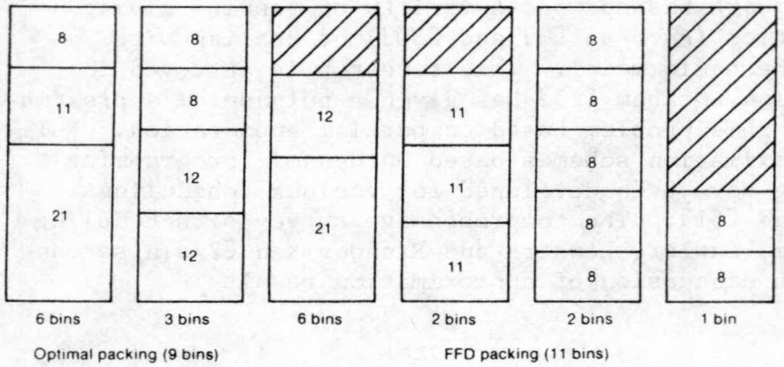

Figure 2. Packings for an example.

In this example, $Z_{FFD}/Z = 11/9$, Johnson [31] showed that asymptotically, this is the worst that FFD can do. Specifically,

$$Z_{FFD} \leq (11/9)Z + 4$$

and examples exist for arbitrarily large Z with $Z_{FFD} = (11/9)Z$. This bound is a little different from (5) because of the constant 4. Note however that this constant is negligible for large Z, which are the problems of principal interest.

Various other bin packing results are given in [32]. For example, the heuristic that assigns items in arbitrary order, assigning each item to the first bin into which it will fit, is shown to have an asymptotic (for large Z) worst-case ratio of $r = 1.7$.

3.3. Parallel Machine Scheduling and Bin Packing

The perceptive reader may have noticed the close connection between the bin packing problem and the parallel machine model discussed above. Specifically, if we equate items with jobs, then the bin packing problem is the problem of minimizing the number of machines required to complete all jobs by time b. This suggests a heuristic for the parallel machine problem in which we apply FFD to a sequence of bin packing problems with $a_j = p_j$ for all j and b varied within lower and upper bounds on the optimal completion time. To illustrate, consider the parallel machine example in figure 1 for which LPT found a schedule with a completion time of 11. The reader can verify that if we apply FFD to this data with b = 10 or 11 we get a schedule that uses 3 machines and completes at time 10. With b = 9 we use 3 machines and complete by time 9.

To use this idea in a parallel machine heuristic we need a systematic procedure for choosing trial values of b. Coffman, Garey and Johnson [3] describe such a procedure and analyze the worst-case performance of the resulting heuristic.

They begin by establishing lower and upper bounds on the values of b it is necessary to consider. The bounds are

$$LB = \max \{\sum_{j=1}^{n} p_j/m \ , \ \max_j p_j\},$$

$$UB = \max \{2 \sum_{j=1}^{n} p_j/m \ , \ \max_j p_j\}.$$

The value LB is an obvious lower bound on the optimal completion time Z and it is shown in [3] that FFD will use at most m machines when b = UB.

The heuristic, called MULTIFIT in [3], consists of repetitive application of FFD using binary search to set b within the interval [LB, UB]. On the first iteration, the value $b = (UB-LB)/2$ is used. If FFD requires more than m machines, the LB is adjusted to (UB-LB)/2. Otherwise, UB is adjusted to (UB-LB)/2. Subsequent iterations proceed in the same fashion.

Let $Z_{MF}(k)$ denote the best feasible completion time found after k applications of FFD. It is shown in [3] that $Z_{MF}(k) \leq (1.22 + (1/2)^k)Z$. Examples are also exhibited where $Z_{MF}(k)/Z = 1.176$ for all k.

3.4. Other Problems

Worst-case studies of heuristics have been conducted for several problems in addition to the knapsack, bin packing and parallel machine models. The most heavily studied models are the uncapacitated location problem and the traveling salesman problem. Important references for the uncapacitated location problem include [5], [11], [43], [44], [47] and [53], and for the traveling salesman problem [2], [6], [10], [11], [15], [29], [39], [48] and [49].

4. CONCLUSIONS AND FUTURE DIRECTIONS

In evaluating the performance of a heuristic, one always has the option of programming the heuristic and testing it on some examples. However, it is also natural to want to gain further insight into the heuristic by thinking about the logic of the heuristic and what it implies about performance. For this reason alone, the formulation of an analytic approach to the study of heuristics is a significant accomplishment. Beyond this, of course, worst-case results have made available information that is of obvious value to a designer or user of heuristics.

What additional work remains to be accomplished? One immediate observation is that most worst-case bounds are rather large compared with observed average performance or with the performance most users would regard as acceptable. Of course, one must keep in mind that the worst-case bound is the guaranteed performance, and most guarantees are worse than what you expect to actually receive. Still, it's always nice to have tighter bounds, and it is possible to refine worst case analysis to provide bounds that are closer to average performance. Whenever a worst case bound is significantly larger than average performance, the worse case performance must be occurring for only a small fraction of the problem instances, with most instances resulting in much better performance. If one can identify those characteristics of the "good" problem instances that enable better performance, then a better performance bound can be obtained subject to those conditions. This can be done either by defining a smaller problem class (an example is the better heuristic and bound obtained by Christofides [2] for the traveling salesman problem when intercity distances are assumed to satisfy the triangle inequality) or by making the performance bound a function of one or more summary statistics of the problem input data. An example of the latter approach is the data dependent bound that was given in § 1 for the greedy heuristic and the knapsack problem: $Z_G/Z \geq m/(m+1)$ where $m = [b/\max_j a_j]$. This bound can be thought of as a worst case analysis for an infinite family of problem classes, one for each specification of the parameter m. When m assumes its worst case value of 1, this bound equals the usual 1/2 bound. However, it's not unlikely that in a real application m will be known to be greater than 1. In this case, greedy will perform better than 1/2, and the data dependent bound will predict this better performance. The potential for using data dependent bounds like this to better predict actual performance has not been adequately exploited.

After reviewing the results surveyed in §§ 2 and 3, most readers will have obtained the impression that the worst-case theory of heuristics is quite fragmented, consisting of a collection of many problem specific results. This impression is absolutely correct. There is a clear need for a concept to put order to this chaos of specific results. The best hope would seem to be a concept like problem reduction that has been fundamental in the development of NP-completeness theory. In terms of optimization problems, an NP-completeness reduction from problem A to problem B consists of a polynomial time transformation that maps any instance of problem A to an instance of B, and a polynomial time procedure for computing an optimal solution to the instance of A from an optimal solution to the instance of B. To be useful in studying heuristics, such a reduction would also need to provide a polynomial time rule for computing a "near-optimal" solution of the instance of A from a "near-optimal" solution to the instance of B.

Such a reduction would have at least two important uses. First, it could be used to translate a heuristic and worst-case result for problem B to an analogous heuristic and worst-case result for problem A. Second, it would be useful in classifying problems as to their difficulty from the viewpoint of approximation. A number of results were reviewed in §§ 2 and 3 (e.g., see references [5], [16], [18], [44], [49]) that make some statement about the difficulty of obtaining approximately optimal solutions for a particular problem. A reduction could be used to translate such a result for problem A to an analogous result for problem B.

Some initial work has been done in this area. References [1], [40], and [46] report exploratory efforts to define natural classes of reductions that have the required properties. Reference [3] develops a new heuristic for parallel machine scheduling by exploiting a problem reduction that is just slightly more complicated than the one outlined above. Finally, Fisher [14] has used a reduction from the set covering problem to the uncapacitated location problem to derive a performance bound on greedy covering as a simple corollary of the performance bound on the greedy heuristic for uncapacitated location given in § 3.2.

Worst-case methods can be employed for broader purposes than the study of heuristics. One natural extension is to use these methods to study the performance of relaxations designed for obtaining super optimal bounds in a branch-and-bound algorithm. Such a result for a linear programming relaxation of the uncapacitated location problem has been reported in [5].

The effect of approximations in models can also be studied using either worst-case or probabilistic analysis in a manner that parallels heuristic analysis both philosophically and technically. For example, in most location models, groups of customers are aggregated together and treated as a single market. This aggregation reduces the computational difficulty of analyzing the model at the expense of introducing possible error. One is interested in knowing how much error is introduced by a given level of aggregation. Geoffrion [22], [23] has studied this question using a worst-case approach.

Finally, it would be useful to study heuristics for other kinds of intractible problems. Dempster, et al [8] have recently shown that various kinds of hierarchical scheduling problems can be nicely modeled as stochastic integer programs. From the viewpoint of optimization, stochastic integer programs appear formidable. However, using a worst case approach, it is shown in [9] that a natural heuristic is asymptotically optimal for several specific hierarchical scheduling problems.

The increased interest in heuristics should prompt a closer look at the modeling process. It is well recognized that modeling is an art, and the construction of "good" models requires judicious assumptions that preserve the essence of the problem while allowing tractible analysis. This process has never been well studied, and what folklore exists is tilted towards making models analytically tractable for optimization methods, not heuristics.

As an example, consider the use of the 0-1 knapsack model to represent a capital budgeting problem, where b is the available budget and x_j is a 0-1 variable that determines whether funds will be allocated to project j. The best optimization algorithms for the 0-1 knapsack problem are branch and bound methods for which it is convenient to have b specified as a fixed number. On the other hand, the budget is usually somewhat flexible, and a very profitable investment program would not be rejected merely because it required a slight increase in the budget. Consider the example $n = 5$, $b = 999$, $a_1 = a_2 = a_3 = 333$, $c_1 = c_2 = c_3 = 500$, $a_4 = a_5 = 500$ and $c_4 = c_5 = 1000$. How many financial managers would prefer the optimal solution $x_1 = x_2 = x_3 = 1$, $x_4 = x_5 = 0$, $z = 1500$ to the "infeasible" solution $x_1 = x_2 = x_3 = 0$, $x_4 = x_5 = 1$, $z = 2000$? It would seem that a better model would be to maximize $\Sigma_{j=1}^{n} c_j x_j / b$, where b is allowed to vary over some range. Provided the range is sufficiently large, this is exactly the problem solved by the greedy heuristic. "Soft" constraints of this type arise in many applications. Paradoxically, allowing for this flexibility seems to result in a model which is both a better representation of the real problem and is more amenable to analysis by a heuristic.

ACKNOWLEDGMENTS

The author is indebted to Abba Krieger and Alexander Rinnooy Kan for their helpful comments on an early draft of this paper. This paper was supported in part by NSF Grant ENG-7826500 to the University of Pennsylvania. With the exception of Sections 3.3 and 3.4, it is reprinted by permission of Marshall L. Fisher, Worst-case analysis of heuristic algorithms, *Management Science*, Volume 26, Number 1, January 1980, Copyright 1980 The Institute of Management Sciences.

REFERENCES

1. G. AUSIELLO, A. D'ATRI (1977) On the structure of combinatorial problems and structure preserving reductions. In: A. SALOMAA, M. STEINBY (eds.) (1977) *Automata, Languages and Programming*, Lecture Notes in Computer Science 52, Springer, Berlin, 45-60.
2. N. CHRISTOFIDES (1977) Worst-case analysis of a new heuristic for the travelling salesman problem. Management Sciences Research Report 388, Carnegie-Mellon University.
3. E.G. COFFMAN, JR., M.R. GAREY, D.S. JOHNSON (1977) An

application of bin packing to multiprocessor scheduling. *SIAM J. Comput.* 7,1-17.
4. S.A. COOK (1971) The complexity of theorem-proving procedures. *Proc. 3rd Annual ACM Symp. Theory of Computing*, 151-158.
5. G. CORNUEJOLS, M.L. FISHER, G.L. NEMHAUSER (1977) Location of bank accounts to optimize float: an analytic study of exact and approximate algorithms. *Management Sci.* 23,789-810.
6. G. CORNUEJOLS, G.L. NEMHAUSER (1978) Tight bounds for Christofides' traveling salesman heuristic. *Math. Programming* 14,116-121.
7. G. CORNUEJOLS, G.L. NEMHAUSER, L.A. WOLSEY (1978) Worst-case and probabilistic analysis of algorithms for a location problem. Operations Research Technical Report 375, Cornell University; *Oper. Res.*, to appear.
8. M.A.H. DEMPSTER, M.L. FISHER, L. JANSEN, B.J. LAGEWEG, J.K. LENSTRA, A.H.G. RINNOOY KAN (1981) Analytical evaluation of hierarchical planning systems. *Oper. Res.* 29, to appear.
9. M.A.H. DEMPSTER, M.L. FISHER, L. JANSEN, B.J. LAGEWEG, J.K. LENSTRA, A.H.G. RINNOOY KAN (1981) Analysis of heuristics for stochastic programming: results for hierarchical scheduling problems. Report BW 142, Mathematisch Centrum, Amsterdam
10. M.L. FISHER, G.L. NEMHAUSER, L.A. WOLSEY (1979) An analysis of approximations for finding a maximum weight Hamiltonian circuit. *Oper. Res.* 27,799-809.
11. M.L. FISHER, G.L. NEMHAUSER, L.A. WOLSEY (1978) An analysis of approximations for maximizing submodular set functions - II. *Math. Programming Stud.* 8,73-87.
12. M.L. FISHER (1978) Worst-case analysis of heuristics. In: J. K. LENSTRA, A.H.G. RINNOOY KAN, P. VAN EMDE BOAS (eds.) (1978) *Interfaces Between Computer Science and Operations Research*, Mathematical Centre Tracts 99, Mathematisch Centrum, 87-108.
13. M.L. FISHER, D.S. HOCHBAUM (1980) Probabilistic analysis of the planar K-median problem. *Math. Oper. Res.* 5,27-34.
14. M.L. FISHER (1979) On the equivalence of greedy covering and greedy location. Decision Sciences Working Paper 79-11-01, University of Pennsylvania.
15. A.M. FRIEZE (1978) Worst-case analysis of two algorithms for travelling salesman problems. Technical Report, Department of Computer Science and Statistics, Queen Mary College, London.
16. M.R. GAREY, D.S. JOHNSON (1976) The complexity of near-optimal graph coloring. *J. Assoc. Comput. Mach.* 23,43-49.
17. M.R. GAREY, D.S. JOHNSON (1976) Approximation algorithms for combinatorial problems: an annotated bibliography. In: J.F. TRAUB (ed.) (1976) *Algorithms and Complexity: New Directions and Recent Results*, Academic Press, New York, 41-52.
18. M.R. GAREY, D.S. JOHNSON (1978) "Strong" NP-completeness results: motivation, examples and implications. *J. Assoc. Comput. Mach.* 25,499-508.
19. M.R. GAREY, R.L. GRAHAM, D.S. JOHNSON (1978) Performance guarantees for scheduling algorithms. *Oper. Res.* 26,3-21.

20. M.R. GAREY, R.L. GRAHAM, D.S. JONSON, A.C. YAO (1976) Resource constrained scheduling as generalized bin packing. *J. Combin. Theory Ser. A* 21,257-298.
21. M.R. GAREY, D.S. JOHNSON (1979) *Computers and Intractability: a Guide to the Theory of NP-Completeness*, Freeman, San Francisco.
22. A.M. GEOFFRION (1977) Objective function approximations in mathematical programming. *Math. Programming* 13,23-27.
23. A.M. GEOFFRION (1977) A priori error bounds for procurement commodity aggregation in logistics planning models. *Naval Res. Logist. Quart.* 24,201-212.
24. A.M. GEOFFRION, G.W. GRAVES (1974) Multi-commodity distribution system design by Benders decomposition. *Management Sci.* 20,822-844.
25. R.L. GRAHAM (1966) Bounds for certain multiprocessing anomalies. *Bell System Tech. J.* 45,1563-1581.
26. R.L. GRAHAM (1969) Bounds on multiprocessing timing anomalies. *SIAM J. Appl. Math.* 17,416-429.
27. R.L. GRAHAM, E.L. LAWLER, J.K. LENSTRA, A.H.G. RINNOOY KAN (1979) Optimization and approximation in deterministic sequencing and scheduling: a survey. *Ann. Discrete Math.* 5, 287-326.
28. O.H. IBARRA, C.E. KIM (1975) Fast approximation algorithms for the knapsack and sum of subset problems. *J. Assoc. Comput. Mach.* 22,463-468.
29. T.A. JENKYNS (1977) The greedy travelling salesman's problem. Technical Report, Brock University.
30. D.S. JOHNSON (1974) Approximation algorithms for combinatorial problems. *J. Comput. System Sci.* 9,256-278.
31. D.S. JOHNSON (1973) Near optimal bin packing algorithms. Report MAC TR-109, Massachusetts Institute of Technology.
32. D.S. JOHNSON, A. DEMERS, J.D. ULLMAN, M.R. GAREY, R.L. GRAHAM (1974) Worst-case performance bounds for simple one-dimensional packing algorithms. *SIAM J. Comput.* 3,299-325.
33. R. KANNAN, B. KORTE (1978) Approximative combinatorial algorithms. Report 78107-OR, Institut für Ökonometrie und Operations Research, Universität Bonn.
34. R.M. KARP (1972) Reducibility among combinatorial problems. In: R.E. MILLER, J.W. THATCHER (eds.) (1972) *Complexity of Computer Computations*, Plenum, New York, 85-103.
35. R.M. KARP (1975) On the computational complexity of combinatorial problems. *Networks* 5,45-68.
36. R.M. KARP (1976) The probabilistic analysis of some combinatorial search algorithms. In: J.F. TRAUB (ed.) (1976) *Algorithms and Complexity: New Directions and Recent Results*, Academic Press, New York, 1-19.
37. R.M.KARP (1977) Probabilistic analysis of partitioning algorithms for the traveling-salesman problem in the plane. *Math. Oper. Res.* 2,209-224.
38. E.L. LAWLER (1978) Fast approximation algorithms for knapsack

problems. In: J.K. LENSTRA, A.H.G. RINNOOY KAN, P. VAN EMDE BOAS (eds.) (1978) *Interfaces Between Computer Science and Operations Research*, Mathematical Centre Tracts 99, Mathematisch Centrum, Amsterdam, 109-139.
39. S. LIN, B.W. KERNIGHAN (1973) An effective heuristic algorithm for the traveling-salesman problem. *Oper. Res.* 21, 498-516.
40. N. LYNCH, R.J. LIPTON (1978) On structure preserving reductions. *SIAM J. Comput.* 7,119-126.
41. T.G. MAIRS, G.W. WAKEFIELD, E.L. JOHNSON, K. SPIELBERG (1978) On a production allocation and distribution problem. *Management Sci.* 24,1622-1630.
42. R.E. MARSTEN, M.R. MULLER, C.L. KILLION (1978) Crew planning at Flying Tiger: a successful application of integer programming. MIS Technical Report 533, University of Arizona.
43. G.L. NEMHAUSER, L.A. WOLSEY, M.L. FISHER (1978) An analysis of approximations for maximizing submodular set functions - I. *Math. Programming* 14,265-294.
44. G.L. NEMHAUSER, L.A. WOLSEY (1978) Best algorithms for approximating the maximum of a submodular set function. *Math. Oper. Res.* 3,177-188.
45. G.L. NEMHAUSER, L.A. WOLSEY (1978) Maximizing submodular set functions: formulations, algorithms and applications. CORE Discussion Paper 7832, University of Louvain.
46. A. PAZ, S. MORAN (1977) Non-deterministic polynomial optimization problems and their approximation. In: A. SALOMAA, M. STEINBY (eds.) (1977) *Automata, Languages and Programming*, Lecture Notes in Computer Science 52, Springer, Berlin, 370-379.
47. S. RABIN, P. KOLESAR (1978) Mathematical optimization of glaucoma visual field screening protocols. *Documenta Ophthalmologica* 45,361-380.
48. D.J. ROSENKRANTZ, R.E. STEARNS, P.M. LEWIS II (1977) An analysis of several heuristics for the traveling salesman problem. *SIAM J. Comput.* 6,563-581.
49. S. SAHNI, T. GONZALEZ (1976) P-Complete approximation problems. *J. Assoc. Comput. Mach.* 23,555-565.
50. S. SAHNI (1975) Approximate algorithms for the 0-1 knapsack problem. *J. Assoc. Comput. Mach.* 22,115-124.
51. S. SAHNI (1976) Algorithms for scheduling independent tasks. *J. Assoc. Comput. Mach.* 23,116-127.
52. C. WESTERBERG, B. BJORKLUND, E. HULTMAN (1977) An application of mixed integer programming in a Swedish steel mill. *Interfaces.*
53. E. ZEMEL (1978) Functions for measuring the quality of approximate solutions to zero-one programming problems. Disscussion Paper 323, Graduate School of Management, Northwestern University.

RECENT DEVELOPMENTS IN DETERMINISTIC SEQUENCING AND SCHEDULING: A SURVEY

E.L. Lawler[1], J.K. Lenstra[2], A.H.G. Rinnooy Kan[3]

[1]University of California, Berkeley
[2]Mathematisch Centrum, Amsterdam
[3]Erasmus University, Rotterdam

ABSTRACT

The theory of deterministic sequencing and scheduling has expanded rapidly during the past years. We survey the state of the art with respect to optimization and approximation algorithms and interpret these in terms of computational complexity theory. Special cases considered are single machine scheduling, identical, uniform and unrelated parallel machine scheduling, and open shop, flow shop and job shop scheduling. This paper is a revised version of the survey by Graham *et al.* (*Ann. Discrete Math.* 5(1979)287-326), with emphasis on recent developments.

1. INTRODUCTION

In this paper we attempt to survey the rapidly expanding area of deterministic scheduling theory. Although the field only dates back to the early fifties, an impressive amount of literature has been created and the remaining open problems are currently under heavy attack. An exhaustive discussion of all available material would be impossible - we will have to restrict ourselves to the most significant results, paying special attention to recent developments and omitting detailed theorems and proofs. For further information the reader is referred to the classic book by Conway, Maxwell and Miller [Conway *et al.* 1967], the introductory textbook by Baker [Baker 1974], the advanced expository articles collected by Coffman [Coffman 1976] and a few survey papers and theses [Bakshi & Arora 1969; Lenstra 1977; Liu 1976; Rinnooy Kan 1976]. This paper itself is a revised and updated version of a recent survey [Graham *et al.* 1979].

M. A. H. Dempster et al. (eds.), Deterministic and Stochastic Scheduling, 35–73.

The outline of the paper is as follows. Section 2 introduces the essential notation and presents a detailed problem classification. Sections 3, 4 and 5 deal with single machine, parallel machine, and open shop, flow shop and job shop problems, respectively. In each section we briefly outline the relevant complexity results and optimization and approximation algorithms. Section 6 contains some concluding remarks.

We shall be making extensive use of concepts from the *theory of computational complexity* [Cook 1971; Karp 1972]. Several introductory surveys of this area are currently available [Karp 1975; Garey & Johnson 1979; Lenstra & Rinnooy Kan 1979] and hence terms like *(pseudo)polynomial-time algorithm* and (*binary* and *unary*) *NP-hardness* will be used without further explanation.

2. PROBLEM CLASSIFICATION

2.1. Introduction

Suppose that n *jobs* J_j ($j = 1,...,n$) have to be processed on m *machines* M_i ($i = 1,...,m$). Throughout, we assume that each machine can process at most one job at a time and that each job can be processed on at most one machine at a time. Various job, machine and scheduling characteristics are reflected by a three-field problem classification $\alpha|\beta|\gamma$, to be introduced in this section.

2.2. Job data

In the first place, the following data can be specified for each J_j:

- a *number of operations* m_j;
- one or more *processing times* p_j or p_{ij}, that J_j has to spend on the various machines on which it requires processing;
- a *release date* r_j, on which J_j becomes available for processing;
- a *due date* d_j, by which J_j should ideally be completed;
- a *weight* w_j, indicating the relative importance of J_j;
- a nondecreasing real *cost function* f_j, measuring the cost $f_j(t)$ incurred if J_j is completed at time t.

In general, m_j, p_j, p_{ij}, r_j, d_j and w_j are integer variables.

2.3. Machine environment

We shall now describe the first field $\alpha = \alpha_1\alpha_2$ specifying the machine environment. Let $\circ$ denote the empty symbol.

If $\alpha_1 \in \{\circ,P,Q,R\}$, each J_j consists of a single operation that can be processed on any M_i; the processing time of J_j on M_i is p_{ij}. The four values are characterized as follows:

- $\alpha_1 = \circ$: *single machine*; $p_{1j} = p_j$;
- $\alpha_1 = P$: *identical parallel machines*; $p_{ij} = p_j$ ($i = 1,...,m$);

- α_1 = Q: *uniform parallel machines*; $p_{ij} = p_j/q_i$ for a given *speed* q_i of M_i ($i = 1,\ldots,m$);
- α_1 = R: *unrelated parallel machines*.

If α_1 = O, we have an *open shop*, in which each J_j consists of a set of operations $\{O_{1j},\ldots,O_{mj}\}$. O_{ij} has to be processed on M_i during p_{ij} time units, but the order in which the operations are executed is immaterial. If $\alpha_1 \in \{F,J\}$, an ordering is imposed on the set of operations corresponding to each job. If α_1 = F, we have a *flow shop*, in which each J_j consists of a chain $(O_{1j},\ldots,O_{mj})$. O_{ij} has to be processed on M_i during p_{ij} time units. If α_1 = J, we have a *job shop*, in which each J_j consists of a chain $(O_{1j},\ldots,O_{m_j j})$. O_{ij} has to be processed on a given machine μ_{ij} during p_{ij} time units, with $\mu_{i-1,j} \neq \mu_{ij}$ for $i = 2,\ldots,m_j$.

If α_2 is a positive integer, then m is *constant* and equal to α_2. If $\alpha_2 = \circ$, then m is assumed to be *variable*. Obviously, $\alpha_1 = \circ$ if and only if $\alpha_2 = 1$.

2.4. Job characteristics

The second field $\beta \subset \{\beta_1,\ldots,\beta_5\}$ indicates a number of job characteristics, which are defined as follows.

1. $\beta_1 \in \{pmtn,\circ\}$
   - $\beta_1 = pmtn$: *Preemption* (job splitting) is allowed: the processing of any operation may be interrupted and resumed at a later time.
   - $\beta_1 = \circ$: No preemption is allowed.
2. $\beta_2 \in \{prec,tree,\circ\}$
   - $\beta_2 = prec$: A *precedence relation* $\rightarrow$ between the jobs is specified. It is derived from a directed acyclic graph G with vertex set $\{1,\ldots,n\}$. If G contains a directed path from j to k, we write $J_j \rightarrow J_k$ and require that J_j is completed before J_k can start.
   - $\beta_2 = tree$: G is a *rooted tree* with either outdegree at most one for each vertex or indegree at most one for each vertex.
   - $\beta_2 = \circ$: No precedence relation is specified.
3. $\beta_3 \in \{r_j,\circ\}$
   - $\beta_3 = r_j$: *Release dates* that may differ per job are specified.
   - $\beta_3 = \circ$: All $r_j = 0$.
4. $\beta_4 \in \{m_j \leq \bar{m},\circ\}$
   - $\beta_4 = m_j \leq \bar{m}$: A *constant upper bound* on m_j is specified (only if α_1 = J).
   - $\beta_4 = \circ$: All m_j are arbitrary integers.
5. $\beta_5 \in \{p_{ij}=1,\circ\}$
   - $\beta_5 = p_{ij}=1$: Each operation has *unit processing time* (if $\alpha_1 \in \{\circ,P,Q\}$, we write $p_j=1$; if α_1 = R, $p_{ij}=1$ will not occur).
   - $\beta_5 = \circ$: All p_{ij} (p_j) are arbitrary integers.

2.5. Optimality criteria

The third field $\gamma \in \{f_{max}, \Sigma f_j\}$ refers to the optimality criterion chosen. Given a schedule, we can compute for each J_j:

- the *completion time* C_j;
- the *lateness* $L_j = C_j - d_j$;
- the *tardiness* $T_j = \max\{0, C_j - d_j\}$;
- the *unit penalty* $U_j = 0$ if $C_j \le d_j$, $U_j = 1$ otherwise.

The optimality criteria most commonly chosen involve the minimization of

$$f_{max} \in \{C_{max}, L_{max}\}$$

where $f_{max} = \max_j\{f_j(C_j)\}$ with $f_j(C_j) = C_j, L_j$, respectively, or

$$\Sigma f_j \in \{\Sigma C_j, \Sigma T_j, \Sigma U_j, \Sigma w_j C_j, \Sigma w_j T_j, \Sigma w_j U_j\}$$

where $\Sigma f_j = \Sigma_{j=1}^{n} f_j(C_j)$ with $f_j(C_j) = C_j, T_j, U_j, w_j C_j, w_j T_j, w_j U_j$, respectively.

It should be noted that $\Sigma w_j C_j$ and $\Sigma w_j L_j$ differ by a constant $\Sigma w_j d_j$ and hence are *equivalent*. Furthermore, any schedule minimizing L_{max} also minimizes T_{max} and U_{max}, but not *vice versa*.

The optimal value of γ will be denoted by γ^*, the value produced by an (approximation) algorithm A by $\gamma(A)$. If a known upper bound ρ on $\gamma(A)/\gamma^*$ is best possible in the sense that examples exist for which $\gamma(A)/\gamma^*$ equals or asymptotically approaches ρ, this will be denoted by a dagger (†).

2.6. Examples

$1|prec|L_{max}$: minimize maximum lateness on a single machine subject to general precedence constraints. This problem can be solved in polynomial time (Section 3.2).

$R|pmtn|\Sigma C_j$: minimize total completion time on a variable number of unrelated parallel machines, allowing preemption. The complexity of this problem is unknown (Section 4.4.3).

$J3|p_{ij}=1|C_{max}$: minimize maximum completion time in a 3-machine job shop with unit processing times. This problem is NP-hard (Section 5.4.1).

2.7. Reducibility among scheduling problems

Each scheduling problem in the class outlined above corresponds to an 7-tuple $(v_0, \ldots, v_6)$, where v_i is a vertex of graph G_i drawn in Figure 1 ($i = 0, \ldots, 6$). For two problems $P' = (v_0', \ldots, v_6')$ and $P = (v_0, \ldots, v_6)$, we write $P' \rightarrow P$ if either $v_i' = v_i$ or G_i contains a directed path from v_i' to v_i, for $i = 0, \ldots, 6$. The reader should verify that $P' \rightarrow P$ implies $P' \propto P$. The graphs thus define elementary reductions among scheduling problems. It follows that

- if $P' \rightarrow P$ and P is well solved, then P' is well solved;
- if $P' \rightarrow P$ and P' is NP-hard, then P is NP-hard.

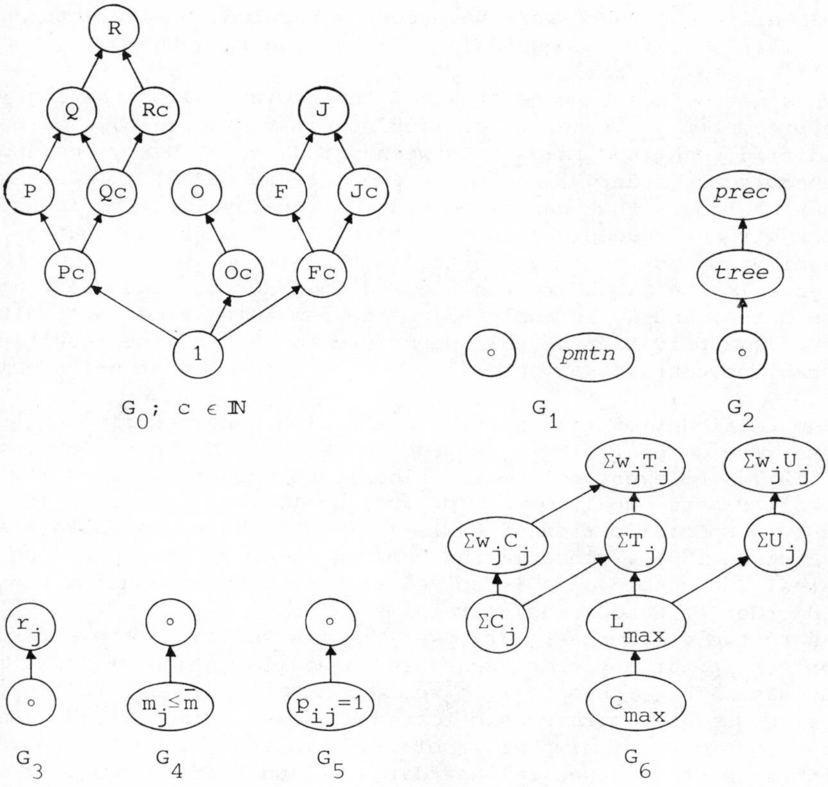

Figure 1

3. SINGLE MACHINE PROBLEMS

3.1. Introduction

The single machine case has been the object of extensive research ever since the seminal work by Jackson [Jackson 1955] and Smith [Smith 1956]. We will give a brief survey of the principal results, classifying them according to the optimality criterion chosen. As a general result, we note that if all $r_j = 0$ we need only consider schedules without preemption and without machine idle time [Conway *et al.* 1967].

3.2. Minimizing maximum cost

A crucial result in this section is an $O(n^2)$ algorithm to solve $1|prec|f_{max}$ for arbitrary nondecreasing cost functions [Lawler 1973]. At each step of the algorithm, let S denote the index set of unscheduled jobs, let $p(S) = \Sigma_{j \in S}\, p_j$, and let $S' \subset S$ indicate the jobs all whose successors have been scheduled. One selects J_k for the last position among $\{J_j | j \in S\}$ by requiring that $f_k(p(S)) \leq f_j(p(S))$ for all $j \in S'$.

This method has been generalized to an $O(n^2)$ algorithm for $1|pmtn,prec,r_j|f_{max}$ [Baker *et al.* 1982]. First, the release dates are modified such that $r_j+p_j \leq r_k$ whenever $J_j \rightarrow J_k$. Next, the jobs are scheduled in order of nondecreasing release dates; this creates a number of *blocks* that can be considered separately. From among the jobs without successors in a certain block, a job J_k that yields minimum cost when put in the last position is selected, the other jobs in the block are rescheduled in order of nondecreasing release dates, and J_k is assigned to the remaining time intervals. By repeated application of this procedure to each of the resulting subblocks, one obtains an optimal schedule with at most n-1 preemptions in $O(n^2)$ time.

The remainder of this section deals with nonpreemptive L_{max} problems. The general $1|r_j|L_{max}$ problem is unary NP-hard [Lenstra *et al.* 1977]. However, polynomial algorithms exist if all r_j are equal, all d_j are equal, or all p_j are equal. The first case is solved by a specialization of Lawler's method, known as *Jackson's rule* [Jackson 1955]: schedule the jobs in order of nondecreasing due dates. The second case is solved similarly by scheduling the jobs in order of nondecreasing release dates.

As to the third case, $1|r_j,p_j=1|L_{max}$ is solved by the *extended Jackson's rule*: at any time, schedule an available job with smallest due date. The problem $1|r_j,p_j=p|L_{max}$, where p is an arbitrary integer, requires a more sophisticated approach [Simons 1978]. Let us first consider the simpler problem of finding a feasible schedule with respect to given release dates r_j and deadlines d_j. If application of the extended Jackson's rule yields such a schedule, we are finished; otherwise, let J_ℓ be the first late job and let J_k be the last job preceding J_ℓ such that $d_k > d_\ell$. If J_k does not exist, there is no feasible schedule; otherwise, the only hope of obtaining such a schedule is to postpone J_k by forcing it to yield precedence to the set of jobs currently between J_k and J_ℓ. This is achieved by declaring the interval between the starting time of J_k and the smallest release date of this set to be a *forbidden region* in which no job is allowed to start and applying the extended Jackson's rule again subject to this constraint. Since at each iteration at least one starting time of the form r_j+hp $(1 \leq j,h \leq n)$ is excluded, at most n^2 iterations will occur and the feasibility question is answered in $O(n^3 \log n)$ time. An improved implementation requires only $O(n \log n)$ time [Garey *et al.* 1981A]. Bisection search over the possible L_{max} values leads to a polynomial algorithm for $1|r_j,p_j=p|L_{max}$.

These three special cases remain well solved in the presence of precedence constraints. It suffices to update release and due dates such that $r_j < r_k$ and $d_j < d_k$ whenever $J_j \rightarrow J_k$ [Lageweg *et al.* 1976].

Various elegant enumerative methods exist for solving

$1|prec,r_j|L_{max}$. Baker and Su [Baker & Su 1974] obtain a lower bound by allowing preemption; their enumeration scheme simply generates all *active schedules*, *i.e.* schedules in which one cannot decrease the starting time of an operation without increasing the starting time of another one. McMahon and Florian [McMahon & Florian 1975] propose a more ingenious approach; a slight modification of their algorithm allows very fast solution of quite large problems [Lageweg *et al.* 1976]. Carlier [Carlier 1980] describes a related method of comparable efficiency.

Very little work has been done on worst-case analysis of approximation algorithms for single machine problems. For $1|r_j|L_{max}$, Potts [Potts 1980B] presents an iterative version of the extended Jackson's rule (IJ) and shows that, if $r_j \geq 0$ and $d_j \leq 0$ $(j = 1,\ldots,n)$,

$$L_{max}(IJ)/L^*_{max} \leq \frac{3}{2}. \qquad (\dagger)$$

3.3. Minimizing total cost

3.3.1. $1|\beta|\Sigma w_jC_j$

The case $1||\Sigma w_jC_j$ can be solved in $O(n \log n)$ time by *Smith's rule*: schedule the jobs according to nonincreasing ratios w_j/p_j [Smith 1956]. If all weights are equal, this amounts to the SPT rule of executing the jobs on the basis of shortest processing time first, a rule that is often used in more complicated situations without much empirical, let alone theoretical, support for its superior quality (*cf.* Section 5.4.2).

This result has been extended to $O(n \log n)$ algorithms that deal with *tree*-like [Horn 1972; Adolphson & Hu 1973; Sidney 1975] and even *series-parallel* [Lawler 1978] precedence constraints; see [Adolphson 1977] for an $O(n^3)$ algorithm covering a slightly more general case. The crucial observation to make here is that, if $J_j \rightarrow J_k$ with $w_j/p_j < w_k/p_k$ and if all other jobs either have to precede J_j, succeed J_k, or are incomparable with both, then J_j and J_k are adjacent in at least one optimal schedule and can effectively be treated as one job with processing time p_j+p_k and weight w_j+w_k. By successive application of this device, starting at the bottom of the precedence tree, one will eventually obtain an optimal schedule. Addition of general precedence constraints results in NP-hardness, even if all $p_j = 1$ or all $w_j = 1$ [Lawler 1978; Lenstra & Rinnooy Kan 1978].

If release dates are introduced, $1|r_j|\Sigma C_j$ is already unary NP-hard [Lenstra *et al.* 1977]. In the preemptive case, $1|pmtn,r_j|\Sigma C_j$ can be solved by an obvious extension of Smith's rule, but, surprisingly, $1|pmtn,r_j|\Sigma w_jC_j$ is unary NP-hard [Labetoulle *et al.* 1979].

For $1|r_j|\Sigma w_jC_j$, several elimination criteria and branch-and-bound algorithms have been proposed [Rinaldi & Sassano 1977; Bianco & Ricciardelli 1981; Hariri & Potts 1981].

3.3.2. $1|\beta|\Sigma w_jT_j$

$1||\Sigma w_jT_j$ is a unary NP-hard problem [Lawler 1977; Lenstra *et al.* 1977], for which various enumerative solution methods have been proposed. Elimination criteria developed for the problem [Emmons 1969; Shwimer 1972] can be extended to the case of arbitrary nondecreasing cost functions [Rinnooy Kan *et al.* 1975]. Lower bounds can be based on a linear assignment relaxation using an underestimate of the cost of assigning J_j to position k [Rinnooy Kan *et al.* 1975], a fairly similar relaxation to a transportation problem [Gelders & Kleindorfer 1974, 1975], and relaxation of the requirement that the machine can process at most one job at a time [Fisher 1976]. In the latter approach, one attaches "prices" (*i.e.*, Lagrangean multipliers) to each unit-time interval. Multiplier values are sought for which a cheapest schedule does not violate the capacity constraint. The resulting algorithm is quite successful on problems with up to 50 jobs, although a straightforward but cleverly implemented dynamic programming approach [Baker & Schrage 1978] offers a surprisingly good alternative.

If all $p_j = 1$, we have a simple linear assignment problem, the cost of assigning J_j to position k being given by $f_j(k)$. If all $w_j = 1$, the problem can be solved by a pseudopolynomial algorithm in $O(n^4\Sigma p_j)$ time [Lawler 1977]; the computational complexity of $1||\Sigma T_j$ with respect to a binary encoding remains an open question.

Addition of precedence constraints yields NP-hardness, even for $1|prec,p_j=1|\Sigma T_j$ [Lenstra & Rinnooy Kan 1978].

If we introduce release dates, $1|r_j,p_j=1|\Sigma w_jT_j$ can again be solved as a linear assignment problem, whereas $1|r_j|\Sigma T_j$ is obviously unary NP-hard.

3.3.3. $1|\beta|\Sigma w_jU_j$

An algorithm due to Moore [Moore 1968] allows solution of $1||\Sigma U_j$ in O(n log n) time: jobs are added to the schedule in order of nondecreasing due dates, and if addition of J_j results in this job being completed after d_j, the scheduled job with the largest processing time is marked to be late and removed. This procedure can be extended to cover the case in which certain specified jobs have to be on time [Sidney 1973]; the further generalization in which jobs have to meet given *deadlines* occurring at or after their due dates is binary NP-hard [Lawler -]. The problem also remains solvable in O(n log n) time if we add *agreeable weights* (*i.e.*, $p_j < p_k \Rightarrow w_j \geq w_k$) [Lawler 1976A] or *agreeable release dates* (*i.e.*, $d_j < d_k \Rightarrow r_j \leq r_k$) [Kise *et al.* 1978]. $1||\Sigma w_jU_j$ is binary NP-hard [Karp 1972], but can be solved by dynamic programming in $O(n\Sigma p_j)$ time [Lawler & Moore 1969].

Again, $1|prec,p_j=1|\Sigma U_j$ is NP-hard [Garey & Johnson 1976], even for *chain*-like precedence constraints [Lenstra & Rinnooy Kan 1980].

Of course, $1|r_j|\Sigma U_j$ is unary NP-hard, but dynamic programming techniques can be applied to solve $1|pmtn,r_j|\Sigma U_j$ in $O(n^6)$ time and $1|pmtn,r_j|\Sigma w_jU_j$ in $O(n^3(\Sigma w_j)^3)$ time [Lawler -].

For $1||\Sigma w_jU_j$, Sahni [Sahni 1976] presents algorithms A_k with $O(n^3k)$ running time such that

$$\Sigma w_j\bar{U}_j(A_k)/\Sigma w_j\bar{U}_j^* \geq 1 - \frac{1}{k},$$

where $\bar{U}_j = 1-U_j$. For $1|tree|\Sigma w_jU_j$, Ibarra and Kim [Ibarra & Kim 1978] give algorithms B_k of order $O(kn^{k+2})$ with the same worst-case error bound.

4. PARALLEL MACHINE PROBLEMS

4.1. Introduction

Recall from Section 2.3 the definitions of *identical*, *uniform* and *unrelated* machines, denoted by P, Q and R, respectively.

Nonpreemptive parallel scheduling problems tend to be difficult. This can be inferred immediately from the fact that $P2||C_{max}$ and $P2||\Sigma w_jC_j$ are binary NP-hard [Bruno *et al.* 1974; Lenstra *et al.* 1977]. If we are to look for polynomial algorithms, it follows that we should either restrict attention to the special case $p_j = 1$, as we do in Section 4.2, or concern ourselves with the ΣC_j criterion, as we do in the first three subsections of Section 4.3. The remaining part of Section 4.3 is entirely devoted to enumerative optimization methods and approximation algorithms for various NP-hard problems.

The situation is much brighter with respect to *preemptive* parallel scheduling. For example, $P|pmtn|C_{max}$ has long been known to admit a simple $O(n)$ algorithm [McNaughton 1959]. Many new results for the ΣC_j, C_{max}, L_{max}, ΣU_j and Σw_jU_j criteria have been obtained quite recently. These are summarized in Section 4.4. With respect to other criteria, $P2|pmtn|\Sigma w_jC_j$ turns out to be NP-hard (see Section 4.4.1). Little is known about $P|pmtn|\Sigma T_j$, but we know from Section 3.3.2 that $1|pmtn|\Sigma w_jT_j$ is already NP-hard.

4.2. Nonpreemptive scheduling: unit processing times

4.2.1. $Q|p_j=1|\Sigma f_j$, $Q|p_j=1|f_{max}$

A simple transportation network model provides an efficient solution method for $Q|p_j=1|\Sigma f_j$ and $Q|p_j=1|f_{max}$.

Let there be n sources j (j = 1,...,n) and mn sinks (i,k) (i = 1,...,m, k = 1,...,n). Set the cost of arc (j,(i,k)) equal to $c_{ijk} = f_j(k/q_i)$. The arc flow x_{ijk} is to have the interpretation:

$$x_{ijk} = \begin{cases} 1 & \text{if } J_j \text{ is executed on } M_i \text{ in the k-th position,} \\ 0 & \text{otherwise.} \end{cases}$$

Then the problem is to minimize

$$\Sigma_{i,j,k}\, c_{ijk} x_{ijk} \quad \text{or} \quad \max_{i,j,k}\{c_{ijk} x_{ijk}\}$$

subject to

$$\Sigma_{i,k}\, x_{ijk} = 1 \quad \text{for all } j,$$

$$\Sigma_{j}\, x_{ijk} \le 1 \quad \text{for all } i,k,$$

$$x_{ijk} \ge 0 \quad \text{for all } i,j,k.$$

The time required to prepare the data for this transportation problem is $O(mn^2)$. A careful analysis reveals that the problem can be solved (in integers) in $O(n^3)$ time. Since we may assume that $m \le n$, the overall running time is $O(n^3)$.

We note that the special case $P|p_j=1|\Sigma U_j$ can be solved in $O(n \log n)$ time [Lawler 1976A]. The problem $P|r_j,p_j=p|L_{max}$ is solvable in polynomial time by an extension of the corresponding single machine algorithm (see Section 3.2) [Simons 1980].

4.2.2. $P|prec,p_j=1|C_{max}$

$P|prec,p_j=1|C_{max}$ is known to be NP-hard [Ullman 1975; Lenstra & Rinnooy Kan 1978]. It is an open question whether this remains true for any constant value of $m \ge 3$. The problem is well solved, however, if the precedence relation is of the *tree*-type or if $m = 2$.

$P|tree,p_j=1|C_{max}$ can be solved in $O(n)$ time by Hu's algorithm [Hu 1961; Hsu 1966; Sethi 1976A]. The *level* of a job is defined as the number of jobs in the unique path to the root of the precedence tree. At the beginning of each time unit, as many available jobs as possible are scheduled on the m machines, where highest priority is granted to the jobs with the largest levels. Thus, Hu's algorithm is a nonpreemptive *list scheduling* algorithm, whereby at each step the available job with the highest ranking on a priority list is assigned to the first machine that becomes available. It can also be viewed as a *critical path* scheduling algorithm: the next job chosen is the one which heads the longest current chain of unexecuted jobs.

If the precedence constraints are in the form of an *intree* (each job has at most one successor), then Hu's algorithm can be adapted to minimize L_{max}; in the case of an *outtree* (each job has at most one predecessor), the L_{max} problem turns out to be NP-hard [Brucker *et al.* 1977]. There are some recent algorithmic and NP-hardness results concerning $P|prec,p_j=1|C_{max}$ for precedence constraints other than intrees or outtrees, such as *opposing forests* (combinations of intrees and outtrees), *level graphs*, and so forth [Dolev 1981; Garey *et al.* 1981B; Warmuth 1980].

$P2|prec,p_j=1|C_{max}$ can be solved by various polynomial algorithms [Fujii *et al.* 1969, 1971; Coffman & Graham 1972; Gabow 1980].

In the approach due to Fujii *et al.*, an undirected graph is constructed with vertices corresponding to jobs and edges {j,k} whenever J_j and J_k can be executed simultaneously, *i.e.*, $J_j \not\to J_k$ and $J_k \not\to J_j$. An optimal schedule is then derived from a maximum cardinality matching in the graph. Such a matching can be found in $O(n^3)$ time [Lawler 1976B].

The Coffman-Graham approach leads to an $O(n^2)$ list algorithm. First the jobs are labelled in the following way. Suppose labels 1,...,k have been applied and S is the subset of unlabelled jobs all of whose successors have been labelled. Then a job in S is given the label k+1 if the labels of its immediate succesors are *lexicographically minimal* with respect to all jobs in S. The priority list is given by ordering the jobs according to decreasing labels. It is possible to execute this algorithm in time almost linear in n plus the number of arcs in the precedence graph, if the graph is given in the form of a *transitive reduction* [Sethi 1976B].

Recently, Gabow developed an algorithm which has the same running time, but which does not require such a representation of the precedence graph.

Garey and Johnson present polynomial algorithm for this problem where, in addition, each job becomes available at its *release date* and has to meet a given *deadline*. In this approach, one processes the jobs in order of increasing modified deadlines. This modification requires $O(n^2)$ time if all $r_j = 0$ [Garey & Johnson 1976] and $O(n^3)$ time in the general case [Garey & Johnson 1977].

We note that $P|prec,p_j=1|\Sigma C_j$ is NP-hard [Lenstra & Rinnooy Kan 1978]. Hu's algorithm does not yield an optimal ΣC_j schedule in the case of intrees, but in the case of outtrees critical path scheduling minimizes both C_{max} and ΣC_j [Rosenfeld -]. The Coffman-Graham algorithm also minimizes ΣC_j [Garey -].

As far as approximation algorithms for $P|prec,p_j=1|C_{max}$ are concerned, the NP-hardness proof given in [Lenstra & Rinnooy Kan 1978] implies that, unless $P = NP$, the best possible worst-case bound for a polynomial-time algorithm would be 4/3. The performance of both Hu's algorithm and the Coffman-Graham algorithm has been analyzed.

When critical path (CP) scheduling is used, Chen and Liu [Chen 1975; Chen & Liu 1975] and Kunde [Kunde 1976] show that

$$C_{max}(CP)/C^*_{max} \le \begin{cases} \frac{4}{3} & \text{for } m = 2, \\ 2 - \frac{1}{m-1} & \text{for } m \ge 3. \end{cases} \qquad (\dagger)$$

Lam and Sethi [Lam & Sethi 1977] use the Coffman-Graham (CG) algorithm to generate lists and show that

$$C_{max}(CG)/C^*_{max} \le 2 - \frac{2}{m} \quad (m \ge 2). \qquad (\dagger)$$

If SS denotes the algorithm which schedules as the next job the one having the greatest number of successors then it can be shown [Ibarra & Kim 1976] that

$$C_{max}(SS)/C^*_{max} \le \frac{4}{3} \text{ for } m = 2. \qquad (\dagger)$$

Examples show that this bound does not hold for $m \ge 3$.

Finally, we mention some results for the more general case in which all $p_j \in \{1,k\}$. Both $P2|prec,p_j\in\{1,2\}|C_{max}$ and $P2|prec,p_j\in\{1,2\}|\Sigma C_j$ are NP-hard [Ullman 1975; Lenstra & Rinnooy Kan 1978]. For $P2|prec,p_j\in\{1,k\}|C_{max}$, Goyal [Goyal 1977] proposes a generalized version of the Coffman-Graham algorithm (GCG) and shows that

$$C_{max}(GCG)/C^*_{max} \le \begin{cases} \frac{4}{3} & \text{for } k = 2, \\ \frac{3}{2} - \frac{1}{2k} & \text{for } k \ge 3. \end{cases} \qquad (\dagger)$$

4.3. Nonpreemptive scheduling: general processing times

4.3.1. $P||\Sigma w_jC_j$

The following generalization of the SPT rule for $1||\Sigma C_j$ (see Section 3.3.1) solves $P||\Sigma C_j$ in $O(n \log n)$ time [Conway *et al.* 1967]. Assume $n = km$ (dummy jobs with zero processing times can be added if not) and suppose $p_1 \le \ldots \le p_n$. Assign the m jobs $J_{(j-1)m+1}$, $J_{(j-1)m+2},\ldots,J_{jm}$ to m different machines ($j = 1,\ldots,k$) and execute the k jobs assigned to each machine in SPT order.

With respect to $P||\Sigma w_jC_j$, Eastman, Even and Isaacs [Eastman *et al.* 1964] show that after renumbering the jobs according to nonincreasing ratios w_j/p_j

$$\Sigma w_jC_j(LS) - \frac{1}{2}\Sigma_{j=1}^n w_jp_j \ge \frac{1}{m}(\Sigma_{j=1}^n \Sigma_{k=1}^j w_jp_k - \frac{1}{2}\Sigma_{j=1}^n w_jp_j). \qquad (\dagger)$$

It follows from this inequality that

$$\Sigma w_jC^*_j \ge \frac{m+n}{m(n+1)} \Sigma_{j=1}^n \Sigma_{k=1}^j w_jp_k.$$

In [Elmaghraby & Park 1974; Barnes & Brennan 1977] branch-and-bound algorithms based on this lower bound are developed.

Sahni [Sahni 1976] constructs algorithms A_k (in the same spirit as his approach for $1||\Sigma w_jU_j$ mentioned in Section 3.3.3) with $O(n(n^2k)^{m-1})$ running time for which

$$\Sigma w_jC_j(A_k)/\Sigma w_jC^*_j \le 1 + \frac{1}{k}.$$

For $m = 2$, the running time of A_k can be improved to $O(n^2k)$.

4.3.2. $Q||\Sigma C_j$

The algorithm for solving $P||\Sigma C_j$ given in the previous section can be generalized to the case of uniform machines [Conway *et al.* 1967]. If J_j is the k-th last job executed on M_i, a cost contribution $kp_{ij} = kp_j/q_i$ is incurred. ΣC_j is a weighted sum of the p_j and is minimized by matching the n smallest weights k/q_i in nondecreasing order with the p_j in nonincreasing order. The procedure can be implemented to run in O(n log n) time [Horowitz & Sahni 1976].

4.3.3. $R||\Sigma C_j$

$R||\Sigma C_j$ can be formulated and solved as an m×n transportation problem [Horn 1973; Bruno *et al.* 1974]. Let

$$x_{ijk} = \begin{cases} 1 & \text{if } J_j \text{ is the k-th last job executed on } M_i, \\ 0 & \text{otherwise.} \end{cases}$$

Then the problem is to minimize

$$\Sigma_{i=1}^{m} \Sigma_{j=1}^{n} \Sigma_{k=1}^{n} kp_{ij} x_{ijk}$$

subject to

$$\Sigma_{i=1}^{m} \Sigma_{k=1}^{n} x_{ijk} = 1 \quad \text{for all } j,$$

$$\Sigma_{j=1}^{n} x_{ijk} \le 1 \quad \text{for all } i,k,$$

$$x_{ijk} \ge 0 \quad \text{for all } i,j,k.$$

This problem, like the similar one in Section 4.2.1, can be solved in $O(n^3)$ time.

4.3.4. *Other cases: enumerative optimization methods*

As we noted in Section 4.1, $P2||C_{max}$ and $P2||\Sigma w_jC_j$ are NP-hard. Hence it seems fruitless to attempt to find polynomial-time optimization algorithms for criteria other than ΣC_j. Moreover, $P2|tree|\Sigma C_j$ is known to be NP-hard, both for intrees and outtrees [Sethi 1977]. It follows that it is also not possible to extend the above algorithms to problems with precedence constraints. The only remaining possibility for optimization methods seems to be implicit enumeration.

$R||C_{max}$ can be solved by a branch-and-bound procedure described in [Stern 1976]. The enumerative approach for identical machines in [Bratley *et al.* 1975] allows inclusion of release dates and deadlines as well.

A general dynamic programming technique [Rothkopf 1966; Lawler & Moore 1969] is applicable to parallel machine problems with the C_{max}, L_{max}, Σw_jC_j and Σw_jU_j optimality criteria, and even to problems with the Σw_jT_j criterion in the special case of a common due date.

Let us define $F_j(t_1,\ldots,t_m)$ as the minimum cost of a schedule without idle time for $J_1,\ldots,J_j$ subject to the constraint that the last job on M_i is completed at time t_i, for $i = 1,\ldots,m$. Then, in the case of f_{max} criteria,

$$F_j(t_1,\ldots,t_m) = \min_{1\le i\le m}\{\max\{f_j(t_i),F_{j-1}(t_1,\ldots,t_i-p_{ij},\ldots,t_m)\}\},$$

and in the case of Σf_j criteria,

$$F_j(t_1,\ldots,t_m) = \min_{1\le i\le m}\{f_j(t_i)+F_{j-1}(t_1,\ldots,t_i-p_{ij},\ldots,t_m)\}.$$

In both cases, the initial conditions are

$$F_0(t_1,\ldots,t_m) = \begin{cases} 0 & \text{if } t_i = 0 \text{ for } i = 1,\ldots,m,\\ \infty & \text{otherwise.}\end{cases}$$

Appropriate implementation of these equations yields $O(mnC^{m-1})$ computations for a variety of problems, where C is an upper bound on the completion time of any job in an optimal schedule. Among these problems are $P|r_j|C_{max}$, $Q||L_{max}$ and $Q||\Sigma w_jC_j$. $P||\Sigma w_jU_j$ can be solved in $O(mn(\max_j\{d_j\})^m)$ time.

Still other dynamic programming approaches can be used to solve $P||\Sigma f_j$ and $P||f_{max}$ in $O(m\cdot\min\{3^n,n2^nC\})$ time.

4.3.5. *Other cases: approximation algorithms*

4.3.5.1. $P||C_{max}$

By far the most studied scheduling model from the viewpoint of approximation algorithms is $P||C_{max}$. We refer to [Garey *et al*. 1978] for an easily readable introduction into the techniques involved in many of the "performance guarantees" mentioned below.

Perhaps the earliest and simplest result on the worst-case performance of list scheduling is given in [Graham 1966]:

$$C_{max}(LS)/C^*_{max} \le 2 - \frac{1}{m}. \qquad (\dagger)$$

If the jobs are selected in LPT order, then the bound can be considerably improved, as is shown in [Graham 1969]:

$$C_{max}(LPT)/C^*_{max} \le \frac{4}{3} - \frac{1}{3m}. \qquad (\dagger)$$

A somewhat better algorithm, called *multifit* (MF) and based on a completely different principle, is given in [Coffman *et al*. 1978]. The idea behind MF is to find (by binary search) the smallest "capacity" a set of m "bins" can have and still accommodate all jobs when the jobs are taken in order of nonincreasing p_j and each job is placed into the first bin into which it will fit. The set of jobs in the i-th bin will be processed by M_i. If k packing attempts are made, the algorithm (denoted by MF_k) runs in time $O(n \log n + knm)$ and satisfies

$$C_{max}(MF_k)/C^*_{max} \le 1.22 + 2^{-k}.$$

We note that if the jobs are not ordered by decreasing p_j then all that can be guaranteed by this method is

$$C_{max}(MF)/C^*_{max} \le 2 - \frac{2}{m+1}. \qquad (\dagger)$$

The following algorithm Z_k was introduced in [Graham 1969]: schedule the k largest jobs optimally, then list schedule the remaining jobs arbitrarily. It is shown in [Graham 1969] that

$$C_{max}(Z_k)/C^*_{max} \le 1 + (1 - \frac{1}{m})/(1 + \left\lfloor\frac{k}{m}\right\rfloor)$$

and that when m divides k, this is best possible. Thus, we can make the bound as close to 1 as desired by taking k sufficiently large. Unfortunately, the best bound on the running time is $O(n^{km})$.

A very interesting algorithm for $P||C_{max}$ is given by Sahni [Sahni 1976]. He presents algorithms A_k with $O(n(n^2k)^{m-1})$ running time which satisfy

$$C_{max}(A_k)/C^*_{max} \le 1 + \frac{1}{k}.$$

For m = 2, algorithm A_2 can be improved to run in time $O(n^2k)$. As in the cases of $1||\Sigma w_jU_j$ (Section 3.3.3) and $P||\Sigma w_jC_j$ (Section 4.3.1), the algorithms A_k are based on a clever combination of dynamic programming and rounding and are beyond the scope of the present discussion.

Several bounds are available which take into account the processing times of the jobs. In [Graham 1969] it is shown that

$$C_{max}(LS)/C^*_{max} \le 1 + (m-1)\max_j\{p_j\}/\Sigma_j p_j.$$

For the case of LPT, Ibarra and Kim [Ibarra & Kim 1977] prove that

$$C_{max}(LPT)/C^*_{max} \le 1 + \frac{2(m-1)}{n} \quad \text{for } n \ge 2(m-1)\max_j\{p_j\}/\min_j\{p_j\}.$$

4.3.5.2. $Q||C_{max}$

In the literature on approximation algorithms for scheduling problems, it is usually assumed that *unforced idleness* (UI) of machines is *not* allowed, *i.e.*, a machine cannot be idle when jobs are available. In the case of identical machines, UI need not occur in an optimal schedule if there are no precedence constraints or if all $p_j = 1$. Allowing UI may yield better solutions, however, in the cases which are to be discussed in Sections 4.3.5.2-5. The optimal value of C_{max} under the restriction of *no* UI will be denoted by

C^*_{max}, the optimum if UI *is* allowed by $C^*_{max}(UI)$.

Liu and Liu [Liu & Liu 1974A, 1974B, 1974C] study numerous questions dealing with uniform machines. They define the algorithm A_k as follows: schedule the k longest jobs first, resulting in a completion time of $C_k(A_k)$, and schedule the remaining tasks for a total completion time of $C_{max}(A_k)$. If $C_{max}(A_k) > C_k(A_k)$, then

$$C_{max}(A_k)/C^*_{max}(UI) \le 1 + \frac{1}{Q} - \frac{1}{Q\Sigma_i q_i}$$

where all $q_i \ge 1$ and

$$Q = \max\{\min_j\{\lceil \frac{k+1}{\Sigma_i \lceil q_i \rceil} \rceil \cdot \frac{\lceil q_j \rceil}{q_j} - \frac{1}{\lceil q_j \rceil q_j}\}, \frac{k+1}{\Sigma_i q_i}\}.$$

This is best possible when the q_i are integers and $\Sigma_i q_i$ divides k.

Gonzalez, Ibarra and Sahni [Gonzalez *et al.* 1977] consider the following generalization LPT' of LPT: assign each job, in order of nonincreasing processing time, to the machine on which it will be completed soonest. Thus, unforced idleness may occur in the schedule. They show

$$C_{max}(LPT')/C^*_{max} \le 2 - \frac{2}{m+1}.$$

Also, examples are given for which $C_{max}(LPT')/C^*_{max}$ approaches 3/2 as m tends to infinity.

4.3.5.3. $R||C_{max}$

Very little is known about approximation algorithms for this model. Ibarra and Kim [Ibarra & Kim 1977] consider five algorithms, typical of which is to schedule J_j on the machine that executes it fastest, *i.e.*, on an M_i with minimum p_{ij}. For all five algorithms A they prove

$$C_{max}(A)/C^*_{max} \le m$$

with equality possible for four of the five. For the special case $R2||C_{max}$, they give an O(n log n) algorithm G such that

$$C_{max}(G)/C^*_{max} \le \frac{1+\sqrt{5}}{2}. \qquad (\dagger)$$

Potts [Potts -] proposes an $R||C_{max}$ algorithm based on linear programming (LP), the running time of which is polynomial only for fixed m. He proves

$$C_{max}(LP)/C^*_{max} \le 2. \qquad (\dagger)$$

4.3.5.4. $P|prec|C_{max}$

In the presence of precedence constraints it is somewhat unexpected [Graham 1966] that the 2-(1/m) bound still holds, *i.e.*,

$$C_{max}(LS)/C^*_{max} \le 2 - \frac{1}{m}.$$

Now, consider executing the set of jobs *twice*: the first time using processing times p_j, precedence constraints, m machines and an arbitrary priority list, the second time using processing times $p'_j \le p_j$, weakened precedence constraints, m' machines and a (possibly different) priority list. Then [Graham 1966]

$$C'_{max}(LS)/C_{max}(LS) \le 1 + \frac{m-1}{m'}. \quad (\dagger)$$

Even when critical path (CP) scheduling is used, examples exist [Graham -] for which

$$C_{max}(CP)/C^*_{max} = 2 - \frac{1}{m}.$$

It is known [Graham -] that unforced idleness (UI) has the following behavior:

$$C_{max}(LS)/C^*_{max}(UI) \le 2 - \frac{1}{m}. \quad (\dagger)$$

Let $C^*_{max}(pmtn)$ denote the optimal value of C_{max} if preemption is allowed. As in the case of UI, it is known [Graham -] that

$$C_{max}(LS)/C^*_{max}(pmtn) \le 2 - \frac{1}{m}. \quad (\dagger)$$

Liu [Liu 1972] shows that

$$C^*_{max}(UI)/C^*_{max}(pmtn) \le 2 - \frac{2}{m+1}. \quad (\dagger)$$

4.3.5.5. $Q|prec|C_{max}$

Liu and Liu [Liu & Liu 1974B] also consider the presence of precedence constraints in the case of uniform machines. They show that, when unforced idleness or preemption is allowed,

$$C_{max}(LS)/C^*_{max}(UI) \le 1+\max_i\{q_i\}/\min_i\{q_i\}-\max_i\{q_i\}/\Sigma_i q_i, \quad (\dagger)$$

$$C_{max}(LS)/C^*_{max}(pmtn) \le 1+\max_i\{q_i\}/\min_i\{q_i\}-\max_i\{q_i\}/\Sigma_i q_i. \quad (\dagger)$$

When all $q_i = 1$ this reduces to the earlier 2-(1/m) bounds for these questions on identical machines.

Suppose that the jobs are executed twice: the first time using m machines of speeds $q_1,\ldots,q_m$, the second time using m' machines of speeds $q'_1,\ldots,q'_{m'}$. Then

$$C'_{max}(LS)/C^*_{max}(UI) \le \max_i\{q_i\}/\min_i\{q'_i\}+(\Sigma_i q_i-\max_i\{q_i\})/\Sigma_i q'_i. \quad (\dagger)$$

Jaffe [Jaffe 1979] develops an algorithm LSi, that uses list scheduling on the fastest i machines for an appropriately chosen value of i. It is shown that

$$C_{max}(LSi)/C^*_{max}(UI) \le \sqrt{m} + O(m^{1/4})$$

and examples are given for which the bound $\sqrt{m-1}$ is approached arbitrarily closely.

4.4. Preemptive scheduling

4.4.1. $P|pmtn|\Sigma C_j$

A theorem of McNaughton [McNaughton 1959] states that for $P|pmtn|\Sigma w_jC_j$ there is no schedule with a finite number of preemptions which yields a smaller criterion value than an optimal nonpreemptive schedule. The finiteness restriction can be removed by appropriate application of results from open shop theory. It therefore follows that the procedure of Section 4.3.1 can be applied to solve $P|pmtn|\Sigma C_j$. It also follows that $P2|pmtn|\Sigma w_jC_j$ is NP-hard, since $P2||\Sigma w_jC_j$ is known to be NP-hard.

4.4.2. $Q|pmtn|\Sigma C_j$

McNaughton's theorem does not apply to uniform machines, as can be demonstrated by a simple counterexample. There is, however, a polynomial algorithm for $Q|pmtn|\Sigma C_j$.

One can show that there exists an optimal preemptive schedule in which $C_j \le C_k$ if $p_j < p_k$ [Lawler & Labetoulle 1978]. Accordingly, first place the jobs in SPT order. Then obtain an optimal schedule by preemptively scheduling each successive job in the available time on the m machines so as to minimize its completion time [Gonzalez 1977]. This procedure can be implemented in $O(n \log n + mn)$ time and yields an optimal schedule with no more than $(m-1)(n-\frac{1}{2}m)$ preemptions. It has been extended to cover the case in which ΣC_j is minimized subject to a common deadline for all jobs [Gonzalez 1977].

4.4.3. $R|pmtn|\Sigma C_j$

Very little is known about $R|pmtn|\Sigma C_j$. This remains one of the more vexing questions in the area of preemptive scheduling.

4.4.4. $P|pmtn,prec|C_{max}$

An obvious lower bound on the value of an optimal $P|pmtn|C_{max}$ schedule is given by

$$\max\{\max_j\{p_j\}, \frac{1}{m}\Sigma_j p_j\}.$$

A schedule meeting this bound can be constructed in $O(n)$ time [McNaughton 1959]: just fill the machines successively, scheduling the jobs in any order and splitting a job whenever the above time bound is met. The number of preemptions occurring in this schedule is at most m-1. It is possible to design a class of problems for which this number is minimal, but the general problem of minimizing the number of preemptions is easily seen to be NP-hard.

In the case of precedence constraints, $P|pmtn,prec,p_j=1|C_{max}$ turns out to be NP-hard [Ullman 1976], but $P|pmtn,tree|C_{max}$ and $P2|pmtn,prec|C_{max}$ can be solved by a polynomial-time algorithm due to Muntz and Coffman [Muntz & Coffman 1969, 1970]. This is as follows.

Define $\ell_j(t)$ to be the level of a J_j wholly or partly unexecuted at time t. Suppose that at time t m' machines are available and that n' jobs are currently maximizing $\ell_j(t)$. If $m' < n'$, we assign m'/n' machines to each of the n' jobs, which implies that each of these jobs will be executed at speed m'/n'. If $m' \geq n'$, we assign one machine to each job, consider the jobs at the next highest level, and repeat. The machines are reassigned whenever a job is completed or threatens to be processed at a higher speed than another one at a currently higher level. Between each pair of successive reassignment points, jobs are finally rescheduled by means of McNaughton's algorithm for $P|pmtn|C_{max}$. The algorithm requires $O(n^2)$ time [Gonzalez & Johnson 1980].

Gonzalez and Johnson [Gonzalez & Johnson 1980] have developed a totally different algorithm that solves $P|pmtn,tree|C_{max}$ by starting at the roots rather than the leaves of the tree and determines priority by considering the total remaining processing time in subtrees rather than by looking at critical paths. The algorithm runs in O(n log m) time and introduces at most n-2 preemptions into the resulting optimal schedule.

Lam and Sethi [Lam & Sethi 1977], much in the same spirit as their work mentioned in Section 4.2.2, analyze the performance of the Muntz-Coffman (MC) algorithm for $P|pmtn,prec|C_{max}$. They show

$$C_{max}(MC)/C^*_{max} \leq 2 - \frac{2}{m} \quad (m \geq 2). \tag{†}$$

4.4.5. $Q|pmtn,prec|C_{max}$

Horvath, Lam and Sethi [Horvath *et al.* 1977] adapt the Muntz-Coffman algorithm to solve $Q|pmtn|C_{max}$ and $Q2|pmtn,prec|C_{max}$ in $O(mn^2)$ time. This results is an optimal schedule with no more than $(m-1)n^2$ preemptions.

A computationally more efficient algorithm due to Gonzalez and Sahni [Gonzalez & Sahni 1978B] solves $Q|pmtn|C_{max}$ in O(n) time, if the jobs are given in order of nonincreasing p_j and the machines in order of nonincreasing q_i. This procedure yields an optimal schedule with no more than 2(m-1) preemptions, which can be shown to be a tight bound.

The optimal value of C_{max} is given by

$$\max\{\max_{1\leq k\leq m-1}\{\Sigma^k_{j=1}\ p_j/\Sigma^k_{i=1}\ q_i\},\ \Sigma^n_{j=1}p_j/\Sigma^m_{i=1}\ q_i\},$$

where $p_1 \geq \ldots \geq p_n$ and $q_1 \geq \ldots \geq q_m$. This result generalizes the one given in Section 4.4.4.

The Gonzalez-Johnson algorithm for $P|pmtn,tree|C_{max}$ mentioned in the previous section can be adapted to the case $Q2|pmtn,tree|C_{max}$.

Jaffe [Jaffe 1980] studies the performance of *maximal usage schedules* (MUS) for $Q|pmtn,prec|C_{max}$, *i.e.*, schedules without unforced idleness in which at any time the jobs being processed are assigned to the fastest machines. It is shown that

$$C_{max}(MUS)/C^*_{max} \leq \sqrt{m} + \frac{1}{2}$$

and examples are given for which the bound $\sqrt{m-1}$ is approached arbitrarily closely.

4.4.6. $R|pmtn|C_{max}$

Many preemptive scheduling problems involving independent jobs on unrelated machines can be formulated as linear programming problems [Lawler & Labetoulle 1978]. For instance, solving $R|pmtn|C_{max}$ is equivalent to minimizing

$$C_{max}$$

subject to

$$\Sigma_{i=1}^{m} x_{ij}/p_{ij} = 1 \quad (j = 1,\ldots,n),$$

$$\Sigma_{i=1}^{m} x_{ij} \leq C_{max} \quad (j = 1,\ldots,n),$$

$$\Sigma_{j=1}^{n} x_{ij} \leq C_{max} \quad (i = 1,\ldots,m),$$

$$x_{ij} \geq 0 \quad (i = 1,\ldots,m,\ j = 1,\ldots,n).$$

In this formulation x_{ij} represents the total time spent by J_j on M_i. The linear program can be solved in polynomial time [Khachiyan 1979], and a feasibie schedule can be constructed in polynomial time by applying the algorithm for $O|pmtn|C_{max}$, discussed in Section 5.2.2.

This procedure can be modified to yield an optimal schedule with no more than about $7m^2/2$ preemptions. It remains an open question as to whether $O(m^2)$ preemptions are necessary for an optimal preemptive schedule.

For fixed m, it seems to be possible to solve the linear program in linear time. Certainly, the special case $R2|pmtn|C_{max}$ can be solved in O(n) time [Gonzalez *et al*. 1981].

We note that a similar linear programming formulation can be given for $R|pmtn,r_j|L_{max}$ [Lawler & Labetoulle 1978].

4.4.7. $P|pmtn,prec,r_j|L_{max}$

$P|pmtn|L_{max}$ and $P|pmtn,r_j|C_{max}$ can be solved by a procedure due to Horn [Horn 1974]. The $O(n^2)$ running time has been reduced to $O(mn)$ [Gonzalez & Johnson 1980].

More generally, the existence of a feasible preemptive schedule with given release dates and deadlines can be tested by means of a network flow model in $O(n^3)$ time [Horn 1974]. A binary search can then be conducted on the optimal value of L_{max}, with each trial value of L_{max} inducing deadlines which are checked for feasibility by means of the network computation. It can be shown that this yields an $O(n^3 \min\{n^2, \log n + \log \max_j\{p_j\}\})$ algorithm [Labetoulle *et al.* 1979].

In the case of precedence constraints, the algorithms of Brucker, Garey and Johnson for $P|intree,p_j=1|L_{max}$, $P2|prec,p_j=1|L_{max}$ and $P2|prec,r_j,p_j=1|L_{max}$ (see Section 4.2.2) have preemptive counterparts. *E.g.*, $P|pmtn,intree|L_{max}$ can be solved in $O(n^2)$ time [Lawler 1980]; see also the next section.

4.4.8. $Q|pmtn,prec,r_j|L_{max}$

In the case of uniform machines, the existence of a feasible preemptive schedule with given release dates and a common deadline can be tested in $O(n \log n + mn)$ time; the algorithm generates $O(mn)$ preemptions in the worst case [Sahni & Cho 1980]. More generally, $Q|pmtn,r_j|C_{max}$ and, by symmetry, $Q|pmtn|L_{max}$ are solvable in $O(n \log n + mn)$ time; the number of preemptions generated is $O(mn)$ [Sahni & Cho 1979B; Labetoulle *et al.* 1979].

The first feasibility test mentioned in the previous section has been adapted to the case of two uniform machines [Bruno & Gonzalez 1976] and extended to a polynomial-time algorithm for $Q2|pmtn,r_j|L_{max}$ [Labetoulle *et al.* 1979].

Most recently, Martel has found a polynomial-time algorithm for $Q|pmtn,r_j|L_{max}$ [Martel 1981]. This method is in fact a special case of a more general algorithm for computing maximal *polymatroidal* network flows [Lawler & Martel 1980].

In the case of precedence constraints, $Q2|pmtn,prec|L_{max}$ and $Q2|pmtn,prec,r_j|L_{max}$ can be solved in $O(n^2)$ and $O(n^6)$ time, respectively [Lawler 1980].

4.4.9. $Q|pmtn|\Sigma w_jU_j$

Binary NP-hardness has been established for $1|pmtn|\Sigma w_jU_j$ (see Section 3.3.3) and $P|pmtn|\Sigma U_j$ [Lawler 1981]. For any *fixed* number of uniform machines, $Qm|pmtn|\Sigma w_jU_j$ can be solved in pseudopolynomial time: $O(n^2(\Sigma w_j)^2)$ if $m = 2$ and $O(n^{3m-5}(\Sigma w_j)^2)$ if $m \geq 3$ [Lawler 1981] Hence, $Qm|pmtn|\Sigma U_j$ is solvable in strictly polynomial time.

5. OPEN SHOP, FLOW SHOP AND JOB SHOP PROBLEMS

5.1. Introduction

We now pass on to problems in which each job requires execution on more than one machine. Recall from Section 2.3 that in an *open shop* (denoted by O) the order in which a job passes through the machines is immaterial, whereas in a *flow shop* (F) each job has the same machine ordering $(M_1,\ldots,M_m)$ and in a *job shop* (J) possibly different machine orderings are specified for the jobs. We survey these problem classes in Sections 5.2, 5.3 and 5.4, respectively.

We shall be dealing exclusively with the C_{max} criterion. Other optimality criteria lead usually to NP-hard problems, such as:

- $O2||L_{max}$ [Lawler *et al.* 1981],
- $O||\Sigma C_j$, $O|pmtn|\Sigma C_j$ [Gonzalez 1979B],
- $F2||L_{max}$ [Lenstra *et al.* 1977], $F2|pmtn|L_{max}$ [Cho & Sahni 1978],
- $F2||\Sigma C_j$ [Garey *et al.* 1976], $F3|pmtn|\Sigma C_j$, $J2|pmtn|\Sigma C_j$ [Lenstra -].

Notable exceptions are $O|pmtn,r_j|L_{max}$, which is solvable in polynomial time by linear programming [Cho & Sahni 1978], and $O2||\Sigma C_j$ and $F2|pmtn|\Sigma C_j$, which are open.

5.2. Open shop scheduling

5.2.1. *Nonpreemptive case*

The case $O2||C_{max}$ admits of an O(n) algorithm [Gonzalez & Sahni 1976]. A simplified exposition is given below.

For convenience, let $a_j = p_{1j}$, $b_j = p_{2j}$. Let $A = \{J_j|a_j \geq b_j\}$, $B = \{J_j|a_j < b_j\}$. Now choose J_r and J_ℓ to be any two distinct jobs (whether in A or B) such that

$$a_r \geq \max_{J_j\in A}\{b_j\}, \quad b_\ell \geq \max_{J_j\in B}\{a_j\}.$$

Let $A' = A-\{J_r,J_\ell\}$, $B' = B-\{J_r,J_\ell\}$. We assert that it is possible to form feasible schedules for $B'\cup\{J_\ell\}$ and for $A'\cup\{J_r\}$ as indicated in Figure 2(a), the jobs in A' and B' being ordered arbitrarily. In each of these separate schedules, there is no idle time on either machine, from the start of the first job on that machine to the completion of the last job on that machine.

Let $T_1 = \Sigma_j a_j$, $T_2 = \Sigma_j b_j$. Suppose $T_1-a_\ell \geq T_2-b_r$ (the case $T_1-a_\ell < T_2-b_r$ being symmetric). We then combine the two schedules as shown in Figure 2(b), pushing the jobs in $B'\cup\{J_\ell\}$ on M_2 to the right. Again, there is no idle time on either machine, from the start of the first job to the completion of the last job.

We finally propose to move the processing of J_r on M_2 to the first position on that machine. There are two cases to consider.

(1) $a_r \le T_2 - b_r$. The resulting schedule is as in Figure 2(c). The length of the schedule is $\max\{T_1, T_2\}$.

(2) $a_r > T_2 - b_r$. The resulting schedule is as in Figure 2(d). The length of the schedule is $\max\{T_1, a_r + b_r\}$.

For any feasible schedule we obviously have that

$$C_{max} \ge \max\{T_1, T_2, \max_j\{a_j + b_j\}\}.$$

Since, in all cases, we have met this lower bound, it follows that the schedules constructed are optimal.

There is little hope of finding polynomial-time algorithms for nonpreemptive open shop problems more complicated than $O2||C_{max}$. $O3||C_{max}$ is binary NP-hard [Gonzalez & Sahni 1976] and $O2|r_j|C_{max}$, $O2|tree|C_{max}$ and $O||C_{max}$ are unary NP-hard [Lawler *et al.* 1981; Lenstra -].

The special case of $O3||C_{max}$ is which $\max_j\{p_{hj}\} \le \min_j\{p_{ij}\}$ for some pair (M_h, M_i) $(h \ne i)$ is likely to be solvable in polynomial time [Adiri & Hefetz 1980].

5.2.2. *Preemptive case*

The result on $O2||C_{max}$ presented in the previous section shows that there is no advantage to preemption for m = 2, and hence $O2|pmtn|C_{max}$ can be solved in O(n) time. More generally, $O|pmtn|C_{max}$ is solvable in polynomial time as well [Gonzalez & Sahni 1976; Lawler & Labetoulle 1978; Gonzalez 1979A]. We had already occasion to refer to this result in Section 4.4.6.

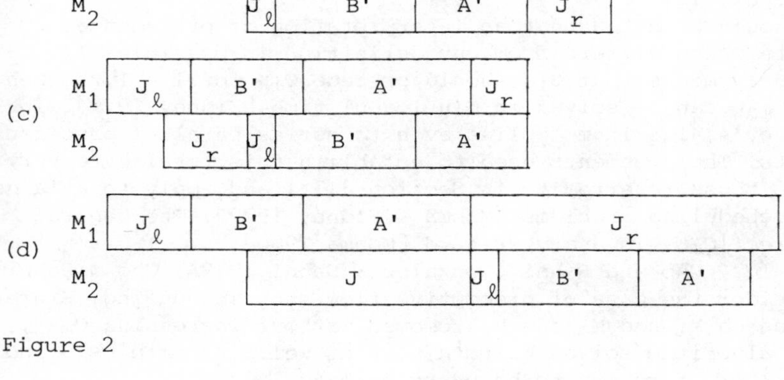

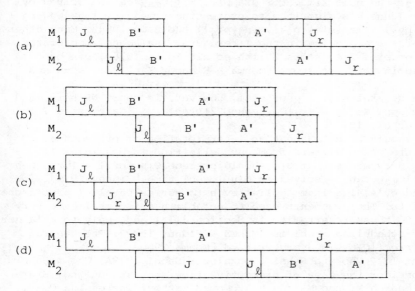

Figure 2

If release dates are introduced, $O2|pmtn,r_j|C_{max}$ is still solvable in O(n) time [Lawler *et al*. 1981]. As mentioned in Section 5.1, even $O|pmtn,r_j|L_{max}$ is well solved [Cho & Sahni 1978].

5.3. Flow shop scheduling

5.3.1. $F2|\beta|C_{max}$, $F3|\beta|C_{max}$

A fundamental algorithm for solving $F2||C_{max}$ is due to Johnson [Johnson 1954]. He shows that there exists an optimal schedule in which J_j precedes J_k if $\min\{p_{1j},p_{2k}\} \le \min\{p_{2j},p_{1k}\}$. It follows that the problem can be solved in O(n log n) time: arrange first the jobs with $p_{1j} \le p_{2j}$ in order of nondecreasing p_{1j} and subsequently the remaining jobs in order of nonincreasing p_{2j}.

Some special cases involve *start lags* ℓ_{1j} and *stop lags* ℓ_{2j} for J_j, that represent minimum time intervals between starting times on M_1 and M_2 and between completion times on M_1 and M_2, respectively [Mitten 1958; Johnson 1958; Nabeshima 1963; Szwarc 1968]. Defining $\ell_j = \min\{\ell_{1j}-p_{1j},\ell_{2j}-p_{2j}\}$ and applying Johnson's algorithm to processing times $(p_{1j}+\ell_j,p_{2j}+\ell_j)$ will produce an optimal *permutation schedule*, *i.e.*, one with identical processing orders on all machines [Rinnooy Kan 1976]. If we drop the latter restriction, the problem is unary NP-hard [Lenstra -].

A fair amount of effort has been devoted to identifying special flow shop problems that can still be solved in polynomial time. The crucial notion here is that of a *nonbottleneck machine*, that can effectively be treated as though it can process any number of jobs at the same time. For example, $F3||C_{max}$ can be solved by applying Johnson's algorithm to processing times $(p_{1j}+p_{2j},p_{2j}+p_{3j})$ if $\max_j\{p_{2j}\} \le \max\{\min_j\{p_{1j}\},\min_j\{p_{3j}\}\}$ [Johnson 1954]. Many other special cases appearing in the literature, including some of the work on *ordered flow shops* [Smith *et al*. 1975, 1976], can be discussed in this framework [Monma & Rinnooy Kan 1981; Achuthan 1980].

The general $F3||C_{max}$ problem, however, is unary NP-hard, and the same applies to $F2|r_j|C_{max}$ and $F2|tree|C_{max}$ [Garey *et al*. 1976; Lenstra *et al*. 1977].

It should be noted that an interpretation of precedence constraints which differs from our definition is possible. If $J_j \to' J_k$ only means that O_{ij} should precede O_{ik} for i = 1,2, then $F2|tree'|C_{max}$ can be solved in O(n log n) time [Sidney 1979]. In fact, Sidney's algorithm applies even to series-parallel precedence constraints. The arguments used to establish this result are very similar to those referred to in Section 3.3.1 and apply to a larger class of scheduling problems [Monma & Sidney 1979]. The general case $F2|prec'|C_{max}$ is unary NP-hard [Monma 1980].

Gonzalez, Cho and Sahni [Gonzalez & Sahni 1978A; Cho & Sahni 1978] consider the case of preemptive flow shop scheduling. Since preemptions on M_1 and M_m can be removed without increasing C_{max}, Johnson's algorithm solves $F2|pmtn|C_{max}$ as well. $F3|pmtn|C_{max}$ and $F2|pmtn,r_j|C_{max}$ turn out to be unary NP-hard.

5.3.2. $F||C_{max}$

As a general result, we note that there exists an optimal flow shop schedule with the same processing order on M_1 and M_2 and the same processing order on M_{m-1} and M_m [Conway *et al.* 1967]. It is, however, not difficult to construct a 4-machine example in which a job "passes" another one between M_2 and M_3 in the optimal schedule. Nevertheless, it has become tradition in the literature to assume identical processing orders on all machines, so that in effect only the best permutation schedule has to be determined.

Most research in this area has focused on enumerative methods. The usual enumeration scheme is to assign jobs to the ℓ-th position in the schedule at the ℓ-th level of the search tree. Thus, at a node at that level a partial schedule $(J_{\sigma(1)},\ldots,J_{\sigma(\ell)})$ has been formed and the jobs with index set $S = \{1,\ldots,n\} - \{\sigma(1),\ldots,\sigma(\ell)\}$ are candidates for the $(\ell+1)$-st position. One then needs to find a lower bound on the value of all possible completions of the partial schedule. It turns out that almost all lower bounds developed so far are generated by the following bounding scheme [Lageweg *et al.* 1978].

Let us relax the capacity constraint that each machine can process at most one job at a time, for all machines but at most two, say, M_u and M_v $(1 \le u \le v \le m)$. We then obtain a problem of scheduling $\{J_j | j \in S\}$ on five machines $N_{*u},M_u,N_{uv},M_v,N_{v*}$ in that order, which is specified as follows. Let $C(\sigma,i)$ denote the completion time of $J_{\sigma(\ell)}$ on M_i. N_{*u}, N_{uv} and N_{v*} have infinite capacity; the processing times on these machines are defined by

$$q_{*uj} = \max_{1\le i\le u}\{C(\sigma,i) + \Sigma_{h=i}^{u-1} p_{hj}\},$$

$$q_{uvj} = \Sigma_{h=u+1}^{v-1} p_{hj},$$

$$q_{v*j} = \Sigma_{h=v+1}^{m} p_{hj}.$$

M_u and M_v have capacity 1 and processing times p_{uj} and p_{vj}, respectively. Note that we can interpret N_{*u} as yielding release dates q_{*uj} on M_u and N_{v*} as setting due dates $-q_{v*j}$ on M_v, with respect to which L_{max} is to be minimized.

Any of the machines N_{*u},N_{uv},N_{v*} can be removed from this problem by underestimating its contribution to the lower bound to be the minimum processing time on that machine. Valid lower bounds are obtained by adding these contributions to the optimal solution value of the remaining problem.

For the case that u = v, removing N_{*u} and N_{u*} from the problem produces the *machine-based bound* used in [Ignall & Schrage 1965; McMahon 1971]:

$$\max_{1\le u\le m}\{\min_{j\in S}\{q_{*uj}\} + \Sigma_{j\in S}\, p_{uj} + \min_{j\in S}\{q_{u*j}\}\}.$$

Removing only N_{*u} results in a $1||L_{max}$ problem on M_u, which can be solved by Jackson's rule (Section 3.2) and provides a slightly stronger bound.

If $u \neq v$, removal of N_{*u}, N_{uv} and N_{v*} yields an $F2||C_{max}$ problem, to be solved by Johnson's algorithm (Section 5.3.1). As pointed out in that section, solution in polynomial time remains possible if N_{uv} is taken fully into account; the resulting bound dominates the *job-based bound* proposed in [McMahon 1971] and is the best one currently available.

All other variations on this theme (*e.g.*, taking $u = v$ and considering the resulting $1|r_j|L_{max}$ problem) would involve the solution of NP-hard problems. The development of fast algorithms or strong lower bounds for these problems thus emerges as a possibly fruitful research area.

An alternative and somewhat more efficient enumeration scheme [Potts 1980A] builds up a schedule from the front and from the back at the same time. The adaptation of the above bounding scheme to this approach is straightforward.

The computational performance of branch-and-bound algorithms for $F||C_{max}$ might be improved by the use of *elimination criteria*. Particular attention has been paid to conditions under which all completions of $(J_{\sigma(1)},\ldots,J_{\sigma(\ell)},J_j)$ can be eliminated because a schedule at least as good exists among the completions of $(J_{\sigma(1)},\ldots,J_{\sigma(\ell)},J_k,J_j)$. If all information obtainable from the processing times of the other jobs is disregarded, the strongest condition under which this is allowed is as follows. Defining $\Delta_i = C(\sigma kj,i)-C(\sigma j,i)$, we can exclude J_j for the $(\ell+1)$-st position if

$$\max\{\Delta_{i-1},\Delta_i\} \leq p_{ij} \quad (i = 2,\ldots,m)$$

[McMahon 1969; Szwarc 1971, 1973]. Inclusion of these and similar dominance rules can be very helpful from a computational point of view, depending on the lower bound used [Lageweg *et al*. 1978]. It may be worthwhile to consider further extensions that, for instance, involve the processing times of the unscheduled jobs [Gupta & Reddi 1978; Szwarc 1978].

Not much has been done in the way of worst-case analysis of approximation algorithms for $F||C_{max}$. It is not hard to see that for any active schedule (AS)

$$C_{max}(AS)/C^*_{max} \leq \max_{i,j}\{p_{ij}\}/\min_{i,j}\{p_{ij}\}. \qquad (\dagger)$$

Gonzalez and Sahni [Gonzalez & Sahni 1978A] show that

$$C_{max}(AS)/C^*_{max} \leq m. \qquad (\dagger)$$

This bound is tight even for LPT schedules, in which the jobs are ordered according to nonincreasing sums of processing times. They also give an O(mn log n) algorithm H based on Johnson's algorithm with

$$C_{max}(H)/C^*_{max} \le \lceil \frac{m}{2} \rceil.$$

It thus appears that, in general, the obvious algorithms can deviate quite substantially from the optimum.

5.3.3. *No wait in process*

In a variation on the flow shop problem, each job, once started, has to be processed without interruption until it is completed. This *no wait* constraint may arise out of certain job characteristics (*e.g.*, the "hot ingot" problem in which metal has to be processed at continuously high temperature) or out of the unavailability of intermediate storage in between machines.

The resulting $F|no\ wait|C_{max}$ problem can be formulated as a *traveling salesman* problem with cities $0,1,\ldots,n$ and intercity distances

$$c_{jk} = \max_{1\le i\le m}\{\Sigma_{h=1}^{i} p_{hj} - \Sigma_{h=1}^{i-1} p_{hk}\} \quad (j,k = 0,1,\ldots,n),$$

where $p_{i0} = 0$ $(i = 1,\ldots,m)$ [Piehler 1960; Reddi & Ramamoorthy 1972; Wismer 1972].

For the case $F2|no\ wait|C_{max}$, the traveling salesman problem assumes a special structure and the results from [Gilmore & Gomory 1964] can be applied to yield an $O(n^2)$ algorithm [Reddi & Ramamoorthy 1972]. $F4|no\ wait|C_{max}$ is unary NP-hard [Papadimitriou & Kanellakis 1980], and the same is true for $O2|no\ wait|C_{max}$ and $J2|no\ wait|C_{max}$ [Sahni & Cho 1979A]. In spite of a challenging prize awarded for its solution [Lenstra *et al.* 1977], $F3|no\ wait|C_{max}$ is still open.

The *no wait* constraint may lengthen the optimal flow shop schedule considerably. It can be shown [Lenstra -] that

$$C^*_{max}(no\ wait)/C^*_{max} < m \quad \text{for } m \ge 2. \qquad (\dagger)$$

5.4. Job shop scheduling

5.4.1. $J2|\beta|C_{max}$, $J3|\beta|C_{max}$

A simple extension of Johnson's algorithm for $F2||C_{max}$ allows solution of $J2|m_j\le 2|C_{max}$ in $O(n \log n)$ time [Jackson 1956]. Let J_i be the set of jobs with operations on M_i only $(i = 1,2)$ and J_{hi} the set of jobs that go from M_h to M_i $(hi = 12,21)$. Order the latter two sets by means of Johnson's algorithm and the former two sets arbitrarily. One then obtains an optimal schedule by executing the jobs on M_1 in the order (J_{12},J_1,J_{21}) and on M_2 in the order (J_{21},J_2,J_{12}).

Another special case, $J2|p_{ij}=1|C_{max}$, is solvable in $O(n \log n)$ time as well [Hefetz & Adiri 1979].

This, however, is probably as far as we can get. $J2|m_j\le 3|C_{max}$ and $J3|m_j\le 2|C_{max}$ are binary NP-hard [Lenstra *et al.* 1977; Gonzalez

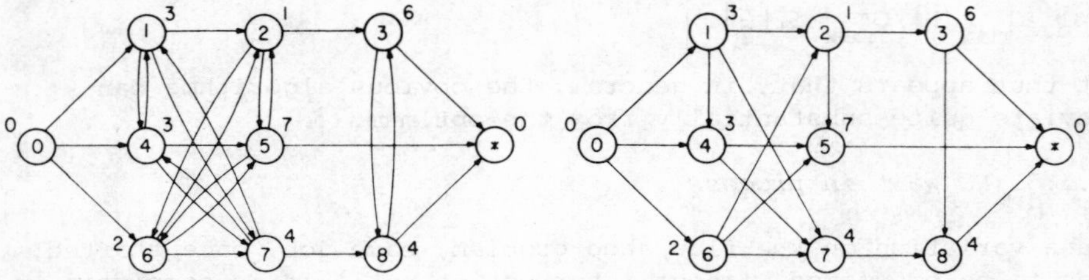

Job shop problem, represented as a disjunctive graph — Job shop schedule, represented as an acyclic directed graph

Figure 3

& Sahni 1978A], $J2|p_{ij}\epsilon\{1,2\}|C_{max}$ and $J3|p_{ij}=1|C_{max}$ are unary NP-hard [Lenstra & Rinnooy Kan 1979], and these results are still true if preemption is allowed.

5.4.2. $J||C_{max}$

The general job shop problem is extremely hard to solve optimally. An indication of this is given by the fact that a 10-job 10-machine problem, formulated in 1963 [Muth & Thompson 1963], still has not been solved.

A convenient problem representation is provided by the *disjunctive graph* model, introduced by Roy and Sussmann [Roy & Sussmann 1964]. Assume each operation O_{ij} being renumbered as O_u with $u = \Sigma_{k=1}^{j-1} m_k + i$ and add two fictitious initial and final operations O_0 and O_* with $p_0 = p_* = 0$. The disjunctive graph is then defined as follows. There is a vertex u with weight p_u corresponding to each operation O_u. The directed *conjunctive arcs* link the consecutive operations of each job, and link O_0 to all first operations and all last operations to O_*. A pair of directed *disjunctive arcs* connects every two operations that have to be executed on the same machine. A feasible schedule corresponds to the selection of one disjunctive arc of every such pair, granting precedence of one operation over the other on their common machine, in such a way that the resulting directed graph is acyclic. The value of the schedule is given by the weight of the maximum weight path from 0 to *. We refer to Figure 3 for an example.

At a typical stage of any enumerative algorithm, a certain subset D of disjunctive arcs will have been selected. We consider the directed graph obtained by removing all other disjunctive arcs. Let the maximum weights of paths from 0 to u and from u to *, excluding p_u, be denoted by r_u and q_u, respectively. In particular, r_* is an obvious lower bound on the value of any feasible schedule obtainable from the current graph [Charlton & Death 1970]. We can get a far better bound in a manner very similar to the development of flow shop bounds in Section 5.3.2 [Lageweg *et al.* 1977].

Let us relax the capacity constraints for all machines except M_i. We then obtain a problem of scheduling the operations O_u on M_i with release dates r_u, processing times p_u, due dates $-q_u$ and precedence constraints defined by the directed graph, so as to minimize maximum lateness. As pointed out in Section 3.2, this $1|prec,r_j|L_{max}$ problem is NP-hard, but there exist fast enumerative methods for its solution on each M_i. Again, almost all lower bounds proposed in the literature appear as special cases of the above one by underestimating the contribution of r_u, q_u or both, by ignoring the precedence constraints, or by restricting the set of machines over which maximization is to take place.

The currently best job shop algorithm [McMahon & Florian 1975] involves the $1|r_j|L_{max}$ bound combined with the enumeration of *active schedules*. Starting from O_0, we consider at each stage the subset S of operations all of whose predecessors have been scheduled and calculate their earliest possible completion times r_u+p_u. It can be shown [Giffler & Thompson 1960] that it is sufficient to consider only a machine on which the minimum value of r_u+p_u is achieved and to branch by successively scheduling next on that machine all O_v for which $r_v < \min_{O_u \in S}\{r_u+p_u\}$. In this scheme, several disjunctive arcs are added to D at each stage. An alternative approach whereby at each stage one disjunctive arc of some *crucial* pair is selected leads to a computationally inferior approach [Lageweg *et al.* 1977].

Surrogate duality relaxations of the job shop problem are investigated in [Fisher *et al.* 1981]. Either the precedence constraints fixing the machine orders for the jobs or the capacity constraints of the machines can be weighted and aggregated to a single constraint. For fixed values of the multipliers, the resulting problems can be solved in (pseudo)polynomial time. Although this approach leads to stronger lower bounds, it appears to be too time consuming to have much computational value.

As far as approximation algorithms are concerned, the performance guarantees due to [Gonzalez & Sahni 1978A] for flow shop algorithms AS and LPT (see Section 5.3.2) also apply to the case of a job shop.

A considerable effort has been invested in the empirical testing of various priority rules [Gere 1966; Conway *et al.* 1967; Day & Hottenstein 1970; Panwalkar & Iskander 1977]. No rule appears to be consistently better than any other and in practical situations one would be well advised to exploit any special structure that the problem at hand has to offer.

6. CONCLUDING REMARKS

If one thing emerges from the preceding survey, it is the amazing success of complexity theory as a means of differentiating between easy and hard problems. Within the very detailed problem classification developed especially for this purpose, surprisingly few open problems remain. For an extensive class of scheduling problems, a computer program has been developed that classifies these problems according to their computational complexity [Lageweg *et al.* 1981A, 1981B]. It employs elementary reductions such as those defined in Section 2.7 in order to deduce the consequences of the development of a new polynomial-time algorithm or a new NP-hardness proof.

As far as polynomial-time algorithms are concerned, the most impressive recent advances have occurred in the area of parallel machine scheduling and are due to researchers with a computer science background, recognizable as such by their use of terms like *tasks* and *processors* rather than *jobs* and *machines*. Single machine, flow shop and job shop scheduling has been traditionally the domain of operations researchers; here, an analytical approach to the performance of approximation algorithms is badly needed.

Several extensions of the problem class considered in this paper appear to be worthy of further study. A quite natural one involves the presence of additional limited *resources*, with the property that each job requires the use of a part of each resource during its execution. These problems turn out to be fairly complicated. We refer to [Davis 1966, 1973] for surveys and extensive bibliographies on resource constrained project scheduling, to [Blazewicz *et al.* 1980] for a partial complexity classification of this problem class, and to [Garey & Johnson 1981] for the famous special case of *bin packing* models.

More fundamentally, the strictly deterministic character of our models represents one of their major shortcomings. The investigation of their *stochastic* counterparts is of obvious interest and forms the subject of several contributions to this volume.

The area of deterministic sequencing and scheduling has emerged as one of the more fruitful interfaces between computer science and operations research. Proper consideration for the practical relevance of further theoretical work should continue to make it a challenging research area for many years to come.

ACKNOWLEDGMENT

The research by the first author was supported by NSF grant MCS78-20054.

REFERENCES

N.R. ACHUTHAN (1980) *Flow-Shop Scheduling Problems*, Ph.D. Thesis, Indian Statistical Institute, Calcutta.

I. ADIRI, N. HEFETZ (1980) Subproblems of openshop - more than two machines - schedule length problem. Operations Research, Statistics and Economics Mimeograph Series 260, Technion, Haifa.

D. ADOLPHSON (1977) Single machine job sequencing with precedence constraints. *SIAM J. Comput.* 6,40-54.
D. ADOLPHSON, T.C. HU (1973) Optimal linear ordering. *SIAM J. Appl. Math.* 25,403-423.
K.R. BAKER (1974) *Introduction to Sequencing and Scheduling*, Wiley, New York.
K.R. BAKER, E.L. LAWLER, J.K. LENSTRA, A.H.G. RINNOOY KAN (1982) Preemptive scheduling of a single machine to minimize maximum cost subject to release dates and precedence constraints. *Oper. Res.*, to appear.
K.R. BAKER, L.E. SCHRAGE (1978) Finding an optimal sequence by dynamic programming: an extension to precedence-related tasks. *Oper. Res.* 26,111-120.
K.R. BAKER, Z.-S. SU (1974) Sequencing with due-dates and early start times to minimize maximum tardiness. *Naval Res. Logist. Quart.* 21,171-176.
M.S. BAKSHI, S.R. ARORA (1969) The sequencing problem. *Management Sci.* 16,B247-263.
J.W. BARNES, J.J. BRENNAN (1977) An improved algorithm for scheduling jobs on identical machines. *AIIE Trans.* 9,25-31.
L. BIANCO, S. RICCIARDELLI (1981) Scheduling of a single machine to minimize total weighted completion time subject to release dates. Istituto di Analisi dei Sistemi ed Informatica, CNR, Rome.
J. BLAZEWICZ, J.K. LENSTRA, A.H.G. RINNOOY KAN (1980) Scheduling subject to resource constraints: classification and complexity. Report BW 127, Mathematisch Centrum, Amsterdam.
P. BRATLEY, M. FLORIAN, P. ROBILLARD (1975) Scheduling with earliest start and due date constraints on multiple machines. *Naval Res. Logist. Quart.* 22,165-173.
P. BRUCKER, M.R. GAREY, D.S. JOHNSON (1977) Scheduling equal-length tasks under tree-like precedence constraints to minimize maximum lateness. *Math. Oper. Res.* 2,275-284.
J. BRUNO, E.G. COFFMAN, JR., R. SETHI (1974) Scheduling independent tasks to reduce mean finishing time. *Comm. ACM* 17,382-387.
J. BRUNO, T. GONZALEZ (1976) Scheduling independent tasks with release dates and due dates on parallel machines. Technical Report 213, Computer Science Department, Pennsylvania State University.
J. CARLIER (1980) Problème à une machine. Manuscript, Institut de programmation, Université Paris VI.
J.M. CHARLTON, C.C. DEATH (1970) A generalized machine scheduling algorithm. *Oper. Res. Quart.* 21,127-134.
N.-F. CHEN (1975) An analysis of scheduling algorithms in multiprocessing computing systems. Technical Report UIUCDCS-R-75-724, Department of Computer Science, University of Illinois at Urbana-Champaign.
N.-F. CHEN, C.L. LIU (1975) On a class of scheduling algorithms for multiprocessors computing systems. In: T.-Y. FENG (ed.) (1975) *Parallel Processing*, Lecture Notes in Computer Science 24,

Springer, Berlin, 1-16.

Y. CHO, S. SAHNI (1978) Preemptive scheduling of independent jobs with release and due times on open, flow and job shops. Technical Report 78-5, Computer Science Department, University of Minnesota, Minneapolis.

E.G. COFFMAN, JR. (ed.) (1976) *Computer and Job-Shop Scheduling Theory*, Wiley, New York.

E.G. COFFMAN, JR., M.R. GAREY, D.S. JOHNSON (1978) An application of bin-packing to multiprocessor scheduling. *SIAM J. Comput.* 7,1-17.

E.G. COFFMAN, JR., R.L. GRAHAM (1972) Optimal scheduling for two-processor systems. *Acta Informat.* 1,200-213.

R.W. CONWAY, W.L. MAXWELL, L.W. MILLER (1967) *Theory of Scheduling*, Addison-Wesley, Reading, Mass.

S.A. COOK (1971) The complexity of theorem-proving procedures. *Proc. 3rd Annual ACM Symp. Theory of Computing*, 151-158.

E.W. DAVIS (1966) Resource allocation in project network models - a survey. *J. Indust. Engrg.* 17,177-188.

E.W. DAVIS (1973) Project scheduling under resource constraints - historical review and categorization of procedures. *AIIE Trans.* 5,297-313.

J. DAY, M.P. HOTTENSTEIN (1970) Review of scheduling research. *Naval Res. Logist. Quart.* 17,11-39.

D. DOLEV (1981) Scheduling wide graphs. Unpublished manuscript.

W.L. EASTMAN, S. EVEN, I.M. ISAACS (1964) Bounds for the optimal scheduling of n jobs on m processors. *Management Sci.* 11, 268-279.

S.E. ELMAGHRABY, S.H. PARK (1974) Scheduling jobs on a number of identical machines. *AIIE Trans.* 6,1-12.

H. EMMONS (1969) One-machine sequencing to minimize certain functions of job tardiness. *Oper. Res.* 17,701-715.

M.L. FISHER (1976) A dual algorithm for the one-machine scheduling problem. *Math. Programming* 11,229-251.

M.L. FISHER, B.J. LAGEWEG, J.K. LENSTRA, A.H.G. RINNOOY KAN (1981) Surrogate duality relaxation for job shop scheduling. Report, Mathematisch Centrum, Amsterdam.

M. FUJII, T. KASAMI, K. NINOMIYA (1969,1971) Optimal sequencing of two equivalent processors. *SIAM J. Appl. Math.* 17,784-789; Erratum. 20,141.

H.N. GABOW (1980) An almost-linear algorithm for two-processor scheduling. Technical Report CU-CS-169-80, Department of Computer Science, University of Colorado, Boulder.

M.R. GAREY (-) Unpublished.

M.R. GAREY, R.L. GRAHAM, D.S. JOHNSON (1978) Performance guarantees for scheduling algorithms. *Oper. Res.* 26,3-21.

M.R. GAREY, D.S. JOHNSON (1976) Scheduling tasks with nonuniform deadlines on two processors. *J. Assoc. Comput. Mach.* 23, 461-467.

M.R. GAREY, D.S. JOHNSON (1977) Two-processor scheduling with start-times and deadlines. *SIAM J. Comput.* 6,416-426.

M.R. GAREY, D.S. JOHNSON (1979) *Computers and Intractability: a Guide to the Theory of NP-Completeness*, Freeman, San Francisco.
M.R. GAREY, D.S. JOHNSON (1981) Approximation algorithms for bin packing problems: a survey. In: G. AUSIELLO, M. LUCERTINI (eds.) (1981) *Analysis and Design of Algorithms in Combinatorial Optimization*, CISM Courses and Lectures 266, Springer, Vienna, 147-172.
M.R. GAREY, D.S. JOHNSON, R. SETHI (1976) The complexity of flowshop and jobshop scheduling. *Math. Oper. Res.* 1,117-129.
M.R. GAREY, D.S. JOHNSON, B.B. SIMONS, R.E. TARJAN (1981A) Scheduling unit-time tasks with arbitrary release times and deadlines. *SIAM J. Comput.* 10,256-269.
M.R. GAREY, D.S. JOHNSON, R.E. TARJAN, M. YANNAKAKIS (1981B) Scheduling opposing forests. Unpublished manuscript.
L. GELDERS, P.R. KLEINDORFER (1974) Coordinating aggregate and detailed scheduling decisions in the one-machine job shop: part I. Theory. *Oper. Res.* 22,46-60.
L. GELDERS, P.R. KLEINDORFER (1975) Coordinating aggregate and detailed scheduling in the one-machine job shop: II - computation and structure. *Oper. Res.* 23,312-324.
W.S. GERE (1966) Heuristics in job shop scheduling. *Management Sci.* 13,167-190.
B. GIFFLER, G.L. THOMPSON (1960) Algorithms for solving production-scheduling problems. *Oper. Res.* 8,487-503.
P.C. GILMORE, R.E. GOMORY (1964) Sequencing a one-state variable machine: a solvable case of the traveling salesman problem. *Oper. Res.* 12,655-679.
T. GONZALEZ (1977) Optimal mean finish time preemptive schedules. Technical Report 220, Computer Science Department, Pennsylvania State University.
T. GONZALEZ (1979A) A note on open shop preemptive schedules. *IEEE Trans. Computers* C-28,782-786.
T. GONZALEZ (1979B) NP-Hard shop problems. Report CS-79-35, Department of Computer Science, Pennsylvania State University, University Park.
T. GONZALEZ, O.H. IBARRA, S. SAHNI (1977) Bounds for LPT schedules on uniform processors. *SIAM J. Comput.* 6,155-166.
T. GONZALEZ, D.B. JOHNSON (1980) A new algorithm for preemptive scheduling of trees. *J. Assoc. Comput. Mach.* 27,287-312.
T. GONZALEZ, E.L. LAWLER, S. SAHNI (1981) Optimal preemptive scheduling of a fixed number of unrelated processors in linear time. To appear.
T. GONZALEZ, S. SAHNI (1976) Open shop scheduling to minimize finish time. *J. Assoc. Comput. Mach.* 23,665-679.
T. GONZALEZ, S. SAHNI (1978A) Flowshop and jobshop schedules: complexity and approximation. *Oper. Res.* 26,36-52.
T. GONZALEZ, S. SAHNI (1978B) Preemptive scheduling of uniform processor systems. *J. Assoc. Comput. Mach.* 25,92-101.
D.K. GOYAL (1977) Non-preemptive scheduling of unequal execution time tasks on two identical processors. Technical Report

CS-77-039, Computer Science Department, Washington State University, Pullman.

R.L. GRAHAM (1966) Bounds for certain multiprocessing anomalies. *Bell System Tech. J.* 45,1563-1581.

R.L. GRAHAM (1969) Bounds on multiprocessing timing anomalies. *SIAM J. Appl. Math.* 17,263-269.

R.L. GRAHAM (-) Unpublished.

R.L. GRAHAM, E.L. LAWLER, J.K. LENSTRA, A.H.G. RINNOOY KAN (1979) Optimization and approximation in deterministic sequencing and scheduling: a survey. *Ann. Discrete Math.* 5,287-326.

J.N.D. GUPTA, S.S. REDDI (1978) Improved dominance conditions for the three-machine flowshop scheduling problem. *Oper. Res.* 26, 200-203.

A.M.A. HARIRI, C.N. POTTS (1981) An algorithm for single machine sequencing with release dates to minimise total weighted completion time. Report BW 143, Mathematisch Centrum, Amsterdam.

N. HEFETZ, I. ADIRI (1979) An efficient optimal algorithm for the two-machines, unit-time, job-shop, schedule-length, problem. Operations Research, Statistics and Economics Mimeograph Series 237, Technion, Haifa.

W.A. HORN (1972) Single-machine job sequencing with treelike precedence ordering and linear delay penalties. *SIAM J. Appl. Math.* 23,189-202.

W.A. HORN (1973) Minimizing average flow time with parallel machines. *Oper. Res.* 21,846-847.

W.A. HORN (1974) Some simple scheduling algorithms. *Naval Res. Logist. Quart.* 21,177-185.

E. HOROWITZ, S. SAHNI (1976) Exact and approximate algorithms for scheduling nonidentical processors. *J. Assoc. Comput. Mach.* 23,317-327.

E.C. HORVATH, S. LAM, R. SETHI (1977) A level algorithm for preemptive scheduling. *J. Assoc. Comput. Mach.* 24,32-43.

N.C. HSU (1966) Elementary proof of Hu's theorem on isotone mappings. *Proc. Amer. Math. Soc.* 17,111-114.

T.C. HU (1961) Parallel sequencing and assembly line problems. *Oper. Res.* 9,841-848.

O.H. IBARRA, C.E. KIM (1976) On two-processor scheduling of one- or two-unit time tasks with precedence constraints. *J. Cybernet.* 5,87-109.

O.H. IBARRA, C.E. KIM (1977) Heuristic algorithms for scheduling independent tasks on nonidentical processors. *J. Assoc. Comput. Mach.* 24,280-289.

O.H. IBARRA, C.E. KIM (1978) Approximation algorithms for certain scheduling problems. *Math. Oper. Res.* 3,197-204.

E. IGNALL, L. SCHRAGE (1965) Application of the branch-and-bound technique to some flow-shop scheduling problem. *Oper. Res.* 13,400-412.

J.R. JACKSON (1955) Scheduling a production line to minimize maximum tardiness. Research Report 43, Management Science Research Project, University of California, Los Angeles.

J.R. JACKSON (1956) An extension of Johnson's results on job lot scheduling. *Naval Res. Logist. Quart.* 3,201-203.

J.M. JAFFE (1979) Efficient scheduling of tasks without full use of processor resources. Report MIT/LCS/TM-122, Laboratory for Computer Science, Massachusetts Institute of Technology, Cambridge.

J.M. JAFFE (1980) An analysis of preemptive multiprocessor job scheduling. *Math. Oper. Res.* 5,415-421.

S.M. JOHNSON (1954) Optimal two- and three-stage production schedules with setup times included. *Naval Res. Logist. Quart.* 1, 61-68.

S.M. JOHNSON (1958) Discussion: sequencing *n* jobs on two machines with arbitrary time lags. *Management Sci.* 5,299-303.

R.M. KARP (1972) Reducibility among combinatorial problems. In: R. E. MILLER, J.W. THATCHER (eds.) (1972) *Complexity of Computer Computations*, Plenum Press, New York, 85-103.

R.M. KARP (1975) On the computational complexity of combinatorial problems. *Networks* 5,45-68.

L.G. KHACHIYAN (1979) A polynomial algorithm in linear programming. *Soviet Math. Dokl.* 20,191-194.

H. KISE, T. IBARAKI, H. MINE (1978) A solvable case of the one-machine scheduling problem with ready and due times. *Oper. Res.* 26,121-126.

M. KUNDE (1976) Beste Schranken beim LP-Scheduling. Bericht 7603, Institut für Informatik und Praktische Mathematik, Universität Kiel.

J. LABETOULLE, E.L. LAWLER, J.K. LENSTRA, A.H.G. RINNOOY KAN (1979) Preemptive scheduling of uniform machines subject to release dates. Report BW 99, Mathematisch Centrum, Amsterdam.

B.J. LAGEWEG, E.L. LAWLER, J.K. LENSTRA, A.H.G. RINNOOY KAN (1981A) Computer aided complexity classification of combinatorial problems. Report BW 137, Mathematisch Centrum, Amsterdam.

B.J. LAGEWEG, E.L. LAWLER, J.K. LENSTRA, A.H.G. RINNOOY KAN (1981B) Computer aided complexity classification of deterministic scheduling problems. Report BW 138, Mathematisch Centrum, Amsterdam.

B.J. LAGEWEG, J.K. LENSTRA, A.H.G. RINNOOY KAN (1976) Minimizing maximum lateness on one machine: computational experience and some applications. *Statist. Neerlandica* 30,25-41.

B.J. LAGEWEG, J.K. LENSTRA, A.H.G. RINNOOY KAN (1977) Job-shop scheduling by implicit enumeration. *Management Sci.* 24,441-450.

B.J. LAGEWEG, J.K. LENSTRA, A.H.G. RINNOOY KAN (1978) A general bounding scheme for the permutation flow-shop problem. *Oper. Res.* 26,53-67.

S. LAM, R. SETHI (1977) Worst case analysis of two scheduling algorithms. *SIAM J. Comput.* 6,518-536.

E.L. LAWLER (1973) Optimal sequencing of a single machine subject to precedence constraints. *Management Sci.* 19,544-546.

E.L. LAWLER (1976A) Sequencing to minimize the weighted number of tardy jobs. *RAIRO Rech. Opér.* 10.5 Suppl.27-33.

E.L. LAWLER (1976B) *Combinatorial Optimization: Networks and Matroids*, Holt, Rinehart and Winston, New York.

E.L. LAWLER (1977) A "pseudopolynomial" algorithm for sequencing jobs to minimize total tardiness. *Ann. Discrete Math.* 1, 331-342.

E.L. LAWLER (1978) Sequencing jobs to minimize total weighted completion time subject to precedence constraints. *Ann. Discrete Math.* 2,75-90.

E.L. LAWLER (1980) Preemptive scheduling of precedence-constrained jobs on parallel machines. Report BW 132, Mathematisch Centrum, Amsterdam.

E.L. LAWLER (1981) Preemptive scheduling of uniform parallel machines to minimize the number of late jobs. Report, Mathematisch Centrum, Amsterdam, to appear.

E.L. LAWLER (-) Unpublished.

E.L. LAWLER, J. LABETOULLE (1978) On preemptive scheduling of unrelated parallel processors by linear programming. *J. Assoc. Comput. Mach.* 25,612-619.

E.L. LAWLER, J.K. LENSTRA, A.H.G. RINNOOY KAN (1981) Minimizing maximum lateness in a two-machine open shop. *Math. Oper. Res.* 6,153-158.

E.L. LAWLER, C.U. MARTEL (1980) Computing "polymatroidal" network flows. Research Memorandum ERL M80/52, Electronics Research Laboratory, University of California, Berkeley.

E.L. LAWLER, J.M. MOORE (1969) A functional equation and its application to resource allocation and sequencing problems. *Management Sci.* 16,77-84.

J.K. LENSTRA (1977) *Sequencing by Enumerative Methods*, Mathematical Centre Tracts 69, Mathematisch Centrum, Amsterdam.

J.K. LENSTRA (-) Unpublished.

J.K. LENSTRA, A.H.G. RINNOOY KAN (1978) Complexity of scheduling under precedence constraints. *Oper. Res.* 26,22-35.

J.K. LENSTRA, A.H.G. RINNOOY KAN (1979) Computational complexity of discrete optimization problems. *Ann. Discrete Math.* 4, 121-140.

J.K. LENSTRA, A.H.G. RINNOOY KAN (1980) Complexity results for scheduling chains on a single machine. *European J. Oper. Res.* 4,270-275.

J.K. LENSTRA, A.H.G. RINNOOY KAN, P. BRUCKER (1977) Complexity of machine scheduling problems. *Ann. Discrete Math.* 1,343-362.

C.L. LIU (1972) Optimal scheduling on multi-processor computing systems. *Proc. 13th Annual IEEE Symp. Switching and Automata Theory*, 155-160.

C.L. LIU (1976) Deterministic job scheduling in computing systems. Department of Computer Science, University of Illinois at Urbana-Champaign.

J.W.S. LIU, C.L. LIU (1974A) Bounds on schedulling algorithms for heterogeneous computing systems. In: J.L. ROSENFELD (ed.) (1974) *Information Processing 74*, North-Holland, Amsterdam, 349-353.

J.W.S. LIU, C.L. LIU (1974B) Bounds on scheduling algorithms for heterogeneous computing systems. Technical Report UIUCDCS-R-74-632, Department of Computer Science, University of Illinois at Urbana-Champaign, 68 pp.

J.W.S. LIU, C.L. LIU (1974C) Performance analysis of heterogeneous multi-processor computing systems. In: E. GELENBE, R. MAHL (eds.) (1974) *Computer Architectures and Networks*, North-Holland, Amsterdam, 331-343.

C. MARTEL (1981) Scheduling uniform machines with release times, deadlines and due times. *J. Assoc. Comput. Mach.*, to appear.

G.B. McMAHON (1969) Optimal production schedules for flow shops. *Canad. Oper. Res. Soc. J.* 7,141-151.

G.B. McMAHON (1971) *A Study of Algorithms for Industrial Scheduling Problems*, Ph.D. Thesis, University of New South Wales, Kensington.

G.B. McMAHON, M. FLORIAN (1975) On scheduling with ready times and due dates to minimize maximum lateness. *Oper. Res.* 23,475-482.

R. McNAUGHTON (1959) Scheduling with deadlines and loss functions. *Management Sci.* 6,1-12.

L.G. MITTEN (1958) Sequencing *n* jobs on two machines with arbitrary time lags. *Management Sci.* 5,293-298.

C.L. MONMA (1980) Sequencing to minimize the maximum job cost. *Oper. Res.* 28,942-951.

C.L. MONMA, A.H.G. RINNOOY KAN (1981) Efficiently solvable special cases of the permutation flow-shop problem. Report 8105, Erasmus University, Rotterdam.

C.L. MONMA, J.B. SIDNEY (1979) Sequencing with series-parallel precedence constraints. *Math. Oper. Res.* 4,215-224.

J.M. MOORE (1968) An *n* job, one machine sequencing algorithm for minimizing the number of late jobs. *Management Sci.* 15,102-109.

R.R. MUNTZ, E.G. COFFMAN, JR. (1969) Optimal preemptive scheduling on two-processor systems. *IEEE Trans. Computers* C-18,1014-1020.

R.R. MUNTZ, E.G. COFFMAN, JR. (1970) Preemptive scheduling of real time tasks on multiprocessor systems. *J. Assoc. Comput. Mach.* 17,324-338.

J.F. MUTH, G.L. THOMPSON (eds.) (1963) *Industrial Scheduling*, Prentice-Hall, Englewood Cliffs, N.J., 236.

I. NABESHIMA (1963) Sequencing on two machines with start lag and stop lag. *J. Oper. Res. Soc. Japan* 5,97-101.

S.S. PANWALKAR, W. ISKANDER (1977) A survey of scheduling rules. *Oper. Res.* 25,45-61.

C.H. PAPADIMITRIOU, P.C. KANELLAKIS (1980) Flowshop scheduling with limited temporary storage. *J. Assoc. Comput. Mach.* 27,533-549.

J. PIEHLER (1960) Ein Beitrag zum Reihenfolgeproblem. *Unternehmensforschung* 4,138-142.

C.N. POTTS (1980A) An adaptive branching rule for the permutation flow-shop problem. *European J. Oper. Res.* 5,19-25.

C.N. POTTS (1980B) Analysis of a heuristic for one machine sequencing with release dates and delivery times. *Oper. Res.* 28,1436-1441.

C.N. POTTS (-) Unpublished.
S.S. REDDI, C.V. RAMAMOORTHY (1972) On the flow-shop sequencing problem with no wait in process. *Oper. Res. Quart.* 23,323-331.
G. RINALDI, A. SASSANO (1977) On a job scheduling problem with different ready times: some properties and a new algorithm to determine the optimal solution. Report R.77-24, Istituto di Automatica, Università di Roma.
A.H.G. RINNOOY KAN (1976) *Machine Scheduling Problems: Classification, Complexity and Computations*, Nijhoff, The Hague.
A.H.G. RINNOOY KAN, B.J. LAGEWEG, J.K. LENSTRA (1975) Minimizing total costs in one-machine scheduling. *Oper. Res.* 23,908-927.
P. ROSENFELD (-) Unpublished.
M.H. ROTHKOPF (1966) Scheduling independent tasks on parallel processors. *Management Sci.* 12,437-447.
B. ROY, B. SUSSMANN (1964) Les problèmes d'ordonnancement avec contraintes disjonctives. Note DS no.9 bis, SEMA, Montrouge.
S. SAHNI (1976) Algorithms for scheduling independent tasks. *J. Assoc. Comput. Mach.* 23,116-127.
S. SAHNI, Y. CHO (1979A) Complexity of scheduling jobs with no wait in process. *Math. Oper. Res.* 4,448-457.
S. SAHNI, Y. CHO (1979B) Nearly on line scheduling of a uniform processor system with release times. *SIAM J. Comput.* 8,275-285.
S. SAHNI, Y. CHO (1980) Scheduling independent tasks with due times on a uniform processor system. *J. Assoc. Comput. Mach.* 27,550-563.
R. SETHI (1976A) Algorithms for minimal-length schedules. In: [Coffman 1976], 51-99.
R. SETHI (1976B) Scheduling graphs on two processors. *SIAM J. Comput.* 5,73-82.
R. SETHI (1977) On the complexity of mean flow time scheduling. *Math. Oper. Res.* 2,320-330.
J. SHWIMER (1972) On the *N*-job, one-machine, sequence-independent scheduling problem with tardiness penalties: a branch-and-bound solution. *Management Sci.* 18,B301-313.
J.B. SIDNEY (1973) An extension of Moore's due date algorithm. In: S.E. ELMAGHRABY (ed.) (1973) *Symposium on the Theory of Scheduling and its Applications*, Lecture Notes in Economics and Mathematical Systems 86, Springer, Berlin, 393-398.
J.B. SIDNEY (1975) Decomposition algorithms for single-machine sequencing with precedence relations and deferral costs. *Oper. Res.* 23,283-298.
J.B. SIDNEY (1979) The two-machine maximum flow time problem with series parallel precedence relations. *Oper. Res.* 27,782-791.
B. SIMONS (1978) A fast algorithm for single processor scheduling. *Proc. 19th Annual IEEE Symp. Foundations of Computer Science*, 246-252.
B. SIMONS (1980) A fast algorithm for multiprocessor scheduling. *Proc. 21st Annual IEEE Symp. Foundations of Computer Science*, 50-53.
M.L. SMITH, S.S. PANWALKAR, R.A. DUDEK (1975) Flow shop sequencing

with ordered processing time matrices. *Management Sci.* 21, 544-549.

M.L. SMITH, S.S. PANWALKAR, R.A. DUDEK (1976) Flow shop sequencing problem with ordered processing time matrices: a general case. *Naval Res. Logist. Quart.* 23,481-486.

W.E. SMITH (1956) Various optimizers for single-stage production. *Naval Res. Logist. Quart.* 3,59-66.

H.I. STERN (1976) Minimizing makespan for independent jobs on non-identical parallel machines - an optimal procedure. Working Paper 2/75, Department of Industrial Engineering and Management, Ben-Gurion University of the Negev, Beer-Sheva.

W. SZWARC (1968) On some sequencing problems. *Naval Res. Logist. Quart.* 15,127-155.

W. SZWARC (1971) Elimination methods in the $m \times n$ sequencing problem. *Naval Res. Logist. Quart.* 18,295-305.

W. SZWARC (1973) Optimal elimination methods in the $m \times n$ sequencing problem. *Oper. Res.* 21,1250-1259.

W. SZWARC (1978) Dominance conditions for the three-machine flow-shop problem. *Oper. Res.* 26,203-206.

J.D. ULLMAN (1975) *NP*-Complete scheduling problems. *J. Comput. System Sci.* 10,384-393.

J.D. ULLMAN (1976) Complexity of sequencing problems. In: [Coffman 1976], 139-164.

M.K. WARMUTH (1980) M Processor unit-execution-time scheduling reduces to M-1 weakly connected components. M.S. Thesis, Department of Computer Science, University of Colorado, Boulder

D.A. WISMER (1972) Solution of the flowshop-scheduling problem with no intermediate queues. *Oper. Res.* 20,689-697.

ON SCHEDULING WITH RELEASE TIMES AND DEADLINES

Barbara Simons

IBM Research K52/282
5600 Cottle Rd.
San Jose, Ca. 95193

Abstract. *Algorithms for the scheduling of unit-time jobs with release times and deadlines on one or many identical machines are presented and generalizations which result in NP-complete problems are discussed.*

Introduction. In this paper we are considering only deterministic scheduling in which the jobs which are to be scheduled are available for only a previously given finite amount of time. In addition, we shall assume that all the machines on which the programs are to be scheduled are identical. Consequently, the machine to which a program is assigned for processing is irrelevant, so long as the program gets processed by some machine and two programs are never scheduled for processing on the same machine at the same time.

Definitions. We shall assume that there is a set of n *jobs:* J(1), J(2), ..., J(n) and m *processors* or *machines:* M(1), M(2), ..., M(m). Each job J(i) will have a *release time,* r(i), a *deadline,* d(i), and a *processing time,* p(i). When we speak of a job J(i) being *released* by time t, we mean that $r(i) \leq t$. The *start time,* s(i) of job J(i) is the time at which J(i) is scheduled to begin in a given schedule. If at some time t there is no job scheduled to be running on a given processor, then we say that the time between the completion of the last job to be scheduled before t and the first job to be scheduled after t on that processor is called *idle time.* In addition, we shall consider a problem which has a partial ordering $\leq$ of the jobs, where $J(i) \leq J(j)$ means that job J(i) must be completed by the time that job J(j) begins execution.

One normally thinks of a *schedule* as being an assignment for each job J(i) of a start time, s(i), and a machine, h(i). More formally, a *feasible schedule* is a mapping $g : \{1,...,n\} \rightarrow R \times \{1,...,m\}$, where R is the reals, and where $g(i) = (s(i),h(i))$, $1 \leq i \leq n$, such that:

M. A. H. Dempster et al. (eds.), Deterministic and Stochastic Scheduling, 75–88.

1. $r(i) \leq s(i) \leq d(i) - p(i)$ (every job is run between its release time and its deadline),
2. if $h(i) = h(j)$, then for $i \neq j$, $s(i) \leq s(j) \Rightarrow s(j) - s(i) \geq p(i)$ (each machine can run at most one job at a time).

If there are some jobs which have not been assigned start times, then we have a *partial schedule.* Finally, in none of the problems will *preemption* be allowed, that is, once a job has begun execution it cannot be interrupted and consequently must run until it is completed.

Lemma 1, which applies to multi-machine problems as well as single-machine ones, is presented below without proof.

Lemma 1. If there exists a feasible schedule for a set of n jobs with release times, deadlines, and unit processing times, then there exists a feasible schedule for those jobs which can be represented as a permutation of the n numbers which are the job indices.

Problem 1: r(i), d(i) integers, p(i)=1. Integer release times and deadlines; unit processing time.

Let us consider initially the case where we have one machine and n jobs. Because all of the constraints are integer, it is clear that if a problem instance has a feasible solution, then there exists a feasible solution in which the start time of each of the jobs is integer. We shall assume schedules obtained for problems with integer release times and deadlines will always have integer start times.

There is an O(n log n) algorithm which either produces a feasible schedule with minimum completion time or determines that none exists. This algorithm has been known for some time and is due to Jackson[9]. We shall be referring to this algorithm as *the earliest deadline algorithm.*

The Earliest Deadline (E.D.) Algorithm. At time $t=0$ run the job with the earliest deadline from among those with release times of 0. If there are no jobs released at time $t=0$, then increase t to the minimum release time of all the jobs and repeat the above procedure. In general, after scheduling a job, set $t=t+1$ and select the job with the minimum deadline from among the unscheduled jobs released at time t or earlier. If there are no unscheduled jobs released at time t or earlier, then increase t to the minimum release time of the unscheduled jobs. Halt when either all the jobs have been scheduled or when the algorithm discovers an unscheduled job which has not been scheduled by its deadline.

Because each job is scheduled at most once and because of the use of priority queues, the running time for the E.D. algorithm is O(n log n). A lower bound for any algorithm which produces a feasible schedule when one exists is $\Omega(n \log n)$. This bound is obtained by reducing the problem of sorting a set of integers to a version of problem 1 as follows. Given a set of integers $A = \{a_1, a_2, ..., a_n\}$, we let $r(i) = a_i$, $d(i) = r(i)+1$, and $p(i) = 1$. By lemma 1 if a feasible schedule exists for a set of jobs, then there is always a feasible schedule which can be represented as a permutation of the job indices. Since the E.D. algorithm will produce such a permutation schedule, a one-machine schedule obtained using the E.D. algorithm will contain the elements of A in sorted order. Consequently $\Omega(n \log n)$, the lower bound for sorting, is a lower bound for the E.D. algorithm.

We note that the E.D. algorithm minimizes C_{max} over all feasible schedules, where C_{max} is defined to be the maximum completion time of all the jobs in the schedule. This follows directly from the observation that the only idle time in a schedule produced by the E.D. algorithm occurs when no unscheduled jobs are available.

If, instead of having one machine at our disposal for scheduling the set of jobs, we have m machines, then we can easily modify the E.D. algorithm to obtain a feasible schedule if one exists. Using the E.D. algorithm to schedule individual jobs, we schedule the first on M(1), the second on M(2), ..., the ith on M(k), where k=m if i is a multiple of m and k=i mod m otherwise. Even though we have extended the problem from one to many machines, the running time of the algorithm remains $O(n \log n)$.

Problem 2: r(i), d(i), p(i) integers. Integer release times, deadlines, and processing times.

We consider only the one-machine version of the problem, for even with one machine, the problem is NP-complete. In fact, it is strongly NP-complete [12]. By strongly NP-complete, we mean that if we were to encode each number in a problem instance using a unary encoding scheme instead of an n-ary one for $n \geq 2$, there would still be no algorithm for solving the problem which was polynomial in the (new) length of the problem instance unless P=NP.

Even if we allow only two different values for the release times and two different values for the deadlines, the problem is still NP-complete [5]. In this case, however, there exists a pseudo-polynomial time dynamic programming algorithm for solving the problem, and so the problem is not strongly NP-complete. The proof that this version of the problem is NP-complete reduces the sum of subsets problem to the scheduling problem. The strong NP-completeness result is obtained by reducing the 3-partition problem to the more general version of problem 2.

A combination of problems 1 and 2 is the version where all the running times are identical, as in problem 1, and the release times and deadlines are integers, as in problem 2. A slightly more general version (real release times and deadlines) is presented below as problem 4. Still more general than problem 4 is problem 3, in which there are multiple release time/deadline intervals for each job. If for some J(i) the number of release time/deadline intervals is greater than one, then J(i) is to be scheduled in one (and only one) of these intervals.

Problem 3: r(i), d(i) real numbers, p(i)=1, m=1 multiple release time/deadline intervals per job. Real release times, deadlines; unit processing time; a job may have several intervals in which it can run, but is to be scheduled to run in only one of them.

The NP-completeness proof, which was obtained jointly with Michael Sipser, involves a reduction of 3-Sat to a restricted form of problem 3 in which the release times and deadlines are integers and p(i)=2. Note that we get a version of problem 3 by setting p(i)=1 and dividing all release times and deadlines by two. For both the construction and the proof see [17].

The problem which we shall discuss in greatest detail is that of scheduling n unit-time jobs, with release times and deadlines which are arbitrary real numbers, on m identical machines. A polynomial time algorithm with time complexity $O(n^2 \log n)$ was first obtained for the single-machine case by Simons [14]. An alternative single-machine algorithm with the same time complexity was subsequently obtained by Carlier [3]. Most recently, a single-machine algorithm with time complexity $O(n \log n)$ has been described in a paper by Garey, Johnson, Simons, and Tarjan [7]. This algorithm, a speeded-up version of an $O(n^2)$ algorithm which is also presented in the paper, eliminates redundancy in the data through the use of clever but somewhat complicated data structures. We shall be presenting the $O(n^2)$ version below. The reader who wishes more information on the faster algorithm is referred to the joint paper.

In the multi-machine case the best known result obtained prior to the one described below was by Carlier, whose algorithm runs in time $O(n^{m+2} \log n)$, where m is the number of machines [3]. The first multi-machine algorithm which is polynomial in *both* m and n, having a running time of $O(n^3 \log n)$, was obtained by Simons [13]. We have recently shown that the running time can be reduced to $O(mn^2 \log \log n)$. We discuss first the multi-machine result and then the single-machine algorithm.

Problem 4: r(i), d(i) real numbers, p(i)=1, m machines. Real release times and deadlines; unit processing time; multiple machines.

The algorithm creates a schedule with the minimum possible completion time if one exists, or determines that the problem instance has no feasible solution. Note that the problem in which all jobs have the same running time p can be made into a version of our problem by dividing all release times and deadlines by p. The process is an iterative one which creates a partial schedule and determines the earliest time, say t, at which the next job can be scheduled. Once t has been determined, the next job is scheduled using the earliest deadline algorithm.

The basic idea behind the algorithm is that *there are times when it is necessary to wait for a crucial job to be released, even though there may be other jobs which are available to be run. But if we do start a job, it should be the available job with the earliest deadline.* This theme also appears in the single machine algorithms, although in the case of the recursive $O(n^2 \log n)$ algorithm of Simons [14], the enforced waiting time is only implicit in the algorithm.

The multi-machine algorithm forces a wait only when a job which cannot be completed by its deadline is discovered. Such an event triggers a call to a "fixup" subroutine, called the crisis subroutine, which determines the point in the schedule at which it is necessary to delay scheduling a job.

The general scheduling strategy is to compute the earliest time at which the next job can be scheduled and then to use the ED algorithm to schedule it. This is combined with techniques for determining the times at which a machine must remain idle, possibly in the presence of jobs which could be scheduled. The result is a fast running algorithm which is relatively simple.

The Barriers Algorithm. We define a v-partial schedule to be a feasible schedule of a set of v jobs, $v \leq n$. (When the size of the v-partial schedule is not of interest, we may speak simply of the partial schedule). The algorithm proceeds by trying to increase the current v-partial schedule to a (v+1)-partial schedule. The next job to be scheduled is said to occupy *slot* v+1. In fact, given any partial schedule obtained by any algorithm, we say that the job occupying the $v+1^{st}$ slot is the job which is the $v+1^{st}$ job scheduled to begin running. So the first job to be scheduled by the algorithm occupies slot 1, the second occupies slot 2, and so forth. A v-partial schedule can be viewed as a permutation of v elements and a slot as a location in that permutation. If at some point in the running of the algorithm the jobs are all removed from the partial schedule, then we begin again with scheduling the first job in slot 1. Note that the job occupying a particular slot can vary from partial schedule to partial schedule. But there will be at most n slots in any partial schedule, and if the nth slot is ever filled in a feasible schedule, then we have succeeded in scheduling the entire set of jobs.

The fundamental notion making this algorithm polynomial is that of "barriers." A *barrier* is an ordered pair (j,r), where j is a slot number and r is a release time. (The manner in which barriers are created is described below). A barrier (j,r) represents the constraint that the jth slot can begin no earlier than time r.

The *earliest deadline with barriers (EDB) subroutine* is called to create a (v+1)-partial schedule from the current v-partial schedule. It first computes h, the number of the machine on which the job in slot v+1 is to be scheduled, by setting $h = m$ if v+1 is a multiple of m and $h = (v+1) \bmod m$ otherwise. It then determines the "earliest allowed time", say t, at which slot v+1 can begin as follows. We first define f, the function mapping slots to start times in the current v-partial schedule, by f(i) = start time of slot i, $1 \leq i \leq v$. Now set $t_1 = 0$ if $v+1 \leq m$; otherwise $t_1 = f(v+1-m) + 1$. (Slot v+1-m is the slot most recently scheduled on machine M(h)). Let t_2 = the minimum release time of the unscheduled jobs, and $t_3 = \max \{\{r : (v+1,r) \text{ is a barrier}\} \cup \{s(i) : i \leq v\} \cup \{0\}\}$. Set $t = \max \{t_1,t_2,t_3\}$. We say that t is the *earliest allowed time respecting barriers* (for slot v+1).

The subroutine calls the earliest deadline algorithm to schedule the job to occupy slot v+1 with a start time of t. Suppose job J(i) has been so selected. If $d(i) \geq t+1$, then J(i) is scheduled in slot v+1, and the subroutine returns to the main routine. Note that if $t > t_1$, there will be idle time preceding slot v+1.

The subroutine will *fail* if $d(i) < t+1$. In this case we say that a *crisis* has occurred and that J(i) is a *crisis job*. The *crisis subroutine* is then called.

The crisis subroutine backtracks over the current partial schedule, searching for the job with the highest slot number with a deadline greater than that of the crisis job. If it does not find such a job, it concludes that there is no feasible schedule and halts. Otherwise, the first such job which it encounters is called the *pull job*. The set of all jobs in the v-partial schedule with a *larger* slot number than that of the pull job together with the crisis job, is called a *restricted set*. (The pull job is not a member of the restricted set). The subroutine then determines the minimum release time of all the jobs in the restricted set, call this time r, and the slot occupied by the pull job, call this j. Finally, it creates a new barrier defined to be the ordered pair (j,r), and returns.

The main routine adds the new barrier to the previous list of barriers, discards the current partial schedule, and calls the E.D.B. subroutine, starting with a 0-partial schedule.

The algorithm halts when either it has succeeded in creating an n-partial schedule or it discovers a crisis job for which there is no pull job. In the former case the output consists of a feasible schedule and in the latter a statement that there is no feasible schedule. We observe that a faster algorithm would be obtained by removing only those jobs occupying slots whose numbers are at least as great as the slot of the pull job. This change does not affect the magnitude of the time complexity of the algorithm, however, and we shall disregard this possible speed-up in the analysis of the algorithm.

We illustrate the working of the algorithm by the following two machine example. The release time is the first number in the ordered pair and the deadline is the second.

A: (0,2) B: (0.2,3) C: (0.5,2) D: (0.5,3)

In the example, which is illustrated in figure 1a, we begin scheduling starting at time 0 on machine 1. The only job which has been released at time 0 is A; using the E.D. algorithm, A is scheduled to begin at time 0. Machine 2 is available to run a job at time 0, but there are no remaining unscheduled jobs which have been released by time 0. So the algorithm determines the job with the minimum release time from among the unscheduled jobs. In our example, this job is job B with a release time of 0.2. Now, machine 1 is available to run a job at time 1, by which time jobs C and D have both been released. The algorithm selects the job with the earlier deadline, C in the example, and schedules it to run on machine 1 starting at time 1. Machine 2 is available at time 1.2, and since only D remains unscheduled, it is scheduled on machine 2 at time 1.2.

In the above example, the jobs were scheduled in one iteration with no complications. Suppose, however, that we change C's deadline from 2 to 1.9. In this case, C would be the crisis job and B would be the pull job. Since B had been scheduled in slot 2 and since C is the only job in the restricted set, the barrier (2, 0.5) is created. In this simple example the one barrier is sufficient to create a feasible schedule. This schedule is shown in figure 1b.

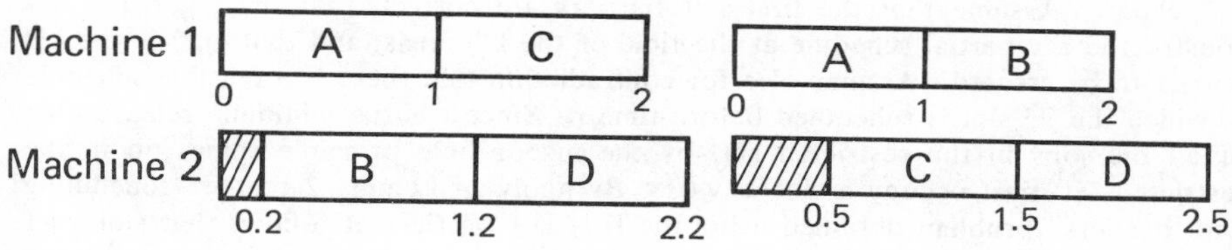

Figure 1a Figure 1b

Figure 1: A 2-machine example

To demonstrate correctness of the algorithm we shall first prove the correctness of the more general problem of "scheduling with barriers." This problem is identical to the originally stated problem, with the additional constraint that for each barrier (j,r) we require that slot j be started no earlier than time r.

Lemma 2. Assume we are given a problem instance of "scheduling with barriers" and that successive calls to the EDB subroutine, starting with an empty partial schedule, produce a v-partial schedule without requiring any calls to the crisis subroutine. Then there is no feasible schedule in which any of the v slots has an earlier start time. Furthermore, slot v+1 can be started no earlier (in any feasible schedule) than the earliest allowed time respecting barriers as computed by the algorithm.

Proof. Assume we are given a v-partial schedule obtained as in the statement of the lemma and assume by induction that the lemma is true for all slots up to and including slot v. Let t = earliest allowed time respecting barriers for slot v+1. If $t = t_1$, then either t = 0 or t = f(v+1-m) + 1. The former case is trivially optimal. In the latter case, f(v+1-m) is by the induction assumption the earliest possible time at which slot v+1-m can start. Since no slot numbered greater than v+1-m can start any earlier and since every machine has scheduled on it a slot which is numbered between v+1-m and v, there is no machine available for running the v+1st slot any earlier than the machine on which slot v+1-m had been scheduled. Therefore, f(v+1-m) + 1 is the earliest possible time for starting slot v+1. If $t = t_2$, then slot v+1 is started at the minimum release time of the unscheduled jobs. Consequently, there are only v jobs with release times less than or equal to t_2. So there is no feasible schedule in which slot v+1 can start earlier than t_2. If $t = t_3$ and $t \neq 0$, then either $t_3 = s(i)$ for some $i \leq v$ and the induction assumption holds, or a barrier constraint is preventing slot v+1 from beginning earlier. □

Lemma 3. The jobs in the restricted set all have release times strictly greater than the start time of the pull job.

Proof. The pull job has a deadline strictly greater than that of any of the jobs in the restricted set. Therefore, if any of them had been released by the time at which the pull job was scheduled, it would have been scheduled at that time. □

Theorem 1. Each barrier that the algorithm creates is correct.

Proof. Assume that the first k-1 barriers are correct, that the algorithm has constructed a v-partial schedule at the time of the kth crisis, and that (j,r) is the kth barrier to be created. Assume also for contradiction that there is a feasible schedule in which the jth slot is scheduled before time r. Since r is the minimum release time of all the jobs in the restricted set, by the pigeon-hole principle some job in the restricted set must occupy a slot $\geq$ v+1. By applying Lemma 2 to the "scheduling with barriers" problem obtained using the first k-1 barriers, it follows that slot v+1 can be started no earlier than the earliest allowed time respecting barriers computed by the EDB subroutine. Since all of the jobs in the restricted set have deadlines less than or equal to that of the crisis job, and since the completion time of slot v+1 must exceed the deadline of the crisis job in any feasible schedule, none of the jobs in the restricted set can be scheduled in slot v+1. Hence in no feasible schedule can slot v+1 begin before time r. □

Lemma 4. If the crisis subroutine does not find a pull job, then there is no feasible schedule.

Proof. Let J(i) be a crisis job for which no pull job is found. It follows from the definition of a pull job that all jobs occupying slots 1 through v in the current v-partial schedule have deadlines no greater than d(i), the deadline of the current crisis job. Therefore, they must all be completed by d(i). Using the same argument as in theorem 1, slot v+1 can begin no earlier than the time computed by the EDB subroutine. Consequently, any job occupying slot v+1 will have a completion time greater than d(i). So either the crisis job or some job which had been scheduled in the v-partial schedule will fail to meet its deadline. □

Since the only way in which the Barriers Algorithm fails to produce a schedule is if the crisis subroutine does not find a pull job, we have the following theorem.

Theorem 2. If the Barriers Algorithm fails to produce a schedule then there is no feasible schedule.

We also note that by lemma 2 if the algorithm produces a schedule, each slot in the schedule, in particular the last, is started at the earliest possible time.

Theorem 3. If the barriers algorithm produces a schedule, then no feasible schedule has an earlier completion time.

Theorem 4. The algorithm runs in time $O(n^3 \log n)$.

Proof. There are at most n distinct release times and n slots. So n^2 is an upper bound on the number of barriers which the algorithm can create. Because of the use of a priority queue, it can take O(log n) time to schedule a job, and at most n jobs can be scheduled before a new barrier is created. □

We note without proof that if more than mn barriers are created that there is no feasible schedule. In addition, it is possible to use priority queues for the release times and deadlines based upon a stratified binary tree which requires only O(log log n) time for insertion or deletion[2]. These two observations give us the lower running time bound of $O(mn^2 \log \log n)$.

If we limit ourselves to only one machine, then there are faster and more general algorithms. We first describe how, given a single-machine problem instance with a partial order, we can modify the problem to eliminate the partial order. We then characterize scheduling algorithms which finds a feasible schedule for the modified problem if and only if one exists for the original problem with the partial order. Finally, we present a one-machine algorithm, belonging to the characterized class, which is joint work with Garey, Johnson, and Tarjan [7]. This algorithm, which runs in $O(n^2)$ time, is the basis for the O(n log n) algorithm.

Problem 5: r(i), d(i) real numbers, p(i)=1, m=1, partial order. Real release times and deadlines; unit processing time; one machine; partial order.

In the one machine case we can eliminate the partial order by revising the release times and deadlines so that when the algorithm presented below is applied to the revised problem instance, it will produce a feasible schedule for the revised problem if and only if there exists a feasible schedule for the original problem instance. Since any schedule so obtained will have the minimum completion time, the modifications do not affect optimality.

Eliminate the partial ordering. We say that a schedule is *normal* if whenever some job is scheduled to begin at time t, that job has the earliest deadline of all the unscheduled jobs which have been released on or before time t. In other words, if we let s(i) be the start time for J(i) in the schedule, then $s(i) < s(j)$ implies that either $d(i) \leq d(j)$ or $s(i) < r(j)$. Clearly, whenever we apply the earliest deadline algorithm to a problem, any schedule which it might produce would be normal.

To modify the release times of the jobs, we make one pass over the DAG which represents the partial order. At an arbitrary point in the modification procedure, we consider only those jobs all of whose predecessors have been dealt with by the procedure. Suppose J(j) is such a job and $J(i) < J(j)$. If $r(j) \geq r(i) + 1$, then r(j) remains unchanged. If, however, $r(j) < r(i) + 1$, then setting $r(j) \leftarrow r(i)+1$ will not transform a set of jobs with a feasible schedule into one without a feasible schedule since J(j) can not be run until J(i) is completed. In particular, we set $r(j) \leftarrow \max \{\{r(j)\} \cup \{r(i)+1 : J(i) < J(j)\}\}$. We make a second pass over the partial order graph setting $d(i) \leftarrow \min \{\{d(i)\} \cup \{d(j)-1 : J(i) < J(j)\}\}$. The modified release times and deadlines are now *consistent* with the partial order, that is $J(i) < J(j)$ implies that $r(i) \leq r(j)-1$ and $d(i) \leq d(j)-1$. The proof of lemma 5 follows fairly directly from the definition of a normal schedule.

Lemma 5. If a set of release times and deadlines is consistent with a given partial order, then any normal one-machine schedule which satisfies the release time and deadline constraints will not violate the partial order.

For the rest of this discussion, we shall assume that there is no partial order. In the one-machine case we can successfully eliminate it because both the multi-machine algorithm (confined to one machine) and the algorithm presented below produce a normal schedule. For the two-machine case with integer release times, arbitrary deadlines, and a partial order there is a polynomial time algorithm [6] and for an arbitrary number of machines the problem is NP-complete even with all r(i) =

0 and all $d(i) = D$ [18]. The boundary at which the problem goes from being polynomial to being NP-complete is currently unknown.

The forbidden regions algorithm. The algorithm determines open intervals of time, called *forbidden regions,* at which no job can *begin* if the schedule is to be feasible. We are going to determine these forbidden regions by using the *backscheduling algorithm.* The backscheduling algorithm schedules a given set of jobs S backwards, starting at a given endpoint and ignoring any information about the jobs, such as their release times or deadlines. Assuming that the last job scheduled by the back-scheduling algorithm begins at time t, the next job to be scheduled will have a start time of t-1 unless t-1 happens to lie in a forbidden region. In the latter case, the next job is scheduled to start at the latest possible time less than t-1 which does not lie in a forbidden region. (Note that this time will be a left end-point of some forbidden region). We say that a forbidden region is *correct* if in every feasible schedule no job can begin in the forbidden region. Suppose that for our set of jobs S, we take all jobs with release times greater than or equal to r(i) and deadlines less than or equal to d(j) for some fixed i and j with $r(i) < d(j)$. Let us also suppose that any forbidden regions within the interval [r(i), d(j)] are correct.

If when we apply the backscheduling algorithm we get some job assigned a start time less than r(i), then clearly S has too many jobs which must be run in the interval [r(i), d(j)], and so S can not have a feasible schedule. Suppose instead that the last job to be scheduled, say J(k), is assigned a start time s(k), $r(i) \leq s(k) < r(i)+1$. Then if some job not in S is started in the interval (s(k)-1, r(i)), there would not be enough time in which to schedule the jobs in S. We note that any job beginning in the interval (s(k)-1, r(i)) could not belong to S since all jobs in S are released no earlier than r(i). Consequently, the open interval (s(k)-1, r(i)) is indeed a forbidden region. In fact, the algorithm determines forbidden regions precisely by applying the backscheduling algorithm to all sets of jobs S(i,j), where S(i,j) is defined to be all jobs released no earlier than r(i) and with deadlines no greater than d(j) for $r(i) < d(j)$.

We have just demonstrated correctness for a new forbidden region when all previously discovered forbidden regions in the interval [r(i), d(j)] are correct. We summarize the above result.

Lemma 6. Forbidden regions obtained by applying the backscheduling algorithm to a set of jobs S(i,j) are correct.

Once all of the forbidden regions as determined by the algorithm are known, then a feasible schedule can be obtained using the E.D. algorithm to schedule the jobs, with the added constraint that no job is scheduled to begin in a forbidden region. This portion of the algorithm has running time of O(n log n), the analysis of which is identical to that of problem 1.

To obtain the relevant forbidden regions we first sort the jobs by their release times. Let us assume that the jobs have been sorted and renamed, so that $r(1) \leq r(2) \leq \ldots \leq r(n)$. We now process the release times in *decreasing* order. When processing release time r(i), we determine for each deadline $d(j) \geq d(i)$ the number of jobs which cannot start before r(i) and which must be completed by d(j). The backscheduling algorithm is then applied going backwards from d(j) to determine the latest time, called c(j), at which the earliest job in S(i,j) can begin. Time c(j) is called the *critical time* for deadline d(j) (with respect to r(i)). Letting c denote the minimum of all defined c(j) with respect to r(i), we declare failure if c < r(i) or declare (c-1, r(i)) to be a forbidden region if r(i) ≤ c < r(i)+1.

By processing release times from largest to smallest, all forbidden regions to the right of r(i) will have been found by the time that r(i) is processed. To increase efficiency and reduce the time bound, we do not recompute the critical time for a deadline when a new release time is processed, but instead update the old value. This is possible because, although the forbidden regions may overlap, the left and right end points of any region are no greater than the corresponding end points of all previously declared regions. So we can merge overlapping forbidden regions, when they occur, in constant time.

We use a stack to store the forbidden regions and for each deadline d(j) we maintain a pointer to the the first forbidden region to the left of c(j) (or d(j) if c(j) has not yet been defined). Since there will never be more than n forbidden regions stored in the stack, the total amount of time to update c(j) for the entire algorithm is O(n). We use this fact to obtain the time bound of the algorithm.

Theorem 5. The forbidden regions algorithm has running time of $O(n^2)$.

Assume there is an interval of idle time (beginning at time t) in a partial schedule which was created using the E.D. algorithm as a subroutine. If this idle time exists because at time t none of the unscheduled jobs had yet been released, then we call this idle time a *division hole.* We note that the jobs which are scheduled before a division hole do not effect the feasibility of the jobs which are scheduled later. The proof of lemma 7 stems directly from this observation.

Lemma 7. Assume that the forbidden regions algorithm using the E.D. algorithm as a subroutine has created a partial schedule with a division hole after the job in slot v. Then the problem can be reduced to the subproblem of scheduling the remaining unscheduled jobs.

The proof of theorem 6 assumes that the forbidden regions algorithm does not determine that a set of jobs is infeasible, but that there is some job, J(i), which is not scheduled to be completed before its deadline by the E.D. algorithm. We then simulate the behavior of the backscheduling algorithm to obtain a contradiction.

Theorem 6. If the forbidden regions algorithm does not declare failure, then it finds a feasible schedule.

Theorem 7. If the forbidden regions algorithm produces a schedule, then no feasible schedule has an earlier completion time.

Minimizing maximum lateness and tardiness. We present a sketch of how to solve the lateness and tardiness problems for the single- and multi-machine cases of scheduling unit length jobs with real release times and deadlines. For a detailed description, see [16].

Let f(i) be the completion time of job J(i) in a schedule. The *lateness,* L(i), of job J(i) is defined by L(i) = f(i) - d(i). In a feasible schedule this value will never be positive. If a problem instance has a feasible schedule, the L_{max} problem consists of finding the minimum L' such that there exists a feasible schedule in which no job has lateness greater than L'.

If a problem instance has no feasible schedule, we relax the constraints by considering due dates instead of deadlines, where due dates are identical to deadlines except that they can be exceeded. The *tardiness* of job J(i) is then defined by T(i) = max{0, f(i) - d(i)}. The T_{max} can be viewed as a special case of the L_{max} problem in which we are dealing with positive values for L_{max}. In the discussion below, we first deal with the T_{max} problem and then mention how to use the same technique to solve the L_{max} problem when there is a feasible schedule.

Assume there is no feasible schedule for the given set of jobs. If for some job J(i), r(i) + 1 > d(i), then we can modify the due dates of all the jobs as follows. Let T = max{r(i) + 1 - d(i): r(i) + 1 > d(i)}. For all jobs J(i) first set $d'(i) \leftarrow d(i) + T$ and then run the appropriate scheduling algorithm to see if the modified problem instance has a feasible schedule. If it does, then T_{max} = T; otherwise proceed as below using the modified due dates and add T to the T_{max} which is computed for the modified problem instance.

In both the barriers and the forbidden regions algorithms, every interval of idle time ends at some release time. So for each job J(i) there are positive integers j and C such that f(i) = r(j) + C for $1 \le j,C \le n$. If J(i) has positive tardiness, we have T(i) = f(i) - d(i) = r(j) + C - d(i). We note that in any schedule which minimizes the maximum tardiness, some job will be completed at its due date. This gives a bound of $O(n^2)$ different values possible for the tardiness any job and $O(n^3)$ different values possible for all n jobs.

We can obtain the integer portion of T_{max}, call it T_{int}, by running the appropriate scheduling algorithm on due dates which have been increased by the integer value being tested. Since a feasible schedule can be produced by adding n-1 to all the due dates and then scheduling the jobs in the order in which they are released, T_{int} is bounded above by n-1. Using binary search, O(log n) iterations are required to find T_{int}. All due dates are then increased by T_{int}.

We now have a problem involving the $O(n^2)$ values of (r(j) - d(i)) (mod 1). To avoid writing down and testing all these values, we use the weighted median algorithm (WMA) [8,10]. This algorithm has as its input two vectors X = (x(1), x(2), ..., x(n)) and Y = (y(1), y(2), ..., y(n)), where the elements of the vectors are in sorted order. The WMA enables us to find the kth largest element in the multiset {x(i) - y(j)} in O(n log n) time and O(n) space.

First the WMA is used to do a binary search over the subproblem in which r(j) (mod 1) $\ge$ d(i) (mod 1). The goal is the minimal fractional value which results in a feasible schedule when added to all the due dates. Recall that the due dates have already been increased by T_{int}. The same procedure is then applied to the subproblem in which r(j) (mod 1) < d(i) (mod 1) by applying the WMA to (1 + r(j) (mod 1)) - d(i) (mod 1). The value of T_{max} is T_{int} plus the minimum of the two subproblem solutions.

If, on the other hand, the given set of jobs has a feasible schedule, we minimize the maximum lateness by decreasing the deadlines of all the jobs using the same binary search technique as for the T_{max} problem. The maximum amount by which the deadlines can all be decreased while still maintaining feasibility is the value for L_{max}. The running time for both the tardiness and the lateness problem exceeds that of the feasibility algorithm being used by only the O(log n) factor required for binary search.

Acknowledgements. I would like to thank Gene Lawler for introducing me to this area. In addition, Dick Karp, my advisor, Manuel Blum, Dan Gusfield, Jan Karel Lenstra, Alexander H. G. Rinnooy Kan, and Ron Fagin were very helpful and encouraging. And most especially, I would like to thank Michael Sipser for many hours of listening to partially formed ideas.

References

[1] A. V. Aho, J. E. Hopcroft, and J. D. Ullman, *The Design and Analysis of Computer Algorithms,* Addison-Wesley, Reading, Mass., 1974.

[2] P. V. E. Boas, "Preserving order in a forest in less than logarithmic time," 16th Annual Symposium on Foundations of Computer Science, IEEE Computer Society, Long Beach, California, 1978, pp. 75-84.

[3] J. Carlier, "Problème à une machine dans le cas où les tâches ont des durées égales," Technical Report Institut de Programmation, Université Paris VI (1979).

[4] S. A. Cook, "The complexity of theorem-proving procedures," Proc. of the Third Annual ACM Symp. on Theory of Computing, pp. 151-158.

[5] M. R. Garey and D. S. Johnson, *Computers and Intractability: A Guide to the Theory of NP-Completeness,* W. H. Freeman, San Francisco, California, 1978.

[6] M. R. Garey and D. S. Johnson, "Two-processor scheduling with start-times and deadlines," *SIAM J. Comput.,* 6 (1977), pp. 416-426.

[7] M. R. Garey, D. S. Johnson, B. B. Simons, and R. E. Tarjan, "Scheduling unit-time tasks with arbitrary release times and deadlines," *SIAM J. Comput.* 10 (1981), pp. 256-269.

[8] Z. Galil and N. Megiddo, "A fast selection algorithm and the problem of optimal distribution of effort," *J. ACM,* 26, 1 (1979), pp. 58-64.

[9] J. R. Jackson, "Scheduling a production line to minimize maximum tardiness," Research Report 43 (1955), Management Science Research Project, UCLA.

[10] D. B. Johnson and T. Mizoguchi, "Selecting the kth element in X + Y and $X_1 + X_2 + ... + X_m$," *SIAM J. Comput.*, 7 (1978), pp. 147-153.

[11] R. M. Karp, "Reducibility among combinatorial problems," *Complexity of Computer Computations,* R. E. Miller and J. M. Thatcher, eds., Plenum Press, New York, 1972, pp. 85-103.

[12] J. K. Lenstra, A. H. G. Rinnooy Kan, and P. Brucker, "Complexity of machine scheduling problems," *Annals of Discrete Mathematics,* 1 (1977), pp. 343-362.

[13] B. B. Simons, "A fast algorithm for multiprocessor scheduling," 21st Annual Symposium on Foundations of Computer Science, IEEE Computer Society, Long Beach, California, 1980, pp. 50-53.

[14] B. B. Simons, "A fast algorithm for single processor scheduling," 19th Annual Symposium on Foundations of Computer Science, IEEE Computer Society, Long Beach, California, 1978, pp. 246-252.

[15] B. B. Simons, "Multiprocessor scheduling of unit-time jobs with arbitrary release times and deadlines," IBM Research Report RJ 3235 (1981), San Jose, Ca.

[16] B. B. Simons, "Minimizing maximum lateness of unit-time jobs with arbitrary release times and deadlines," IBM Research Report RJ 3236 (1981), San Jose, Ca.

[17] B. B. Simons and M. F. Sipser, "On scheduling unit time jobs with multiple release time/deadline intervals," IBM Research Report RJ 3236 (1981), San Jose, Ca.

[18] J. D. Ullman, "NP-complete scheduling problems," *Journal of Computer and System Sciences,* 10 (1975), pp. 384-393.

SCHEDULING UNIFORM MACHINES WITH RELEASE TIMES, DEADLINES AND DUE TIMES

Charles Martel

University of California, Davis

ABSTRACT

Given n jobs each of which has a release time, a deadline, and a processing requirement, we examine the problem of determining whether there exists a preemptive schedule on m uniform machines which completes each job in the time interval between its release time and its deadline. An $O((m+\log n)(m^2n^3+n^4))$ algorithm is presented which uses a polymatroidal flow network to construct such a schedule whenever one exists. This algorithm is then used with search techniques to find a schedule which minimizes maximum lateness.

1. INTRODUCTION

In this paper we show that polymatroidal network flows can be used to schedule jobs with release times and deadlines on uniform machines. A description of polymatroidal network flows can be found in (7), and proofs of several properties can be found in (8). We summarize some of these results and use them to design an efficient implementation of the scheduling algorithm.

As part of the general problem formulation we assume that there are m processors indexed i=1,2,...,m, with speeds $s_1 \geq s_2 \ldots \geq s_m$, and n jobs indexed j=1,2,...,n, each with a processing requirement p_j. A job run on machine i for t time units completes $s_i \cdot t$ units of processing. Thus if

Supported in part by NSF grant MCS78-20054

M. A. H. Dempster et al. (eds.), Deterministic and Stochastic Scheduling, 89–99.

t_{ij} = the time job j runs on machine i,

it is necessary that

$$\sum_{i=1}^{i=m} s_i \cdot t_{ij} = p_j$$

in order for job j to be completed. Processors with speed factors of this type are said to be uniform.

A processor can work on only one job at a time, and a job can be worked on by only one processor at a time. The processing of a job can be interrupted at any time and resumed immediately on a new processor or at a later time on any processor. There is no cost or time loss associated with such an interruption or preemption.

We first consider the feasibility problem in which every job has a release time r_j and a deadline d_j. We want to find a schedule which completes every job in the interval between its release time and its deadline or to determine that no such schedule exists. Several special cases of this problem have been previously solved. When the processors are identical, i.e., all speeds are equal, Horn (4) shows that the problem can be converted to a network flow problem which can be solved in $O(n^3)$ time. Gonzalez and Sahni (3) have an O(n+mlogn) algorithm that solves the problem when all release times are equal, and all the deadlines are equal. Sahni and Cho (10) also solve the problem in O(nlogn+mn) time when the jobs have variable release times but all jobs have a common deadline.

After solving the feasibility problem, we use the solution to solve a scheduling problem in which jobs have due times rather than deadlines, and we extend the model to allow jobs which have arbitrary availabilities and processors whose speeds vary.

2. SOLUTION OF THE FEASIBILITY PROBLEM

In the case when each job has a release time and a deadline we can use these release times and deadlines to divide the time line into 2n-1 intervals. Let t_i be the ith smallest value among the n release times and n deadlines (for i=1,2,...,2n). The ith interval is the time period from t_i to t_{i+1} and let $\Delta_i = t_{i+1} - t_i$. The jth job is available within the ith interval iff $r_j \le t_i$ and $d_j \ge t_{i+1}$. Within each interval the set of available jobs does not change; so, given the amount of processing to be done on each job within an interval, scheduling these jobs within the interval is an instance of the problem where all jobs have the same release time and all jobs have a common deadline. Thus it is sufficient to find the amount of processing to be done on each job within each interval, and then use the Gonzalez-Sahni algorithm to construct the actual schedule.

Horvath, Lam and Sethi (5) show that if $q_1 \geq q_2 \geq \ldots \geq q_k$ are the amounts of processing to be done on k jobs then there is a schedule which completes these processing amounts within an interval of length Δ iff:

$$\sum_{i=1}^{j} q_i \leq \left(\sum_{i=1}^{j} s_i \cdot \Delta \right) \quad \text{for } j=1,2,\ldots,m-1, \tag{2.1}$$

$$\sum_{i=1}^{k} q_i \leq \left(\sum_{i=1}^{m} s_i \right) \cdot \Delta . \tag{2.2}$$

Thus we can easily determine if the processing amounts within an interval can be scheduled.

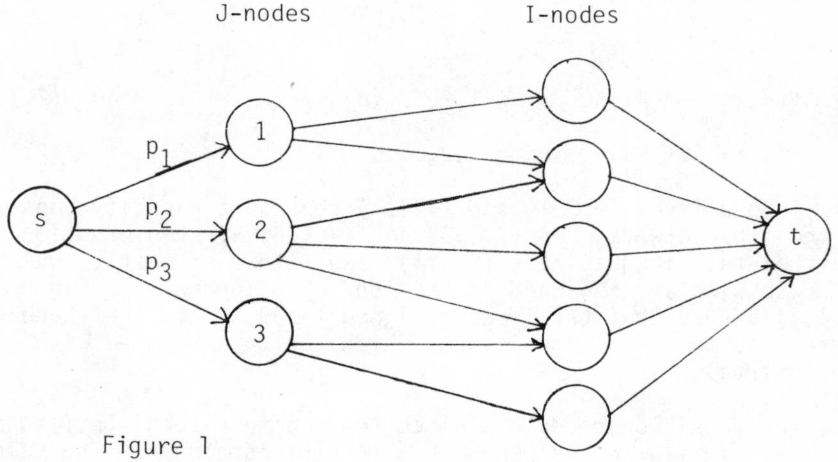

Figure 1

2.1 Converting the Feasibility Problem to a Flow Problem

In order to find the amount of processing to be done on each job within an interval a flow network of the form shown in Figure 1 is used. There is a <u>J-node</u> for each job and in <u>I-node</u> for each interval. An arc connects a J-node to an I-node if the job corresponding to the J-node is available in the corresponding interval. The set of arcs directed out of the jth J-node is A_j and B_i is the set of arcs directed into the ith I-node. There is also a source node s with an arc from s to each J-node, and a sink node t with an arc from each I-node to t. We will denote the arc from s to the jth J-node as <u>(s,j)</u>, the arc from the jth I-node to t as <u>(j,t)</u>, and the arc from the jth J-node to the ith I-node as <u>(j,i)</u>. We will assign flow values to the arcs using a function f which maps the arcs into the non-negative reals. We will apply f to sets of arcs where

$$f(X) = \sum_{x \in X} f(x) \text{ .}$$

A flow function f is <u>feasible</u> iff

$$f((s,j)) \leq p_j \text{ ,} \quad j = 1,2,\ldots n; \tag{2.1.1}$$

$$\text{for all } X \subseteq B_j \text{ , } f(X) \leq \rho(X) \text{ ,} \quad j = 1,2,\ldots,2n-1, \tag{2.1.2}$$

$$\text{where if } X \subseteq B_j \text{ , } \rho(X) = \begin{cases} \Delta_j x \sum_{i=1}^{|X|} s_i & \text{if } |X| < m, \\ \Delta_j x \sum_{i=1}^{m} s_i & \text{if } |X| \geq m; \end{cases}$$

$$f(s,j) = f(A_j) \text{ ,} \quad j = 1,2,\ldots,n; \tag{2.1.3}$$

$$f(j,t) = f(B_j) \text{ ,} \quad j = 1,2,\ldots,2n-1. \tag{2.1.4}$$

Inequalities (2.1.1) and (2.1.2) represent capacity constraints on the flow network, and (2.1.3) and (2.1.4) are conservation constraints. Inequalities (2.1.1) assure that at most p_j units of processing are assigned to the jth job. Inequalities in (2.1.2) assure that the flows assigned to arcs into an I-node will correspond to processing amounts which can be scheduled within the interval.

Any feasible schedule will correspond to a feasible flow and any feasible flow corresponds to a feasible schedule. The value of a flow is the amount of processing completed in the corresponding schedule. Thus a schedule which completes all jobs exists iff the maximum flow is equal to the sum of the processing amounts.

2.2 Finding the Maximum Flow

The constraints (2.1.1), (2.1.3) and (2.1.4) on f, describe a classical network flow problem. The addition of the constraints in (2.1.2) creates a polymatroidal flow problem. However, we have shown in (7) that a maximum flow can be found using techniques very similar to those used for classical flow networks. Starting with any feasible flow, this flow is successively increased using augmenting paths until a maximum flow is achieved. A modified max flow-min cut theorem is used to show that if a flow admits no augmenting path then the flow is maximum. With respect to a flow, an augmenting path is an undirected path from s to t for which there is an amount $\delta>0$, such that increasing the flow in all forward arcs by δ and decreasing the flow in all backward arcs by δ results in a feasible flow. The maximum such value of δ is the path capacity.

Each augmenting path starts with a forward arc (s,j) s.t. $f((s,j))<p_j$. The second arc will be from the jth J-node to an I-node, i. If this second arc, (j,i) can have its flow increased without violating any constraints in (2.1.2), then $(s,j)\to(j,i)\to(i,t)$ is an augmenting path. If increasing the flow in arc (j,i) violates a constraint in (2.1.2), then we can try to offset this by decreasing the flow in an arc (k,i). The augmenting path now contains arcs: (s,j)(j,i)(k,i) where (k,i) is a backwards arc. The next arc in the path will be a forward arc (k,ℓ). If increasing the flow in (k,ℓ) violates no constraints, then $(s,j)(j,i)(k,i)(k,\ell)(\ell,t)$ is an augmenting path. If it does violate a constraint then the path must be extended further. We will now introduce some terminology to allow us to formally describe which paths are augmenting paths. All of the descriptions are with respect to a given flow f.

If y can follow x on an augmenting path and both are incident to the same I-node then (x,y) are an admissible pair.

If for $A \subseteq B_i$, $f(A) = \rho(A)$ then A is saturated and all arcs in A are saturated arcs. When a saturated arc is directed forward in an augmenting path it must be followed by a backwards arc. Also, if $q_1 \geq q_2 \geq \ldots \geq q_k$ are the flows in a saturated set, then:

$$\sum_{j=1}^{k} q_k = \Delta_i \cdot S_k \qquad \text{if } k \leq m,$$
$$= \Delta_i \cdot S_m \qquad \text{if } k > m,$$

where

$$S_k = \sum_{i=1}^{k} s_i .$$

2.3 Finding Augmenting Paths

We will increase the flow by successively finding shortest augmenting paths, using the following labeling algorithm.

```
L:=1;
Label L all arcs (j,i) such that f(s,j) < p_j;
WHILE there is an arc with label L and all arcs with label
L are saturated DO
BEGIN  {scan forward arcs}
    FOR each arc x with label L, give label L+1 to all
        unlabeled arcs y, such that (x,y) is an admissible
        pair;
    L:=L+1;   {we now scan backwards arcs}
    FOR each arc (j,i) with label L, give label L+1 to all
    unlabeled arcs incident from J-node j;
    L:=L+1
END;
FOR each unsaturated arc (j,i) with label L, give label L+1
to (i,t).
```

Note that each arc gets at most one label, and that all arcs directed forward have odd labels and all arcs directed backwards have even labels.

This labeling procedure will find all shortest augmenting paths. To get a shortest path from the labeling we start with an arc incident to t with label L+1 and work backwards successively choosing arcs which have label one less than the current arc, and which could precede the current arc.

In general there may be several arcs which could precede the current arc. If the current arc has an odd label then the arc to precede it is chosen arbitrarily; however, if the current arc has an even label then we will always choose the arc with maximum flow which could precede it. A shortest augmenting path chosen in this way will be called a <u>minimal augmenting path</u>.

If we always augment the flow using a minimal augmenting path then no new minimal augmenting paths as short or shorter than the path chosen will ever be created (8). Thus it will only be necessary for us to label the network once for each path length. We will now look more closely at the details of finding paths and augmenting the flow.

2.4 Complexity Analysis and Implementation

A high level outline of the scheduling algorithm is:

```
form the network from the scheduling problem;
assign an initial feasible flow;
FOR each path length DO
BEGIN
    label the network;
    while an augmenting path of the current length exists DO
```

```
7          BEGIN
8               find a minimal augmenting path;
9               compute the path capacity;
10              augment the flow
     END   END;
11   FOR each interval use the Gonzalez-Sahni algorithm to form
     the schedule.
```

Line 1 requires $O(n^2)$ time to create adjacency lists for each J-node and I-node. Line 2 can be done in $O(n^2)$ if we start with the zero flow. Line 11 requires at most 2n calls to the Gonzalez-Sahni algorithm which runs in O(n+mlogn); so, the total time is $O(n^2+mnlogn)$.

We have shown in (8) that a shortest augmenting path has at most 2n arcs, and that for each path length $O(m^2n+n^2)$ augmentations are needed.

Thus the loop at line 3 is executed O(n) times, and the inner loop at line 6 is executed $O(m^2n^2+n^3)$ times. These loops form the bottleneck of the program and we will now show how to execute lines 5, 8, 9 and 10 efficiently.

In the labeling algorithm of Section 2.3 we have to label arcs, and for each odd labeled arc, x, determine all arcs y such that (x,y) is an admissible pair. To determine this we precompute several quantities. Take the arcs at each I-node and sort them by their flow values. Now for each arc x there is an arc y such that (x,y) is admissible and for all arcs z, (x,z) is admissible iff $f(z) \geq f(y)$ (or there are no arcs y such that (x,y) is admissible). Thus for each arc x we create a pointer to the arc y in the sorted list of arcs.

Thus when we give an odd label to an arc x, we can give label L+1 to all arcs in the list which are above the arc that x points to. We now remove all these arcs from the list; so, they will never be relabeled.

There are $O(n^2)$ arcs in the network, but at most n at each I-node or J-node. The sorted lists at each I-node and the pointers to the proper place in the list can be computed in $O(n^2logn)$ time. Once these are constructed each arc is labeled once, and finding an arc's successors takes constant time. Thus the labeling is $O(n^2logn)$.

Once we have labeled the network we will construct some new data structures. At each I-node we will create a sorted linked list of arcs for each odd label, and at each J-node we will create a linked list of arcs for each even label. These lists are used to extract a minimal augmenting path from the labeling. These lists of arcs are called label lists.

We will also create a binary search tree at each interval node which contains all arcs incident to the node. We will use this to maintain a sorted list of the m arcs at each I-node with the largest flow values. This list of the m largest arcs is used to determine which pairs of arcs are admissible and to compute the capacity of an augmenting path.

Constructing the label lists and the binary search trees will require O(nlogn) time per interval; so, $O(n^2 \log n)$ total time. However, this is only done once per labeling; so, it can be considered part of the labeling. Using these data structures we can backtrack through the network to find a minimal augmenting path. If the current arc has an odd label, L, then we find the arc to precede it by finding list L-1 at the correct J-node and picking the first arc in the list. If the current arc has an even label, L, then we find list L-1 at the correct I-node and pick the largest arc in the list which can precede the current arc.

The arc to precede an odd labeled arc is found in constant time. The arc to precede an arc with even label L can be found in O(m) time since only the m largest arcs with label L-1 need to be considered. The proper arc is chosen using the list of the m largest arcs at the I-node. Since the length of a minimal augmenting path is O(n), the total time to extract a minimal augmenting path is O(mn).

We asserted earlier that using a minimal augmenting path of length L would never create new paths of length L or less. Thus we only relabel when no paths of the current length exist. However, we do need to remove paths which are destroyed when we augment along a minimal augmenting path.

We will remove these paths as we detect them. While trying to extract a minimal path from the network if we find that the current arc, x, with label L has no arc which can precede it, then we remove x from its list of arcs with label L, and attempt to find a new arc from this list. We spent O(m) time to find x and to determine that x had no predecessor, and for each labeling this happens at most once per arc; so, the total time spent removing arcs is $O(mn^2)$ per labeling.

Once we have found the minimal path we need to compute the path capacity. For an admissible pair (x,y) there is an amount σ which is the maximum amount that can be added to x's flow, subtracted from y's flow and still result in a feasible flow. We show in (8) that σ depends only on x,y and the m-largest flow values at the I-node. Thus it can be computed in O(m) time.

Let a minimal path P be (s,j)(j,k)...(ℓ,i)(i,t). The capacity of P is the minimum of

$p_j - f(s,j)$, (2.4.1)

min{σ: σ is a value for an admissible pair in P}, (2.4.2)

the amount we add to (ℓ,i) to make it saturated. (2.4.3)

Evaluating (2.4.2) requires O(n) computations each of which requires O(m) time; so, O(mn) total time is required.

Once we have found the path capacity we can update the flow values. Only O(n) arcs change their flows; so, changing the flow is O(n). However we also need to update the binary search trees and the label lists at each I-node. Updating the binary search trees requires deleting the O(n) arcs in the path, changing their flows, and then inserting them. Thus the total update time for binary search trees is O(nlogn). Only the lists for odd labeled arcs need to be updated. In these label lists only the m largest arcs in each list could have changed and all odd labeled arcs had their flows increased. Thus each changed arc can be moved to its correct position in O(m) time; so, updating the label lists requires O(mn).

We can now give the algorithm's complexity. The labeling of the network, including building label lists and binary search trees, requires $O(n^3 \log n)$ total time.

The inner loop requires O(mn) to find a minimal augmenting path and to compute its capacity. Updating the flow, including data structures, is O(n(m+logn)). This inner loop is executed $O(m^2n^2+n^3)$ times; so, the total complexity is $O((m^2n^3+n^4)(m+\log n))$.

2.5 Number of Preemptions Used

The Gonzalez-Sahni algorithm which constructs the schedule for each interval produces at most 2(m-1) preemptions within each interval (3). However, since a job may appear in O(n) intervals, a schedule may have $O(n^2)$ preemptions. In (8) we show that this schedule can be modified in $O(n^4)$ time to get a schedule with at most 6mn preemptions.

3. THE Lmax PROBLEM

The feasibility algorithm can also be used to solve another scheduling problem. Suppose that each job has a due time rather than an absolute deadline. The lateness of a job is defined to be its completion time minus its due time. The Lmax problem is to find a schedule which minimizes the maximum value of lateness. To solve this problem we find the smallest value, L such that adding L to all the due times and treating the sums as deadlines, yields a feasible scheduling problem. Call this the scheduling problem induced by L. The minimum value of L, which we will denote L*, can be found to within any desired precision using binary search, but if all the data are integers, it is possible to find this value exactly.

There are $O(n^2)$ critical values of L such that

$$d_i + L = r_j$$

for some release time and due time. The intervals within which a job can be processed only change when a critical value of L is reached. Thus we can determine which arcs connect J-nodes to I-nodes by doing a binary search over the $O(n^2)$ critical values. This search will find two consecutive critical values L_0 and L' such that L_0 induces an infeasible scheduling problem and L' induces a feasible one.

The values L_0 and L' allow us to determine the topology of the flow network for the scheduling problem induced by L*. From this we can formulate the problem as a linear program and using this linear program we show in (8) that $L^* = L_0+x$, where $x = y/z$ is a rational with

$$\max\{y,z\} \le M = 3^{O(n^2)} \cdot s_1^n \cdot \mathrm{bmax} ,$$

where bmax is the largest due date plus the sum of all processing requirements. Papadimitriou has shown that x can be found with at most logM tests (9); so L* can be found with $O(n^2+n\cdot\log(s_1)+\log(\mathrm{bmax}))$ calls to the feasibility routine.

4. EXTENSIONS TO THE MODEL

The feasibility algorithm can be used to solve several scheduling problems which are more general than the one considered. Rather than having a release time and a deadline, each job has associated with it a list of time intervals within which it can be processed. Also rather than having the processors available at all times, each processor has associated with it a list of the the time intervals during which it can be used. Using these lists of time intervals we can divide the time line into intervals such that within each interval the jobs and the processors which are available do not change. It is then easy to construct a flow network to solve this problem. The Lmax problem can also be solved when jobs have release times and due times, and each processor has a list of intervals during which it is available.

REFERENCES

1. J. BRUNO, T. GONZALEZ (1976) Scheduling independent tasks with release dates and due dates on parallel machines. Technical Report 213, Computer Science Department, Pennsylvania State University.
2. T. GONZALEZ, S. SAHNI (1976) Open shop scheduling to minimize finish time. *J. Assoc. Comput. Mach.* 23,665-679.
3. T. GONZALEZ, S. SAHNI (1978) Preemptive scheduling of uniform processor systems. *J. Assoc. Comput. Mach.* 25,92-101.
4. W. HORN (1974) Some simple scheduling algorithms. *Naval Res. Logist. Quart.* 21,177-185.
5. E.C. HORVATH, S. LAM, R. SETHI (1977) A level algorithm for preemptive scheduling. *J. Assoc. Comput. Mach.* 25,32-43.
6. J. LABETOULLE, E.L. LAWLER, J.K. LENSTRA, A.H.G. RINNOOY KAN (1979) Preemptive scheduling of uniform machines subject to release dates. Report BW 99, Mathematisch Centrum, Amsterdam.
7. E.L. LAWLER, C.U. MARTEL (1980) Computing maximal "polymatroidal" network flows. Research Memorandum ERL M80/52, Electronics Research Laboratory, University of California, Berkeley.
8. C. MARTEL (1981) Preemptive scheduling with release times, deadlines and due times. *J. Assoc. Comput. Mach.*, to appear.
9. C. PAPADIMITRIOU (1979) Efficient search for rationals. *Inform. Process. Lett.* 8,1-9.
10. S. SAHNI, Y. CHO (1980) Scheduling independent tasks with due times on a uniform processor system. *J. Assoc. Comput. Mach.* 27,550-563.

PREEMPTIVE SCHEDULING OF PRECEDENCE-CONSTRAINED JOBS ON PARALLEL MACHINES

E.L. Lawler

University of California, Berkeley

ABSTRACT

Polynomial time-bounded algorithms are presented for solving three problems involving the preemptive scheduling of precedence-constrained jobs on parallel machines: the "intree problem", the "two-machine problem with equal release dates", and the "general two-machine problem". These problems are preemptive counterparts of problems involving the nonpreemptive scheduling of unit-time jobs previously solved by Brucker, Garey and Johnson and by Garey and Johnson. The algorithms and proofs (and the running times of the algorithms) closely parallel those presented in their papers. These results improve on previous results in preemptive scheduling and also suggest a close relationship between preemptive scheduling problems and problems in nonpreemptive scheduling of unit-time jobs.

1. INTRODUCTION

In this paper we present polynomial time-bounded algorithms for solving three problems involving the preemptive scheduling of precedence-constrained jobs on parallel machines. These three problems, which we call the "intree problem", the "two-machine problem with equal release dates", and the "general two-machine problem", are preemptive counterparts of problems involving the nonpreemptive scheduling of unit-time jobs previously solved by Brucker, Garey and Johnson [1] and by Garey and Johnson [2,3]. The algorithms and proofs we present closely parallel those given in their papers, and the running times of the algorithms are nearly comparable. Problems involving nonpreemptive scheduling of unit-time jobs are sometimes formulated as approximations of preemptive scheduling

M. A. H. Dempster et al. (eds.), Deterministic and Stochastic Scheduling, 101–123.

problems. The results presented in this paper suggest that there is an even closer algorithmic relation between preemptive scheduling and nonpreemptive scheduling of unit-time jobs than had previously been appreciated.

2. PROBLEM DEFINITION

We shall define a general scheduling problem and then indicate how each of the three problems dealt with in this paper is a specialization of this problem.

There are n *jobs* to be scheduled for processing. For each job j, $j = 1,2,\ldots,n$, there are specified a *processing requirement* $p_j > 0$, a *release date* $r_j \geq 0$, prior to which the job is unavailable for processing, and a *due date* $d_j \geq 0$.

The jobs are to be scheduled subject to *precedence constraints* "$\rightarrow$" in the form of a partial order induced by a given acyclic digraph on nodes $j = 1,2,\ldots,n$. If $i \rightarrow j$ then job i must be completed before the processing of job j is begun.

The jobs are to be scheduled on m parallel *machines*. At least m-1 of the machines are identical. One machine is permitted to have the same speed or a strictly slower speed than the others. More specifically let s_i denote the *speed* of machine i, $i = 1,2,\ldots,m$, and assume that $s_1 = s_2 = \ldots = s_{m-1} = 1$, $s_m = s \leq 1$. The *processing capacity* of machine i in a time interval $[t,t']$ is equal to $s_i(t'-t)$. In order for a job to be completed, the job must be allocated sufficient processing capacity to satisfy its processing requirement.

We make the usual assumptions which apply to the scheduling of parallel machines. A machine can process at most one job at a time and a job can be processed by at most one machine at a time. The schedules we consider are *preemptive*, in that processing of a job can be interrupted at any time and processing resumed at the same time on another machine or at a later time on any machine. There is no penalty for such an interruption or "preemption".

A schedule is *feasible* if no job is processed prior to its release date, if all jobs are completed, and if precedence constraints are observed. A feasible schedule *meets all due dates* if each job is completed no later than its due date.

If in a given feasible schedule the *completion time* of a job j is C_j, then its *lateness* with respect to its specified due date d_j is

$$L_j = C_j - d_j.$$

Our objective is to find a schedule which minimizes

$$L_{max} = \max_j \{L_j\}.$$

(Note that there exists a schedule which meets all due dates if and only if there exists a schedule for which $L_{max} \le 0$.)

We now indicate the three special cases of the general problem which are dealt with in this paper.

The Intree Problem: All release dates are zero and the precedence constraints are in the form of a forest of intrees. That is, each job has at most one immediate successor. In the notation of [5], this problem is essentially $P|pmtn,intree|L_{max}$.

The Two-Machine Problem with Equal Release Dates: All release dates are zero and there are exactly two machines. This problem is denoted $Q2|pmtn,prec|L_{max}$.

The General Two-Machine Problem: There are exactly two machines. This is $Q2|pmtn,prec,r_j|L_{max}$.

It should be noted that, by symmetry of release dates and due dates, other specifications are equivalent to the first and second problems above. The "outtree problem" with arbitrary release dates and equal due dates (which may be assumed to be zero) is equivalent to the intree problem. Similarly, the "two-machine problem with equal due dates" (and arbitrary release dates) is equivalent to the two-machine problem with equal release dates. Minimizing maximum lateness for each of these symmetric problems is equivalent to minimizing

$$C_{max} = \max_j\{C_j\}.$$

(This is the same as minimizing "makespan".) The algorithms presented in this paper apply equally well to these symmetric problems, with certain obvious modifications.

3. PREVIOUS RESULTS

A more specialized version of the intree problem was considered by Muntz and Coffman [11] and by Gonzalez and Johnson [4]. In this version, release dates are equal, due dates are equal and the objective is to minimize makespan, *i.e.* the problem is $P|pmtn,intree|C_{max}$. The Muntz-Coffman algorithm requires $O(n^2)$ time and the Gonzalez-Johnson algorithm requires $O(n \log m)$ time. Our procedure solves the more general problem of $P|pmtn,intree|L_{max}$ in $O(n^2)$ time. (In both [11] and [4] it is also assumed that the m machines are strictly identical.)

Muntz and Coffman [10] also solved a more specialized version of the two-machine problem in which release dates are equal, due dates are equal and in which the two machines are identical, *i.e.* $P2|pmtn,prec|C_{max}$. Horvath, Lam and Sethi [6] extended the Muntz-Coffman algorithm to allow the two machines to have different speeds, *i.e.* to $Q2|pmtn,prec|C_{max}$. The running time required for both the Muntz-Coffman and the Horvath-Lam-Sethi algorithms is $O(n^2)$, which is the same as we require for $Q2|pmtn,prec|L_{max}$. (This time bound does not take into account the time required to obtain the transitive closure of the precedence constraints.)

It follows that the intree problem and the general two-machine problem are strictly more general than those for which polynomial-bounded algorithms have previously been obtained. However, for the more special case of the intree problem considered by Gonzalez and Johnson, their time bound is preferable.

The problems solved by Brucker, Garey and Johnson [1] and by Garey and Johnson [2,3] differ from our problems in that jobs with unit processing times ($p_j = 1$ for all j) are to be nonpreemptively scheduled on strictly identical machines. Our algorithms, theorems and proofs are modelled closely after those in [1,2,3], and we achieve fairly similar running time bounds: For $P|intree,p_j=1|L_{max}$ a time bound of $O(n \log n)$ is achieved in [1]. (This can be reduced to $O(n)$ time, as shown by Monma [9]). For $P|pmtn,intree|L_{max}$ we require $O(n^2)$ time. In [2], $Q2|prec,p_j=1|L_{max}$ is solved in $O(n^2)$ time, the same as we require for $Q2|pmtn,prec|L_{max}$. (Time for transitive closure not included.) In [3], $Q2|prec,p_j=1,r_j,d_j|-$ is solved in $O(n^3)$ time, the same as we require for $Q2|pmtn,prec,r_j,d_j|-$. However, Garey and Johnson require only $O(n^3 \log n)$ time for $Q2|prec,p_j=1,r_j|L_{max}$, whereas the best time we have achieved for $Q2|pmtn,prec,r_j|L_{max}$ is $O(n^6)$.

4. PLAN OF ATTACK

In the next section we shall describe a priority scheduling algorithm which can be applied to any instance of the three problems we propose to deal with. This priority scheduling algorithm plays the same role as, but is necessarily more complex than, the list scheduling procedure employed for nonpreemptive scheduling of unit-time jobs in [1,2,3].

In general, there is no assurance that the priority scheduling algorithm yields an optimal schedule, or even finds a schedule in which all jobs meet their due dates, if such a schedule exists. Hence we consider special conditions under which the algorithm yields such a schedule. For the problems $P|pmtn,intree|L_{max}$, $Q2|pmtn,prec|L_{max}$, and $Q2|pmtn,prec,r_j|L_{max}$ we obtain successively stronger "consistency" conditions on the problem data and show that if these conditions are met, then the priority scheduling algorithm finds an optimal schedule (or, in the case of the last problem, one in which all due dates are met).

We than describe procedures for modifying due dates so that the appropriate consistency conditions are satisfied. We prove that the priority scheduling algorithm finds a schedule meeting the modified due dates if and only if there exists such a schedule with respect to the original due dates.

The general procedure for solving each of the three scheduling

problems is then:
(1) Compute the transitive closure of the precedence constraints. (Unnecessary for $P|pmtn,intree|L_{max}$.)
(2) Modify the due dates. (More than one application of the due date modification procedure required for $Q2|pmtn,prec,r_j|L_{max}$.)
(3) Apply the priority scheduling algorithm to determine a sequence of time intervals and the amount of each job to be processed in each interval.
(4) Construct an optimal schedule from the output of the priority scheduling algorithm.

5. THE PRIORITY SCHEDULING ALGORITHM

The priority scheduling algorithm schedules jobs in successive time intervals. Having scheduled intervals $[t_1,t_2],\ldots,[t_{k-1},t_k]$, the algorithm determines a schedule for the next interval $[t_k,t_{k+1}]$ from data for the jobs available for scheduling at t_k. A job j is said to be *available* at time t_k if:
(1) $r_j \leq t_k$,
(2) the processing of each of the predecessors of j (with respect to precedence constraints $\rightarrow$) has been completed, and
(3) there is a nonzero amount of processing $p_j^{(k)} > 0$ remaining to be done on j.

The principal differences between our priority scheduling algorithm for preemptive scheduling and the list scheduling algorithm for nonpreemptive scheduling of unit-time jobs in [1,2,3] are:
(1) Priorities are assigned to jobs according to the values $b_j^{(k)} = d_j - p_j^{(k)}$ (where the smallest value has highest priority). In the case of unit-time jobs, priorities are determined by due dates d_j.
(2) Priorities are dynamic. That is, the priority of a job relative to other jobs is not necessarily known in advance of the time the job becomes available. (However, relative priorities of jobs do not change, as long as they remain available.) In the case of unit-time jobs, priorities are static.
(3) The time intervals $[t_k,t_{k+1}]$ vary in length. In the case of unit-time jobs, each interval has unit length.
(4) Typically only a fraction of the total processing required for a job is done in a single interval. In the case of unit-time jobs, the processing of a job is "all or none".

Suppose without loss of generality that there are n jobs available at t_k and that these jobs are indexed in order of priority, *i.e.*

$$b_1^{(k)} \leq b_2^{(k)} \leq \ldots \leq b_n^{(k)}.$$

(Recall $b_j^{(k)} = d_j - p_j^{(k)}$, where $p_j^{(k)} > 0$ is the amount of processing remaining to be done on job j.) Let $x_j^{(k)}$ be the amount of processing of job j to be done in the interval $[t_k,t_{k+1}]$, where

$\Delta = t_{k+1} - t_k$. We propose to determine the values $x_j^{(k)}$ by solving the following optimization problem:

lexicographically maximize $(b_1^{(k+1)}, b_2^{(k+1)}, \ldots, b_n^{(k+1)})$

subject to

(5.1) $b_1^{(k+1)} \le b_2^{(k+1)} \le \ldots \le b_n^{(k+1)}$,

(5.2) $\max_j \{x_j^{(k)}\} \le \Delta$,

(5.3) $\Sigma_j x_j^{(k)} \le (m-1+s)\Delta$,

(5.4) $0 \le x_j^{(k)} \le p_j^{(k)}$, for all j.

In words, what we propose to do, subject to constraints (5.1)-(5.4), is to process as much as possible of the highest priority job. Then, subject to this amount of processing of the highest priority job, to process as much as possible of the job with second-highest priority, and so forth. Inequalities (5.1) require us to maintain the relative priorities of the jobs which remain uncompleted at time t_{k+1}. Inequalities (5.2) and (5.3) are well-known necessary and sufficient conditions for the processing amounts $x_j^{(k)}$ to be schedulable within an interval of length Δ. Moreover, given values $x_j^{(k)}$, $j = 1,2,\ldots,n$, satisfying (5.2),(5.3), a feasible schedule for the interval $[t_k, t_{k+1}]$ can be constructed in O(n) time [5]. Inequalities (5.4) simply require us to do no more processing of any job than is required to complete it.

The interval length Δ is chosen to be the minimum of the next release date (if any), *i.e.*

$$\min_j \{r_j \mid r_j > t_k\},$$

and the smallest value necessary to complete at least one available job, when the $x_j^{(k)}$ values are chosen as above.

Let us now consider the form of a solution to the optimization problem posed above. We begin with the case in which Δ is fixed and the values $p_j^{(k)}$ are suitably large, *e.g.* $p_j^{(k)} > \Delta$. If there are m-1 or fewer available jobs, the choice of the $x_j^{(k)}$ values is simple:

$$x_j^{(k)} = \Delta, \quad \text{for all j.}$$

This solution is indicated in Figure 1(a), in which each job j is represented by a bar extending from $b_j^{(k)}$ to d_j. The shaded portion of bar j represents $x_j^{(k)}$.

Now suppose that there are at least m jobs available at time t_k. We assert that the x_j values are determined by two parameters: an integer u, $1 \le u \le m$, and a real number T. For given u and T,

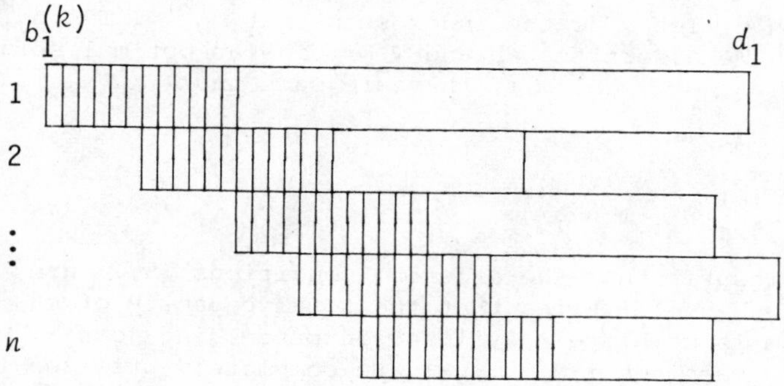

(a) $n \le m-1$

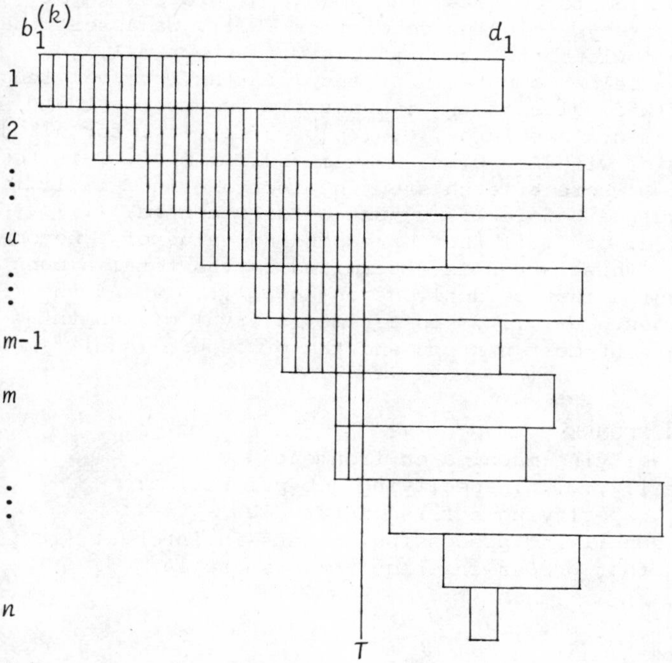

(b) $n \ge m$

Figure 1. Scheduling when t_{k+1} determined by release date.

$$x_j^{(k)} = \begin{cases} \Delta, & j < u, \\ \max\{0, T-b_j^{(k)}\}, & j \ge u. \end{cases}$$

A solution of this form is illustrated in Figure 1(b).

For convenience, let v be the index such that $b_v^{(k)} \le T < b_{v+1}^{(k)}$ (where $b_{n+1}^{(k)} = +\infty$). We assert that u and T provide an optimal solution to our optimization problem if they are such that

$$(5.5) \qquad b_{u-1}^{(k)}+\Delta \le T,$$

$$(5.6) \qquad \Sigma_{j=u}^{v} (T-b_j^{(k)}) = (m-u+s)\Delta.$$

Inequality (5.5) insures that the ordering conditions (5.1) are satisfied. Equation (5.6) implies that the total capacity of machines u,u+1,...,m is completely utilized in processing jobs u,u+1,...,v (while machines 1,2,...,u-1 are completely utilized in processing jobs 1,2,...,u-1). There is thus no idle time on any of the m machines; in order to increase the amount of processing of any job, we would need to decrease the amount of processing of some other job, thereby violating conditions (5.1). We assert that it is intuitively obvious that such a solution is optimal.

The procedure below computes u, v and T for an interval length determined by a fixed value of t_{k+1}, under the assumption that processing requirements are large, *e.g.* $p_j^{(k)} \ge t_{k+1}-t_k$, for all j. The procedure begins with $\Delta = 0$, u = v = m and $T = b_m^{(k)}$. T is then increased in steps, where at each step the increase in T is limited by the amount required before (i) u must be decremented, (ii) v must be incremented, or (iii) the corresponding value of Δ becomes equal to $t_{k+1}-t_k$. (This latter is determined by the total amount of processing P which must be done on jobs u,u+1,...,v and the capacity of machines u,u+1,...,m in a time interval of length $t_{k+1}-t_k$.) There are at most n steps and the procedure requires at most O(n) time.

FIXED INTERVAL PROCEDURE

Input: m,s, specifying machine environment;
b_j, j = 1,2,...,n, specifying job priorities;
t_k, t_{k+1}, specifying a fixed interval.

Output: u,v,T, specifying processing amounts in interval (it is assumed that processing requirements are large).

```
begin procedure
    Δ := 0;
    u := m;
    v := m;
    T := b_m;
    P := 0;
    while Δ < t_{k+1}-t_k
    do  while T = b_{u-1}+Δ
        do  u := u-1; [If u = 1 then let b_0 = -∞.]
            P := P+Δ
        od;
        while T = b_{v+1}
        do  v := v+1 [If v = n then let b_{n+1} = +∞.]
        od;
```

```
        T1 := ((b_{u-1}+Δ)(m-u+s)-(v-u+1)T)/((m-u+s)-(v-u+1));
        T2 := b_{v+1};
        T3 := T + (t_{k+1}-t_k-Δ)(m-u+s)/(v-u+1);
        [T1,T2,T3 are limits on the new size of T, as described
        in conditions (i),(ii),(iii) in text above.]
        T' := min{T1,T2,T3};
        P := P + (v-u+1) (T'-T);
        Δ := P/(m-u+s); T := T'
    od
end procedure.
```

Now let us consider how to determine Δ in the case that the interval length is to be just large enough so that one or more jobs are completed. Clearly, when Δ is properly determined and u,v,T are found as above, we should have

(5.7) $\quad b_j+\Delta \le d_j, \quad j = 1,2,\ldots,u-1,$

(5.8) $\quad T \le d_j, \quad j = u,u+1,\ldots,v,$

with equality in at least one case.

Our strategy is as follows. We first carry out a computation very much as in the fixed interval procedure, but with two modifications:

(1) When T is stepped, T is not permitted to violate conditions (5.8). When $T = d_j$, for some $j \ge u$, the computation stops.

(2) When u is decremented, the condition $b_{u-1}+\Delta > d_{u-1}$ is checked. When this is found to hold, the computation stops.

We shall call the modified procedure the "variable interval procedure". (We leave implementation of the modifications to the reader.) At the conclusion of the variable interval procedure we shall have either:

(1) determined u,v,T for the interval $[t_k,t_{k+1}]$, without violating (5.7) or (5.8);

(2) found values u,v,T and $\Delta < t_{k+1}-t_k$ such that (5.7),(5.8) are satisfied, and one of the constraints (5.8) is satisfied with equality;

(3) found values u,v,T and $\Delta < t_{k+1}-t_k$ such that conditions (5.8) are satisfied, but at least one constraint (5.7) is violated.

In the first two cases we are done: the value $t_{k+1} = t_k+\Delta$ found by the procedure is correct. In the third case, the correct value of t_{k+1} is simply

$$t_{k+1} = t_k + \min\{p_j^{(k)} \mid 1 \le j \le u-1\}.$$

Running the procedure one more time with the correct value of t_{k+1} completes the computation.

We are now ready to indicate the complete priority scheduling procedure.

```
PRIORITY SCHEDULING PROCEDURE
Input:    m,s, specifying machine environment;
          triples (p_j,r_j,d_j), j = 1,2,...,n, specifying jobs;
          an acyclic digraph specifying precedence constraints.
Output:   intervals [t_k,t_{k+1}], k = 1,2,...,N-1, where N ≤ 2n-1;
          values x_j^(k) ≥ 0, indicating the amount of processing of
          job j to be performed in interval [t_k,t_{k+1}].
begin procedure
    compute the in-degree of each job;
    create a priority queue Q (by release date) containing all
    jobs with in-degree zero;
    create a list A of available jobs in nondecreasing order of
       b_j (job priority) values; initially A is empty;
    k := 0;
    while Q is nonempty
    do    k := k+1;
          t_k := min{r_j|j ∈ Q};
          add to A and remove from Q all jobs in Q such that
          r_j = t_k;
          while A is nonempty
          do    t_{k+1} := min{r_j|j ∈ Q};
                apply the variable interval procedure to the jobs
                in A; if the resulting values of Δ,u,v,T violate
                conditions (5.7), rerun the procedure with t_{k+1} =
                t_k + min{p_j^(k)|1 ≤ j ≤ u-1}; [This determines u,v,T,
                and x_j^(k), for all j.]
                remove from A all jobs completed in the interval
                [t_k,t_{k+1}]; for each such job decrement the in-degree
                of each of its successors and place in Q each job
                whose in-degree is reduced to zero;
                add to A and remove from Q all jobs in Q such that
                r_j ≤ t_{k+1};
                k := k+1
          od
    od
end procedure.
```

We assert that the priority scheduling procedure can be implemented to run in $O(n^2)$ time overall. As we commented above, the output of the procedure can also be transformed into an actual schedule in $O(n^2)$ time.

6. A USEFUL LEMMA

We wish to consider the effect of applying the priority scheduling procedure to jobs whose due dates satisfy the *consistency conditions*

(6.1) $\quad d_j \le b_k \quad$ whenever $j \to k$

and

(6.2) $r_j \le b_j$ for all j.

The following is a self-evident property of the priority scheduling procedure.

PROPOSITION. *The procedure fails to schedule all jobs to meet due dates if and only if at some interval* $[t_{\ell-1}, t_\ell]$ *there is a job* j *(available at* $t_{\ell-1}$*) for which* $b_j^{(\ell)} < t_\ell$.

LEMMA 6.1. *Suppose the priority scheduling algorithm is applied to a set of jobs for which* (6.1) *and* (6.2) *hold. Let* $[t_{\ell-1}, t_\ell]$ *be the earliest interval at which there is a job* j *with* $b_j^{(\ell)} < t_\ell$. *Then* $b_j^{(\ell)} = T < t_\ell$ *for all jobs scheduled for processing in the interval and there is no idle time on any of the machines in the interval.*

Proof. Assume the jobs are numbered so that $b_1^{(\ell-1)} \le b_2^{(\ell-1)} \le \ldots \le b_n^{(\ell-1)}$, $b_1^{(\ell)} \le b_2^{(\ell)} \le \ldots \le b_n^{(\ell)}$. Then $b_1^{(\ell)} < t_\ell$. If $n \ge m$ and $u = 1$ the result follows immediately.

So let us suppose that either $n \le m-1$ or $n \ge m$ and $u \ge 2$. Then we would have $b_1^{(\ell)} = b_1^{(\ell-1)} + t_\ell - t_{\ell-1}$ and $b_1^{(\ell-1)} < t_{\ell-1}$. If $\ell \ge 2$ and job 1 was available at $t_{\ell-2}$, then $[t_{\ell-1}, t_\ell]$ would not be an earliest interval at which there is a job j with $b_j^{(\ell)} < t_\ell$. If $\ell = 1$ or if job 1 was not available at $t_{\ell-2}$, then job 1 first became available at $t_{\ell-1}$. This implies that either $r_1 = t_{\ell-1} > b_1^{(\ell-1)} = b_1$, in contradiction to (6.2), or that some job j, with $j \to 1$, was completed in the interval $[t_{\ell-2}, t_{\ell-1}]$. But if this latter were the case, then, by (6.1), $b_j^{(\ell-1)} = d_j \le b_1 = b_1^{(\ell-1)} < t_{\ell-1}$ and $[t_{\ell-1}, t_\ell]$ would not be the earliest interval with a $b_j^{(\ell)} < t$. It follows that $n \ge m$ and $u = 1$, and the lemma is proved. □

7. THE INTREE PROBLEM

We now turn to the in-tree problem, $P|pmtn, intree|L_{max}$. Each job has at most one immediate successor and $r_j = 0$ for all jobs.

We first consider the question of whether or not there exists a schedule meeting all due dates. If a job k is to meet its due date d_k, then the latest possible time at which its processing can begin is $b_k = d_k - p_k$. If $j \to k$, job j must be completed by time b_k, else k will be late. Thus not only must job j be completed by time d_j, it must be completed by time b_k. Since the processing of job j must obey both constraints, we can replace due date d_j by a new due date $d_j' = \min\{d_j, b_k\}$ without changing the problem in any essential way. A schedule meets all due dates in the new problem if and only if it meets all due dates in the original problem. This type of due date modification can be applied repeatedly until we finally obtain an equivalent problem having *modified due dates* d_j' satisfying the consistency conditions

(7.1) $d'_j \le b'_k = d'_k - p_k$, whenever $j \rightarrow k$.

The following simple algorithm constructs such a set of modified due dates and can be implemented to run in time $O(n)$.

DUE DATE MODIFICATION PROCEDURE

Input: ordered pairs (p_j, d_j), $j = 1,2,\dots,n$, specifying jobs; an acyclic digraph in which each node has out-degree at most one.

Output: modified due dates d'_j, $j = 1,2,\dots,n$.

begin procedure
  for each job j which has no successor
  do $d'_j := d_j$
  od;
  while there is a job which has not been assigned its modified due date and whose immediate successor k has had its due date modified, select such a job j and
  do $d'_j := \min\{d_j, b'_k = d'_k - p_k\}$
  od
end procedure.

The arguments above establish the following result.

LEMMA 7.1. *There exists a schedule meeting the original due dates if and only if there exists a schedule meeting the modified due dates.*

THEOREM 7.2. *The schedule obtained by applying the priority scheduling algorithm to the problem with modified due dates meets all original due dates if and only if such a schedule exists.*

Proof. Since $d'_j \le d_j$, for all j, if the schedule obtained by applying the priority scheduling algorithm meets the modified due dates then it meets the original due dates. So suppose the schedule obtained does not meet the modified due dates. We shall show that this implies that there exists no schedule meeting the modified due dates. Then, by Lemma 7.1, there is no schedule meeting the original due dates.

So suppose the schedule does not meet the modified due dates and let us apply Lemma 6.1. Let $[t_{\ell-1}, t_\ell]$ be the earliest interval at which there is a job j with $b_j^{(\ell)} = T < t_\ell$. We assert that in each earlier interval $[t_{k-1}, t_k]$, $k = 2,\dots,\ell-1$, there is no idle time on any machine and $b_j^{(k)} \le T$ for each job processed in the interval. For suppose this were not the case. Then at least one job j processed in $[t_{\ell-1}, t_\ell]$ was unavailable at t_{k-1}, else the algorithm would have scheduled it (in preference to idle time or to a job i with $b_i^{(k)} > T$). But j could only have been unavailable at t_{k-1} because one or more of its predecessors was not yet completed. But for each predecessor i of j, $d'_i \le b'_j$, so necessarily

$b_i^{(k)} \leq T$. No two jobs processed in $[t_{\ell-1},t_\ell]$ have a common predecessor (a property of intree orders), so we have established our assertion.

Let

$$p_j(t) = \begin{cases} p_j, & t \geq d'_j, \\ \max\{0,p_j-(d'_j-t)\}, & t \leq d'_j. \end{cases}$$

Then $p_j(t)$ denotes an amount of processing that must be done on job j prior to time t, if its due date is to be met. The above analysis shows that

$$\Sigma_j\ p_j(T) \geq (m+s-1)t_\ell > (m+s-1)T.$$

In other words, there is more processing which must be performed before time T than the m machines have processing capacity. Hence there can be no schedule meeting the modified due dates and the theorem is proved. □

As it turns out, not only have we solved the problem of constructing a schedule meeting all due dates, if such a schedule exists, but we have also solved the problem of minimizing maximum lateness.

THEOREM 7.3. *The schedule obtained by applying the priority scheduling algorithm to the problem with modified due dates minimizes maximum lateness with respect to the original due dates.*

Proof. The problem of minimizing L_{max} is clearly equivalent to the problem of determining the smallest possible value of L such that when the original due dates d_j are replaced by due dates d_j+L, $j = 1,2,\ldots,n$, there exists a schedule meeting the new due dates. But for any such L the modified due dates are simply changed by the same constant L. That is, if modified due dates d'_j are obtained from the original due dates d_j and d''_j are obtained from due dates d_j+L, then $d''_j = d'_j+L$, $j = 1,2,\ldots,n$. (This can be proved by straightforward induction.) Moreover, the priority scheduling algorithm yields precisely the same schedule when applied to due dates d''_j and d'_j. (We also omit proof of this fact, considering it to be obvious.) It follows that the schedule constructed by the algorithm when applied to the problem with modified due dates d'_j minimizes maximum lateness. □

Notice that if all the original due dates are the same, our procedure constructs a schedule which minimizes makespan. However, as we pointed out earlier, the algorithm of Gonzalez and Johnson [4] provides a running time of $O(n \log m)$, whereas ours is $O(n^2)$.

8. THE TWO-MACHINE PROBLEM WITH EQUAL RELEASE DATES

Now let us consider the two-machine problem with equal release dates, $Q2|pmtn,prec|L_{max}$.

We first consider the question of whether or not there exists a schedule which meets all due dates. As in the case of the intree problem, we note that due dates can be modified so as to observe condition (6.1). However, additional modifications are necessary, as well as more notation.

As in the proof of Theorem 7.2, let us define

$$p_j(t) = \begin{cases} p_j, & t \geq d_j, \\ \max\{0, p_j-(d_j-t)\}, & t < d_j. \end{cases}$$

In words, $p_j(t)$ is a lower bound on the amount of processing that must be done on job j prior to time t, if job j is to meet its due date. Let S(j) denote the set of all successors (immediate or not) of job j. Then in any schedule in which each successor k of j meets its due date, it must be the case that

$$C_j \leq t - \frac{1}{1+s} \Sigma_{k \in S(j)}\, p_k(t), \quad \text{for all } t \geq d_j.$$

The reason for this is that the processing capacity of the two machines in the interval $[C_j,t]$ is $(1+s)(t-C_j)$ and this must not be less than the total amount of processing of the successors of j which must be performed prior to t.

This suggests the additional consistency condition

$$(8.1) \qquad d_j \leq t - \frac{1}{1+s} \Sigma_{k \in S(j)}\, p_k(t), \quad \text{for all } j \text{ and } t \geq d_j.$$

Fortunately, in order to modify due dates so that they conform with this consistency condition, it is not necessary to check (8.1) for all values of $t \geq d_j$.

LEMMA 8.0. *If for given* j, *conditions* (8.1) *are satisfied for all*

$$d \in \{d_k | j \to k\},$$

then condition (8.1) *is satisfied for all* $t \geq d_j$.

Proof. Let $t \geq d_j$ and let

$$\bar{d} = \min\{d_k | d_k \geq t,\ j \to k\},$$

$$\underline{d} = \max\{d_k | d_k \leq t,\ j \to k\}.$$

If both sets above are empty, then j has no successors and (8.1) is clearly satisfied. If the first set is empty, but not the second, $\underline{d} = \max\{d_k | j \to k\}$ and it is easy to verify that satisfaction of (8.1) for $\underline{d}$ implies satisfaction for t. There is a similar argument if the second set is empty. So assume neither set is empty and $\underline{d},\bar{d}$ are both well defined.

If $\Sigma p_k(\underline{d}) \geq \Sigma p_k(t) - (1+s)(t-\underline{d})$ or if $\Sigma p_k(\bar{d}) \geq \Sigma p_k(t) + (1+s)(\bar{d}-t)$, then satisfaction of (8.1) for $\underline{d},\bar{d}$, respectively, implies satisfaction for t. Suppose that $\Sigma p_k(\underline{d}) < \Sigma p_k(t) - (1+s)(t-\underline{d})$ and $\Sigma p_k(\bar{d}) < \Sigma p_k(t) + (1+s)(\bar{d}-t)$. Examination of the definition of $p_j(t)$ reveals that

$$\frac{\Sigma p_k(t) - \Sigma p_k(\underline{d})}{t-\underline{d}} \leq \frac{\Sigma p_k(\bar{d}) - \Sigma p_k(t)}{\bar{d}-t},$$

which yields a contradiction. □

The following algorithm constructs a set of modified due dates satisfying both (6.1) and (8.1). It can be implemented to run in $O(n^2)$ time, provided precedence constraints are already in transitively closed form.

DUE DATE MODIFICATION PROCEDURE

```
Input:   s, 0 < s ≤ 1, specifying speed of second machine;
         ordered pairs (p_j,d_j), j = 1,2,...,n, specifying jobs;
         a transitively closed acyclic digraph specifying prece-
         dence constraints →.
Output:  modified due dates d'_j, j = 1,2,...,n.
begin procedure
    for each job j which has no successors
    do   d'_j := d_j
    od;
    create two ordered lists of jobs whose due dates have been
    modified, one ordered by modified due dates d'_j, the other
    ordered by the values b'_j = d'_j-p_j;
    while there is a job which has not been assigned its modified
    due date and all of whose successors have had their due dates
    modified, select such a job j and
    do   scan the two lists of d'_j and b'_j values, extracting all
         the successors of j;
         merge the two sublists and use the merged list to compute
         the sum Σ_{k∈S(j)} p_k(d'), for each modified due date d'
         assigned to a successor of j; [All such sums can be com-
         puted in O(n) time.]
         d'_j := min{d_j,min_d'{d' - Σp_k(d')/(1+s)}};
         for each successor k of j
         do   d'_j := min{d'_j,b'_k = d'_k-p_k}
         od;
         insert d'_j, b'_j = d'_j-p_j into ordered lists
    od
end procedure.
```

LEMMA 8.1. *There exists a schedule meeting the original due dates if and only if there exists a schedule meeting the modified due dates.*

THEOREM 8.2. *The schedule obtained by applying the priority scheduling algorithm to the problem with modified due dates meets all the original due dates if and only if such a schedule exists.*

Proof. As in the proof of Theorem 7.2, if the schedule obtained by applying the priority scheduling algorithm meets the modified due dates, then it meets the original due dates. So, as before, suppose the schedule obtained does not meet the modified due dates and let us show that this implies that there exists no schedule meeting the modified due dates.

Let $[t_{\ell-1}, t_\ell]$ be the first interval in which $b_j^{(\ell)} < t_\ell$ for some job j. By Lemma 6.1, there is no idle time on any machine in this interval and $b_j^{(\ell)} = T < t_\ell$ for all jobs j processed in $[t_{\ell-1}, t_\ell]$.

Case 1. In each earlier interval $[t_{k-1}, t_k]$ there is no idle time on any machine and $b_j^{(\ell)} \leq T$ for each job processed in the interval. Then we have

$$\Sigma_j\, p_j(T) \geq (1+s)t_\ell > (1+s)T,$$

and there can be no schedule meeting the modified due dates.

Case 2. There is an earlier interval $[t_{k-1}, t_k]$ in which there is either idle time or a job j is processed with $b_j^{(\ell)} > T$. Let $[t_{k-1}, t_k]$ be the latest such interval. It must be the case that the jobs processed in $[t_{\ell-1}, t_\ell]$ are unavailable for processing in $[t_{k-1}, t_k]$, from which it follows that some job j is processed and completed in $[t_{k-1}, t_k]$ which is a predecessor of all such jobs. Let j be this job. By (8.1),

$$d'_j \leq T - \frac{1}{1+s}\, \Sigma_{k\in S(j)}\, p_k(T).$$

But

$$\Sigma_{k\in S(j)}\, p_k(T) \geq (1+s)(t_\ell - t_k) > (1+s)(T - t_k),$$

which implies $d'_j < t_k$. But $d'_j = b_j^{(k)} \geq t_k$, which yields a contradiction. Hence Case 1 must apply and there can exist no schedule meeting the modified due dates. □

THEOREM 8.3. *The schedule obtained by applying the priority scheduling algorithm to the problem with modified due dates minimizes maximum lateness with respect to the original due dates.*

Proof. Similar to that for Theorem 7.3. □

9. THE GENERAL TWO-MACHINE PROBLEM

Now consider the general two-machine problem, $Q2|pmtn,prec,r_j|L_{max}$: there are arbitrary release dates and precedence constraints, but of course only two machines.

Once again we first consider the question of whether or not there exists a schedule which meets all due dates. Consistency conditions (6.1) and (8.1) apply as in the previous section, but are not strong enough for our purposes. More restrictive conditions can be formulated as follows.

Let $p_j(t)$ be some known lower bound on the amount of processing of job j which must be done prior to time t. A function defining lower bounds of this type will be required to satisfy the conditions

$$(9.0) \qquad p_j(t) \begin{cases} = p_j, & t \geq d_j, \\ \geq \max\{0, p_j(t')-(t'-t)\}, & t \leq t' \leq d_j. \end{cases}$$

For given j,r,t, with $r_j \leq r \leq t$, define

$$P(j,r,t) = \Sigma_{k \in S(j)}\, p_k(t) + \Sigma_{k \notin S(j), r_k \geq r, k \neq j}\, p_k(t).$$

If $P(j,r,t) > (1+s)(t-r)$, the amount of processing to be done prior to time t and after the completion of job j exceeds the total processing capacity of the two machines in the interval [r,t]. It follows that if all due dates are to be met we must have

$$C_j \leq t - \frac{1}{1+s} P(j,r,t).$$

Similarly, if $P(j,r,t) \leq (1+s)(t-r)$, $(1+s)(t-r)-P(j,r,t)$ is an upper bound on the amount of processing of job j which can be done in the interval [r,t]. It follows that we should have

$$p_j(r) \geq p_j(t)-(1+s)(t-r)+P(j,r,t).$$

Motivated by these observations, we frame the following definition.

Definition. Release dates r_j and due dates d_j, $j = 1,2,\ldots,n$, are *internally consistent* if there exist functions $p_j(t)$ satisfying (9.0), such that for all j,r,t, with $r_j \leq r \leq t$,

$$(9.1) \qquad r_j \geq r_i+p_i \qquad \text{whenever } i \rightarrow j,$$

$$(9.2) \qquad d_j \leq t-p_k(t) \qquad \text{whenever } j \rightarrow k \text{ and } p_k(t) > 0,$$

$$(9.3) \qquad p_j(r_j) = 0 \qquad \text{for all } j,$$

$$(9.4) \qquad d_j \leq t - \frac{1}{1+s} P(j,r,t) \quad \text{if } P(j,r,t) > (1+s)(t-r),$$

(9.5) $$p_j(r) \geq p_j(t)-(1+s)(t-r)+P(j,r,t) \quad \text{if } P(j,r,t) \leq (1+s)(t-r).$$

The following lemma shows that in order to establish internal consistency of due dates, it is unnecessary to verify conditions (9.0)-(9.5) for all j,r,t, with $r_j \leq r \leq t$. It is sufficient to consider only $r \in R$, $t \in R \cup D$, where $R = \{r_k | k = 1,2,\ldots,n\}$, $D = \{d_k | k = 1,2,\ldots,n\}$.

LEMMA 9.0. *Let* $p_j(t)$ *be functions with domain* $R \cup D$. *If these functions satisfy* (9.0)-(9.5) *for* $r \in R$, $t \in R \cup D$, *then the due dates are internally consistent.*

Proof. Extend the functions to $t \notin R \cup D$ by the rule

$$p_j(t) = \begin{cases} p_j, & t > d_j, \\ \max\{0, p_j(t')-(t'-t)\}, & t < d_j. \end{cases}$$

The lemma follows by reasoning similar to that used in the proof of Lemma 8.0. □

Lemma 9.0 suggests that it is possible to construct an algorithm for modifying due dates, based on the construction of functions $p_j(t)$ over the domain $R \cup D$. We shall employ three loops in this algorithm. The outer loop considers values of $t \in R \cup D$, in decreasing order. The middle loop considers values of j, in arbitrary order, and the inner loop considers values of $r \in R$ in increasing order.

For a given triple t,j,r, the algorithm compares $P(j,r,t)$ with $(1+s)(t-r)$ and modifies either d_j or $p_j(r)$, in accordance with (9.4),(9.5), as necessary. When the processing of j is completed for a given value of t, the values $p_j(u)$, $u \leq t$, are revised to conform with condition (9.0). The algorithm then halts if $p_j(r_j) > 0$, in violation of (9.3). The due dates of predecessors of j are then modified in accordance with (9.2).

Each value t considered in the outer loop is either a (fixed) release date or a due date which remains unchanged by any further processing. Because of the maintenance of conditions (9.2), if $t = d_j$ then $p_k(t) = 0$ for all successors k of j. Hence if $P(j,r,t) > (1+s)(t-r)$ (which is the only other condition under which d_j can be changed), then $P(k,r,t) > (1+s)(t-r)$ for any job k such that $r = r_k$, and $p_k(r_k) > 0$, in violation of (9.3). It follows that at most $2n$ values of t are considered by the outer loop.

We next note that $P(j,r,t)$ is determined only by the values of $p_k(t)$, $k = 1,2,\ldots,n$. Since values of t are processed in decreasing order, the values $p_k(t)$ are not changed by further repetitions of the loop. It follows that if the due date modification algorithm runs to termination without halting because of violation of condition (9.3), the functions $p_j(t)$ constructed by the algorithm satisfy conditions (9.0)-(9.5). By Lemma 9.0, the modified due dates are then internally consistent.

DUE DATE MODIFICATION PROCEDURE

```
Input:   m,s, specifying machine environment;
         triples (p_j,r_j,d_j), j = 1,2,...,n, specifying jobs;
         a transitively closed acyclic digraph specifying prece-
         dence constraints →.
         [We assume that

              r_j ≥ max{r_i+p_i | i → j},

              d_j ≤ min{d_i-p_i | j → i};

         if this is not so, release and due dates should be modi-
         fied accordingly.]
Output:  modified due dates d_j, j = 1,2,...,n, or indication that
         release dates and due dates are not internally consistent.
begin procedure
     R := {r_j | j = 1,2,...,n};
     D := {d_j | j = 1,2,...,n};
     p_j(t) := max{0,min{p_j-(d_j-t),p_j}}, t ∈ R∪D; t := +∞;
     loop if all t ∈ R∪D have been considered
          then stop
          fi; [Modified due dates are internally consistent.]
          set t to the largest value in R∪D less than current t;
          for all j with d_j ≤ t
          do   r := r_j;
               P(j,r,t) := Σ_{k∈S(j)} p_k(t) + Σ_{k∉S(j),r_k≥r,k≠j} p_k(t);
               while there is an r' in R such that r < r' ≤ t, let
               r' be the smallest value in R greater than r and
               do   P(j,r,t) := P(j,r,t) - Σ_{k∉S(j),r≤r_k<r'} p_k(t);
                    r := r';
                    if P(j,r,t) ≤ (1+s)(t-r)
                    then p_j(r) := max{p_j(r),p_j(t)-(1+s)(t-r)+P(j,r,t)}
                    else d_j := min{d_j,t - P(j,r,t)/(1+s)};
                         update D, as necessary
                    fi
               od;
               let r_j = u_1 < u_2 < ... < u_ℓ = t be the different
               values in R∪D between r_j and t;
               if p_j(u_i) has not previously been computed
               then p_j(u_i) := 0
               fi;
               for i = ℓ-1,ℓ-2,...,1
               do   p_j(u_i) := max{p_j(u_i),p_j(u_{i+1})-(u_{i+1}-u_i)}
               od;
               if p_j(r_j) > 0
               then stop
               fi; [Due dates are not internally consistent.]
               t' := min{u ∈ R∪D | p_j(u) > 0},
               for each k such that k → j
```

do $d_k := \min\{d_k, t'-p_j(t')\}$; update D
od
od
forever
end procedure.

Notice that the for loop on j is executed $O(n^2)$ times and that the computation of $P(j,r,t)$ requires only $O(n)$ time. The while loop on r is executed $O(n^2)$ times and the revision of $P(j,r,t)$ requires at most one reference to a given job k, $k \notin S(j)$, for each combination of t and j. Hence only $O(n^3)$ time is required for the revision of $P(j,r,t)$. A similar analysis of other details confirms that the algorithm runs in $O(n^3)$ time.

We now wish to show that if due dates are internally consistent, then there exists a schedule meeting all due dates. We shall do this, not by applying the priority scheduling algorithm directly to the n jobs with modified due dates, but by applying the priority scheduling algorithm to a new problem with an expanded set of $O(n^2)$ jobs.

Suppose at the termination of the due date modification algorithm there is a job j and a $t \in R \cup D$, $t < d_j$, such that $p_j(t)$, as computed by the algorithm, is such that

$$p_j(t) > p_j-(d_j-t),$$

where d_j is the modified due date. If we were simply to apply the priority scheduling algorithm to the original set of jobs with modified due dates, it might fail to schedule a sufficient amount of processing for job j prior to time t, thereby constructing a schedule which fails to meet due dates, though such a schedule exists. To overcome this difficulty, we replace each job j by a chain of smaller jobs, as follows.

Let $p_j(t)$, $j = 1,2,\ldots,n$, $t \in R \cup D$, be the values computed by the due date modification algorithm. For each job j, let

$$\{u | u \in R \cup D,\ p_j(u) > 0,\ u \le d_j\} = \{u_1, u_2, \ldots, u_\ell\},$$

with $u_1 < u_2 < \ldots < u_\ell = d_j$. Create ℓ new jobs, $j(1), j(2), \ldots, j(\ell)$. Set $p_{j(1)} = p_j(u_1)$, $d_{j(1)} = u_1$, $r_{j(1)} = r_j$, and set $p_{j(i+1)} = p_j(u_{i+1})-p_j(u_i)$, $d_{j(i+1)} = u_{i+1}$, $r_{j(i+1)} = r_{j(i)}+p_{j(i)}$, for $i = 2,3,\ldots,\ell$. Replace job j by the ℓ new jobs, and modify the precedence constraints so that $h \to j(1)$, for all $h \to j$, $j(\ell) \to k$, for all $j \to k$, and $j(i) \to j(i+1)$, for $i = 1,2,\ldots,\ell-1$.

PROPOSITION. *There exists a schedule meeting the modified due dates of the* n *jobs if and only if there exists a schedule meeting the due dates of the expanded set of jobs, as obtained above.*

We also note that any feasible schedule for the $O(n^2)$ jobs is easily transformed into a feasible schedule for the original n jobs.

THEOREM 9.1. *If due dates and release dates are internally consistent, then there exists a schedule meeting all due dates.*

Proof. Apply the priority scheduling algorithm to the expanded set of $O(n^2)$ jobs obtained as above. Suppose it fails to obtain a schedule meeting all due dates, and let $[t_{\ell-1},t_\ell]$ be the first interval in which there is a job j with $b_j^{(\ell)} = T < t_\ell$.

Case 1. In all earlier intervals $[t_{k-1},t_k]$ there is no idle time and there is no job i with $b_i^{(k)} > T$. Then

$$\Sigma_i\, p_i(T) \geq (1+s)(t_\ell - r_1) > (1+s)(T-r_1),$$

where r_1 is the earliest release date. But this is not possible, else $p_1(r_1) > 0$.

Case 2. There is an earlier interval $[t_{k-1},t_k]$ in which there is either idle time or some job i is processed with $b_i^{(k)} > T$. Let $[t_{k-1},t_k]$ be the latest such interval.

(a) Suppose that a job h is completed at time t_k. Then each job processed between time t_k and time t_ℓ is either a successor of job h or is a job with release date t_k or later. (Note that condition (9.1) insures that each successor of a job with release date t_k or later has a release date later than t_k.) It follows that

$$P(h,t_k,T) \geq (1+s)(t_\ell - t_k) > (1+s)(T-t_k).$$

But this is not possible, else $d_h < t_k$ and $[t_{\ell-1},t_\ell]$ would not be an earliest interval with a job j such that $b_j^{(\ell)} < t_\ell$.

(b) Suppose that no job is completed at time t_k, but t_k is the release date r_h of a job h. By an argument similar to the above, it follows that $p_h(r_h) > 0$, which is not possible by internal consistency. □

We note that although the priority scheduling algorithm is applied to $O(n^2)$ jobs, the time required for the computation is bounded by $O(n^3)$. The precedence constraints are such that at most n jobs are available at any given time and the number of intervals is $O(n^2)$. It follows that $O(n^3)$ time is sufficient to determine whether or not there exists a schedule meeting all due dates.

Unfortunately, it is not possible to minimize maximum lateness as easily as in the previous two sections. It is clearly *not* the case that if L is added to all due dates then all modified due dates are changed by the same constant L.

One way to minimize maximum lateness is to apply the technique of Megiddo [8]. In this approach, the due date modification algorithm is applied to due dates d_j+L, where L is maintained as a symbolic variable. Each time a numerical comparison is made whose outcome depends upon the value of L, there is an easily computed "critical" value L^* such that one outcome occurs if $L \leq L^*$ and another if $L > L^*$. If the minimum value of L yielding an internally consistent set of due dates is already known to be larger or smaller than L* there is no ambiguity about the outcome of the comparison. Otherwise, the due date modification algorithm is applied to due dates d_j+L^*. If the resulting due dates are internally consistent, the desired value of L is known to be no greater than L*; otherwise, the desired value is known to be larger. Proceeding in this way, one can determine the minimum value of maximum lateness in $O(n^3)$ applications of the due date modification algorithm, or $O(n^6)$ time overall.

10. EXTENSIONS AND GENERALIZATIONS

The possibilities for finding polynomial time algorithms for more general cases of the problems considered in this paper appear to be limited. NP-hardness proofs exist for several more general problems.

In the context of nonpreemptive scheduling of unit jobs, Brucker, Garey and Johnson [1] showed that the intree problem becomes NP-hard if outtree constraints replace intree constraints (but release dates remain equal and due dates are arbitrary). The problem $P|pmtn, outtree|L_{max}$ is NP-hard [7], since there is no advantage to preemption for the subclass of problem instances of $P|outtree, p_j=1|L_{max}$ shown to be NP-hard in [1].

One question which remains unresolved is the status of the intree problem when generalized to allow the m machines to have different speeds instead of m-1 identical machines and one possibly slower one. The algorithmic approach taken here appears to be not quite adequate for this case.

Also with reference to nonpreemptive scheduling of unit jobs, Ullman [13] proved that it is NP-hard to schedule jobs to minimize makespan, when scheduling jobs for an arbitrary number of machines subject to arbitrary precedence constraints. The preemptive version of this problem is also NP-hard [14].

Ullman's result leaves open the question of whether NP-hardness holds for any *fixed* number of machines and, in particular, for three machines. It is reasonable to believe that if a polynomial time algorithm can be found for nonpreemptive scheduling of unit jobs on three machines, a similar algorithm can be found for the preemptive version of the three-machine problem.

ACKNOWLEDGMENTS

This research was supported in part by NSF grant MCS78-20054. The author wishes to acknowledge the help of Harold Gabow, in pointing out an error in [3]. (It is necessary to require that release dates be consistent, as in (9.1).) He also wishes to thank Jan Karel Lenstra for his advice, patience and encouragement in the preparation of this paper, and Ben Lageweg for his careful reading of the manuscript. Any errors which remain are of course the author's responsibility.

REFERENCES

1. P. BRUCKER, M.R. GAREY, D.S. JOHNSON (1977) Scheduling equal length tasks under tree-like precedence constraints to minimize maximum lateness. *Math. Oper. Res.* 2,275-284.
2. M.R. GAREY, D.S. JOHNSON (1976) Scheduling tasks with nonuniform deadlines on two processors. *J. Assoc. Comput. Mach.* 23,461-467.
3. M.R. GAREY, D.S. JOHNSON (1977) Two-processor scheduling with start times and deadlines. *SIAM J. Comput.* 6,416-426.
4. T.F. GONZALEZ, D.B. JOHNSON (1980) A new algorithm for preemptive scheduling of trees. *J. Assoc. Comput. Mach.* 27, 287-312.
5. R.L. GRAHAM, E.L. LAWLER, J.K. LENSTRA, A.H.G. RINNOOY KAN (1979) Optimization and approximation in deterministic sequencing and scheduling: a survey. *Ann. Discrete Math.* 5, 287-326.
6. E.C. HORVATH, S. LAM, R. SETHI (1977) A level algorithm for preemptive scheduling. *J. Assoc. Comput. Mach.* 24,32-43.
7. J.K. LENSTRA. Private communication.
8. N. MEGIDDO (1979) Combinatorial optimization with rational objective functions. *Math. Oper. Res.* 4,414-423.
9. C.L. MONMA (1979) Linear-time algorithm for scheduling equal-length tasks with due dates subject to precedence constraints. Technical Memorandum 79-1712, Bell Laboratories, Holmdel, NJ.
10. R.R. MUNTZ, E.G. COFFMAN, JR. (1969) Optimal preemptive scheduling on two-processor systems. *IEEE Trans. Comput.* C-18, 1014-1020.
11. R.R. MUNTZ, E.G. COFFMAN, JR. (1970) Preemptive scheduling of real-time tasks on multiprocessor systems. *J. Assoc. Comput. Mach.* 2,324-338.
12. S. SAHNI, Y. CHO (1979) Nearly on line scheduling of a uniform processor system with release times. *SIAM J. Comput.* 8, 275-285.
13. J.D. ULLMAN (1975) *NP*-Complete scheduling problems. *J. Comput. System Sci.* 10,384-393.
14. J.D. ULLMAN (1976) Complexity of sequencing problems. In: E.G. COFFMAN, JR. (ed.) (1976) *Computer & Job/Shop Scheduling Theory*, Wiley, New York, 139-164.

FORWARDS INDUCTION AND DYNAMIC ALLOCATION INDICES

J. C. Gittins

Keble College, Oxford

ABSTRACT

The principle of forwards induction is that jobs, or parts of a job, are scheduled in decreasing order of the expected reward per unit time which they yield. Alternatively, each job, or part of a job, may have an associated priority index, and be scheduled accordingly. This paper explores the wide class of stochastic scheduling problems for which forwards induction leads to optimal policies, and the priority indices which are also optimal for these problems.

1. INTRODUCTION

These ideas are probably best communicated by means of specific instances.

Problem 1 (Single Machine Scheduling)

A set of n jobs is to be processed by a single machine. For job i the processing time τ_i is independent of the processing times for other jobs, and the completion (or hazard) rate is an increasing function of the time already spent on the job. If job i is completed at time t the resulting reward is $V_i\exp(-\beta t)$ $(i = 1,2\ldots,n)$. A schedule is required which maximises the expected total reward. Switching between jobs may take place at any stage, with no switching costs or delays. The answer is that the jobs should be scheduled in decreasing order of the indices

$$\nu_i \triangleq V_i E e^{-\beta\tau_i}/(1-Ee^{-\beta\tau_i}),$$

M. A. H. Dempster et al. (eds.), Deterministic and Stochastic Scheduling, 125–156.

and there should be no switches between jobs except at completion times. Note that $V_i Ee^{-\beta\tau_i}$ is the expected reward from job i if it is placed first in the schedule, and that

$$\beta^{-1}(1-Ee^{-\beta\tau_i}) = E\int_0^{\tau_i} e^{-\beta t}dt \quad (=E\tau_i \text{ when } \beta = 0),$$

so that the denominator in the expression for ν_i may be regarded, apart from the constant β, as the expectation of the *discounted* time required to complete the job. Thus the schedule defined by the indices ν_i has the property that at each completion time the next job processed is one for which the expected discounted reward per unit of discounted processing time is maximal, in the set of as yet uncompleted jobs. It is in this sense that the solution conforms to the forwards induction principle. The assumption of increasing completion rates ensures that switches between jobs occur only at completion times, for a forwards induction policy as well as for an optimal policy.

Problem 2 (Search)

A stationary object is hidden in one of n boxes. The probability that a search of box i finds the object if it is in box i is q_i $(i = 1,2,..,n)$. The probability that the object is in box i is p_i, and changes by Bayes' Theorem if this, or any other box, is searched. The cost of a single search of box i is c_i.

The boxes are to be searched one at a time until the object is found. At each stage the box to be searched next may be chosen so as to depend on the numbers of times each of the boxes have so far been searched. The problem is to choose the sequence in which the boxes are searched so as to minimise the expected cost of finding the object.

The answer is that an optimal policy is one which at each stage searches a box j such that

$$\nu_j \overset{\Delta}{=} p_j q_j / c_j = \max_i\{p_i q_i / c_i\}.$$

In this problem there is a single reward of unity which occurs when the object is found. Thus at any stage before this happens $p_i q_i$ is the expected reward if box i is searched next. Again the schedule defined by the indices ν_i conforms to the forwards induction principle. This time there is no discounting, and instead of a time, or expected time, we now have a cost in the denominator. This last change is, however, a trivial one; c_i could perfectly well be interpreted as the time taken to search box i, and the problem reinterpreted as that of finding the object in the minimum expected time.

Problem 3 (Chemical Synthesis)

There are n different routes by which a chemist might try to synthesise a certain compound. Each route consists of a fixed sequence of stages. The probability that the chemist will succeed in synthesising the compound at the j'th stage of route i, conditional on his not succeeding in any of the first j-1 stages, is p_{ij}, and the cost of that stage is c_{ij}. For each i, p_{ij}/c_{ij} is a decreasing function of j. At any particular time the chemist works on just one stage of one of the routes, which he completes before proceeding to the next stage in the sequence, either for that route, or for one of the other routes. Once he has succeeded in synthesising the compound he does no more work on any of the routes. How should the chemist schedule the stages of the various routes, consistent with the fixed orders of the stages within each route, so as to minimise the expected cost of synthesising the compound?

The answer is that on reaching that point in the schedule when j(i) stages of route i (i = 1, 2, .., n) have been completed without success, the chemist should use

$$\nu_i(j(i)) \triangleq p_{ij(i)+1}/c_{ij(i)+1}$$

as a priority index. This scheduling rule once again conforms to the forwards induction principle, as is clear on comparison with the index for Problem 2. This conformity, and the optimality of the priority index rule, depend on the assumption that $\nu_i(j)$ is decreasing in j for all i. Without this assumption it might have been possible to achieve a higher expected reward per unit cost by first scheduling two or more stages of a route k for which $\nu_k(j)$ is increasing, although $\nu_k(0) \neq \max_i \nu_i(0)$.

Problem 4 (Multi-attribute Testing)

In a project designed to find a chemical formulation to serve a certain commercial purpose each formulation passes through screening tests for n different attributes before proceeding to large scale tests. The screening tests for attribute i are divided into m_i stages (i=1, 2, .., n) which occur in a fixed order, and unless a formulation passes successfully through every stage of testing for all n attributes it is eliminated from further consideration. The probability that a formulation passes the j'th stage (j=1, 2, .., m_i) of testing for attribute i, conditional on it passing the first (j-1) stages and irrespective of its success or failure so far as the other attributes are concerned, is p_{ij}, and the cost of this stage of testing is c_{ij}. For each i, $(1-p_{ij})/c_{ij}$ is a decreasing function of j.

Thus the total cost of the screening tests for a formulation which passes all of them is

$$\sum_{i=1}^{n} \sum_{j=1}^{m_i} c_{ij} ,$$

and this is independent of the scheduling of the tests for the different attributes. For a formulation which fails one or more stages of screening tests, on the other hand, the scheduling of the tests does make a difference, since there is no need to incur the costs of the remaining tests, once the formulation has failed at one of the stages. The problem is to schedule the stages of testing for the various attributes, consistent with the fixed orders of the stages for each attribute, so as to minimise the expected cost of putting a formulation through the screening procedure.

The answer is that when $j(i)$ stages of testing for attribute i $(i=1, 2, .., n)$ have been scheduled, the next stage in the schedule should be selected using

$$\nu_i(j(i)) \triangleq \begin{cases} (1-p_{ij(i)+1})/c_{ij(i)+1}, & j(i) < m_i, \\ 0, & j(i) = m_i \end{cases}$$

as a priority index. As has been noted, screening costs differ between scheduling rules only for those formulations which fail at some stage. For formulations which fail it is an advantage to know this at an early stage, so as to avoid the costs of further testing. Perversely as it may seem, we may therefore regard the sole reward as equal to unity and occurring when a formulation fails. We therefore have an exact analogy with Problem 3 except that the roles of success and failure are reversed. Since the analogous monotonicity property has been assumed the index rule again conforms with the forwards induction principle.

Problem 5 (A Gold-mining Problem)

A man owns n gold mines and a gold-mining machine. Each day he must assign the machine to one of the mines. When the machine is assigned to mine i $(i=1, 2, .., n)$ for the j'th time there is, if the machine has not already broken down, a probability p_{ij} that it extracts an amount r_{ij} of gold, and a probability $1 - p_{ij}$ that it extracts no gold and breaks down permanently. For each i, $p_{ij}r_{ij}/(1-p_{ij})$ is decreasing in j. The man's problem is to schedule the mines to which the machine is assigned on successive days so as to maximise the expected total amount of gold mined before the machine breaks down.

The answer is that when mine i has been scheduled $j(i)$ times $(i = 1, 2, .., n)$, the next mine to appear in the schedule should be chosen using

$$\nu_i(j(i)) \triangleq p_{ij(i)+1} r_{ij(i)+1} / (1 - p_{ij(i)+1})$$

as a priority index. Again this is a scheduling rule which conforms to the forwards induction principle, except that the principle must now be redefined in terms of the ratio of expected reward to probability of failure. This conformity, and the optimality of the rule, depend on the assumption that $\nu_i(j)$ is decreasing in j. Otherwise it might have been possible to achieve a higher ratio of expected reward to probability of failure by first scheduling for two or more days a mine k for which $\nu_k(j)$ is increasing, although $\nu_k(0) \neq \max_i \nu_i(0)$.

This problem is discussed by Bellman [1].

Section 2 sets out the concepts and theorems whereby these problems may be solved. The theorems are not proved, but the reasons why they hold are discussed, and these may be used to provide proofs. The proofs themselves, and a generally more complete exposition, will appear in a book to be published by John Wiley and Sons. A preliminary version of the main theoretical chapters is already available (Gittins [7], as is a review (Gittins [6] which complements the present paper in that the emphasis there is on sequential problems, such as the multi-armed bandit problem, with a statistical flavour.

Section 3 provides the solutions of Problems 1 to 5, setting these in the context of other similar problems.

Finally, in an appendix Theorem 2 is proved. This is a somewhat more general version of a result due to Whittle [33], and is important because it establishes a necessary and sufficient condition for the solution of a wide class of scheduling problems to be expressible in terms of priority indices.

2. CONCEPTS AND THEOREMS

The theorems justifying the optimality of the forwards induction principle, and of the equivalent use of dynamic allocation indices, are set in the context of *semi-Markov decision processes* with discounted rewards. Such a process is defined on a state-space, Θ, which is a Borel subset of some complete separable metric space, together with a σ-algebra $\mathcal{X}$ of subsets of Θ which includes every subset consisting of just one element of Θ. Controls are applied to the process at a succession of *decision times*, the intervals between which are random variables.

The first decision time is at time zero. Transitions between states, and rewards, occur instantaneously at these times, and at no other times. It will prove convenient to regard the control applied at a decision time as being continuously in force up to, but not including, the next decision time.

When the process is in state x at decision time t the set of controls which may be applied is $\Omega(x)$. Application of control u at time t with the process in state x yields a reward $e^{-\beta t} r(x,u) (0 < \beta)$. $P(A|x,u)$ is the probability that the state y of the process immediately after time t belongs to $A(\in X)$, given that at time t the process is in state x and control u $(\in\Omega(x))$ is applied. $F(B|x,y,u)$ is the probability that the duration of the interval until the next decision time belongs to the Borel set B, given that at time t the process is in state x, and control u is applied, leading to a transition to state y.

A *Markov* decision process is a semi-Markov decision process with the property that, for some $c(>0)$,

$$F(\{c\}|x,y,u) = 1, \forall\ x,y, \text{ and } u.$$

Thus the interval between decision times takes the constant value c. Without loss of generality we shall assume that $c = 1$, since this is just a question of choosing the units in which time is measured.

A *policy* for a semi-Markov decision process is any rule, satisfying the obvious measurability requirements, including randomised rules, which for each decision time t specifies the control to be applied at time t as a function of t, the state at time t, the set of previous decision times, and of the states of the process and the controls applied at those times. Thus the control applied at time t depends on the entire previous history of the process, a situation which we shall describe by saying that it is *sequentially determined*. This definition is to be understood to include the natural requirement that the control chosen does not depend on what happens to the process after time t. A policy which maximises the expected total reward over the set of all policies for every initial state will be termed an *optimal* policy. *Deterministic* policies are those which involve no randomisation. *Stationary* policies are those which involve no explicit time-dependence. *Markov* policies are those for which the control chosen at time t depends only on t and the state at time t.

A *bandit process* is a semi-Markov decision process for which the control set $\Omega(x)$ consists, for every state x, of two elements 0 and 1. The control 0 *freezes* the process in the sense that when it is applied no reward accrues and the state of the process does not change. Application of control 0 also initiates a time interval in which every point is a decision time. In contrast, control 1 is termed the *continuation* control. Apart from measurability and regularity conditions the functions $r(\cdot,1)$, $P(\cdot|\cdot,1)$, and $F(\cdot|\cdot,\cdot,1)$ may take any form.

At any stage, the total time during which the process has not been frozen by applying control 0 is termed the *process time*. The process time when control 1 is applied for the $(i+1)$th time is denoted by $t_i (i=0,1,2...)$. The state at process time t is denoted by $x(t)$. The sequences of times t_i and states $x(t_i)(i=0,1,2,...)$ constitute a *realisation* of the bandit process, which thus does not depend on the sequence of controls applied. The reward at decision time t if control 1 has been applied at every previous decision time, so that process time coincides with real time, is $e^{-\beta t}r(x(t), 1)$, which we abbreviate to $e^{-\beta t}r(t)$. A *standard* bandit process is a bandit process with just one state, and for which every point in time is a decision time. Thus a policy for a standard bandit process is simply a random Lebesgue-measurable function $I(t)$ from the positive real line to the set $\{0,1\}$, specifying the control to be applied at each point in time. The total reward yielded by such a policy is $\lambda\int_0^\infty e^{-\beta t}I(t)dt$, the parameter depending on the particular standard bandit process.

An arbitrary policy for any bandit process is termed a *freezing rule*. Given any freezing rule f for a non-standard bandit process, the random variables $f_i (i=0,1,2...)$ are sequentially determined, where $f_i (\geq f_{i-1})$ is the total time for which control 0 is applied before the $(i+1)$th application of control 1. We also define $f_{-1} = 0$. A policy is said to be a *stopping rule* if it divides the state space Θ into a *continuation set*, on which control 1 is applied, and a *stopping set*, on which control 0 is applied. Stopping rules are the deterministic stationary Markov policies for bandit processes. For a stopping rule the time τ when the process first reaches the stopping set is the corresponding *stopping time*. Thus $f_i = 0$ if $t_i < \tau$ and $f_i = \infty$ if $t_i \geq \tau$, for all i. We note that τ may take the value infinity with positive probability. Stopping rules have been extensively studied, for the most part in the context of stopping problems (e.g. see Chow et al. [4]), which may be regarded as being defined by stoppable bandit processes (see Section 3.5).

The following notation will be used in conjunction with an arbitrary bandit process D. $R_f(D)$ denotes the expected total reward under the freezing rule f. Thus

$$R_f(D) = E \sum_{i=0}^{\infty} e^{-\beta(t_i+f_i)} r(t_i).$$

$R(D)$ denotes the expected total reward under the freezing rule, termed the null freezing rule, for which $f_i = 0 (i=0,1,2,\ldots)$. Also

$$W_f(D) = E \sum_{i=0}^{\infty} e^{-\beta f_i} \int_{t_i}^{t_{i+1}} e^{-\beta t} dt, \ \nu_f(D) = R_f(D)/W_f(D), \text{ and}$$

$$\nu'(D) = \sup_{\{f:f_0=0\}} \nu_f(D).$$

Similarly, for stopping rules,

$$R_\tau(D) = E \sum_{t_i<\tau} e^{-\beta t_i} r(t_i), \ W_\tau(D) = E\int_0^\tau e^{-\beta t} dt,$$

$$\nu_\tau(D) = R_\tau(D)/W_\tau(D), \text{ and } \nu(D) = \sup_{\{\tau>0\}} \nu_\tau(D).$$

All these quantities naturally depend on the initial state $x(0)$ of the bandit process D. When necessary $R_f(D,x)$ and $W_f(D,x)$, for example, will be used to indicate the values of $R_f(D)$ and $W_f(D)$ when $x(0) = x$.

The quantities $\nu_f(D)$ and $\nu_\tau(D)$ are thus weighted average rewards per unit time under f and τ respectively, the weights being the discount factors. The conditions $f_0 = 0$ and $\tau > 0$ in the definitions of $\nu'(D)$ and $\nu(D)$ mean that the policies considered are all such that at time zero control 1 is applied. This restriction is required to rule out zero denominators $W_f(D)$ or $W_\tau(D)$. In the case of $\nu'(D)$ it also has the effect of removing a common factor from the numerator of $\nu_f(D)$ for those f for which $f_0 \neq 0$, and otherwise implies no loss of generality.

Given any semi-Markov decision process M, together with a policy g, a bandit process may be defined by introducing the freeze control 0 with the same properties as for a bandit process, and requiring that at each decision time either the control 0 or the control given by g be applied. This bandit process is termed the *superprocess* (M,g). Thus, if g is deterministic, stationary, and Markov, application of the continuation control 1 to (M,g) when M is in state x, is equivalent to applying control g(x) to M. The quantities $t_i (i=0,1,2,\ldots), x_g(t), r_g(t)$ and $R_g(M)$ have the same definitions as $t_i (i=0,1,2,\ldots), x(t), r(t)$ $R(D)$, but with the superprocess (M,g) in place of the ordinary bandit process D. Also $R(M) = \sup_g R_g(M)$, $R_{g\tau}(M) = R_\tau((M,g))$, $W_{g\tau}(M) = W_\tau((M,g))$, $\nu_{g\tau}(M) = R_{g\tau}(M)/W_{g\tau}(M)$, $\nu_g(M) = \sup_{\tau>0} \nu_{g\tau}(M)$, and $\nu(M) = \sup_g \nu_g(M)$. Similarly $R_{gf}(M) = R_f((M,g))$, etc. As for bandit processes, $\nu(M,x)$, for example, denotes the value of $\nu(M)$ when M is initially in state x. The useful concept of a superprocess is due to Nash [28].

This notation enables us to give a formal definition of a forwards induction policy for a semi-Markov decision process M with initial state x_0.

The first step is to find a deterministic stationary Markov policy γ_1 such that $\nu_{\gamma_1}(M,x_0) = \nu(M,x_0)$. Then let σ_1 be the stopping time for the superprocess (M,γ_1) defined by the stopping set $\{x:\nu(M,x) < \nu(M,x_0)\}$. It may be shown that γ_1 always exists with these properties, and that $\nu_{\gamma_1\sigma_1}(M,x_0) = \nu(M,x_0)$.

Let x_1 be the (random) state of the superprocess (M,γ) at time σ_1, γ_2 a deterministic stationary Markov policy such that $\nu_{\gamma_2}(M,x_1) = \nu(M,x_1)$, and σ_2 the stopping time for the superprocess (M,γ_2) defined by the stopping set $\{x:\nu(M,x) < \nu(M,x_1)\}$ when the initial state is x_1. The state x_i, policy γ_{i+1}, and stopping time σ_{i+1} (i=2,3,...) are defined inductively by replacing $x_1,\gamma_1,\sigma_1,\gamma_2$ and σ_2 by $x_i,\gamma_i,\sigma_1 + \sigma_2 + \ldots \sigma_i$, γ_{i+1}, and $\sigma_1 + \sigma_2 + \ldots + \sigma_{i+1}$, respectively, in the previous sentence. The policy for M which starts by applying policy γ_1 up to time σ_1, then applies policy γ_2 from σ_1 up to $\sigma_1 + \sigma_2$, γ_3 from $\sigma_1 + \sigma_2$ up to $\sigma_1 + \sigma_2 + \sigma_3$, and so on, is said to be a *forwards induction* policy for M. The successive intervals during which policies γ_1, γ_2, γ_3, and so on, are applied will be described as the *stages* of such a policy.

We now come to the central concept of a *family F of alternative components*. This is a semi-Markov decision process formed by bringing together a set of n more basic semi-Markov decision processes called components, for each of which the control set is augmented at every decision time by inclusion of the freeze control 0. Time zero is a decision time for every component, and at time zero control 0 is applied to every component except one, component M_1 say, to which some other control is applied, which we describe by saying that M_1 is *continued*. These controls remain in force until the next decision time for M_1 is reached. Again, this is a decision time for every component and just one of them, M_2 say, is continued, a state of affairs which persists until the next decision time for M_2 is reached. Once again, this is a decision time for every component, and all except one of them is frozen.

Realisations of F are constructed by proceeding in this way from one decision time to the next. The state space for F is the product of the states for the individual components. The reward which accrues at each decision time is the one yielded by the component which is then continued. The control set at a decision time is the union, excluding control 0, of the control sets of the components of F. If control u is selected from the control set for component M_i this means that component M_i is selected for continuation and control u is applied to M_i.

The quantity $\nu(M_i,x_i)$ is termed the *dynamic allocation index* (*DAI*) for component M_i when it is in state x_i. For a family of alternative components $\{M_i: i=1, 2, .., n\}$ a class of policies may be defined in terms of these indices. These policies, termed *DAI policies*, are those which, when F is in state $(x_1, x_2, .., x_n)$ at a decision time, select for continuation a component M_j for which $\nu(M_j,x_j) = \max_i \nu(M_i,x_i)$ and apply to M_j a control $\gamma_j(x_j)$, where γ_j is a deterministic stationary Markov policy for M_j for which $\nu(M_j,x_j) = \nu((M_j,\gamma_j),x_j)$.

An important special case of a family of alternative components occurs when each of the components is a bandit process. For this case we shall sometimes limit the set of bandit processes which may be selected for continuation, so as to represent precedence constraints, and jobs which arrive after time zero. The absence of such constraints will be indicated by calling F a *simple* family of alternative bandit processes (*SFABP*).

For a single bandit process the sequence of states through which it passes is independent of the freezing rule used, which serves only to control the time intervals between transitions from one state to the next. The sequence of rewards yielded by a bandit process is also independent of the freezing rule, which does, however, control the discount factor associated with each reward by fixing the time at which it occurs. For a SFABP the situation is similar. Each bandit process generates a stream of rewards which is independent of the policy followed. The role of a policy is to splice together portions of these different reward streams to form a single reward stream for the SFABP.

The total discounted reward from a SFABP thus depends on the policy followed insofar as this affects the time at which each individual reward occurs, and hence the discount factor associated with it, but in no other way. Good policies are clearly those which place reward stream portions with a high average reward per unit time early in the spliced stream.

Now for a bandit process D the expected total discounted reward per unit of discounted time for the initial portion of the reward stream is maximised by terminating this portion at time σ, where $\nu_\sigma(D) = \nu(D)$. A good policy for a SFABP $\{D_i: i=1,2,..,n\}$ should therefore be as follows.

For any initial set of states $x_1, x_2, .., x_n$ for $D_1, D_2, .., D_n$ find a bandit process D_i such that $\nu(D_i,x_i) \geq \nu(D_j,x_j)$ $(j=1,2,..,n)$. The first stage of the policy is to continue D_i

up to a stopping time σ_i for which $\nu_{\sigma_i}(D_i,x_i) = \nu(D_i,x_i)$. At time σ_i there will be another vector of states for $D_1,D_2,..,D_n$; it will, o course, be the same as the initial state vector except for the state of D_i. Select a bandit process D_j for continuation up to time $\sigma_i + \sigma_j$ exactly as before, but now with respect to the new vector. This leads to another new state vector at time $\sigma_i + \sigma_j$, at which point we select a bandit process D_k for continuation up to time $\sigma_i + \sigma_j + \sigma_k$, and so on. Call this policy γ.

Policy γ is in fact a forwards induction policy for a SFABP. The reason is that each stage of a forwards induction policy is defined so as to maximise the expected reward rate during that stage, conditional on the state at the end of the previous stage, and this maximum expected reward rate is achieved by the initial portion of the reward stream for the bandit process with the largest DAI. No increase in expected reward rate can be achieved by splicing mixtures of reward streams from more than one bandit process.

Next note that if $\nu_\sigma(D,x(0)) = \nu(D,x(0))$ for a bandit process D, then $\nu(D,x(t)) \leq \nu(D,x(0))$ for any decision time $t < \sigma$. Otherwise we could achieve a higher expected reward rate from D by stopping the reward stream at t instead of σ, giving $\nu_t(D,x(0)) > \nu(D,x(0))$, contrary to the defininition of $\nu(D,x(0))$. It follows that policy γ is also a DAI policy.

In the light of these observations Theorem 1 should cause no surprise.

Theorem 1 (The DAI Theorem for a SFABP). For a simple family of alternative bandit processes a policy is optimal if and only if it is a DAI policy, and forwards induction policies are optimal.

For a family of alternative components which are not necessarily bandit processes a more general version of policy γ may be defined. This starts by choosing a component M_i such that $\nu(M_i,x_i) \geq \nu(M_j,x_j)(j=1,2,..,n)$. The policy then continues M_i up to time σ_i using policy γ_i, where $\nu_{\gamma_i\sigma_i}(M_i,x_i) = \nu(M_i,x_i)$. At time σ_i this procedure is repeated, using the new state vector, and so on. The policy defined in this way is again a fowards induction policy; the argument showing this is virtually the same as for a SFABP. However, forwards induction policies and DAI policies are not always optimal for families of alternative components.

Suppose, for example, that there is just one component M_1 with two continuation controls, 1 and 2, and that M_1 yields zero rewards after the second decision time, whichever control is applied at time zero. Let the second (deterministic) decision time be t_i if control i ($i=1,2$) is applied at time zero. Clearly it is optimal to apply control i if $r(x_1,i) \geq r(x_1,3-i)$, whereas under a forwards induction policy or a DAI policy control i is used only if

$$\beta r(x_1,i)/(1-e^{-\beta t_i}) = \nu(M,x_1) \geq \beta r(x_1,3-i)/(1-e^{-\beta t_{3-i}}).$$

These two conditions are not always compatible.

One way of obtaining a DAI theorem for a family of alternative components $\{M_i : i=1,2,..,n\}$ is to impose conditions which ensure that an optimal policy which continues component M_i when it is in state x_i must use the same control $u_i(x_i)$, irrespective of the other components. This reduces each component to a superprocess, and hence Theorem 1 applies.

One might suppose that very restrictive conditions would be required, but it turns out to be sufficient to ensure that a component M reduces to the same superprocess when there is just one alternative component Λ, which is a standard bandit process with arbitrary parameter λ. We now proceed to state this more precisely. Note that in general an optimal deterministic stationary Markov policy for the family of alternative components $\{M,\Lambda\}$ is defined by a policy g for M which is applied up to some stopping time σ, after which Λ is continued permanently, where both g and σ may depend on λ.

Condition A. A semi-Markov decision process M will be said to satisfy Condition A if (i) there is a deterministic stationary Markov policy g for M, independent of λ, which, together with a suitable stopping time for the superprocess (M,g), defines an optimal policy for the family of alternative components $\{M,\Lambda\}$, where Λ is a standard bandit process with parameter λ, and (ii) for no λ is there an optimal deterministic stationary Markov policy for $\{M,\Lambda\}$ which is not of this form, for some g with the stated property. A family of alternative components satisfies Condition A if the same is true for each of its components.

Theorem 2 (The General DAI Theorem). (i) For a family of alternative components satisfying Condition A, a policy is optimal if and only if it is a DAI policy, and forwards induction policies are optimal. (ii) The classes of DAI policies and optimal policies coincide for the family of alternative components formed by merging an arbitrary SFABP with a fixed component M if and only if M satisfies Condition A.

Part (i) of Theorem 2 for the discrete time case is due to Whittle [33], whose elegant proof involves conjecturing the form of the maximum expected total reward function, and showing that this satisfies the appropriate dynamic programming recurrence equation. The appendix to the present paper gives an alternative, and perhaps more intuitive and direct, proof of the theorem as a whole.

Condition A, then, as also noted by Glazebrook [18], comes close to being necessary and sufficient for a DAI policy to be optimal. However, it is not always easy to check whether it holds, and alternative conditions consequently may also be useful. One such condition holds for a family of alternative bandit processes (FABP) representing jobs subject to precedence and arrival constraints.

The term *job* here means a bandit process with a completion state C, a wider meaning than the conventional one. Once a job reaches state C it never leaves it again, and generates no further rewards or costs. A precedence constraint is a requirement of the form 'job i must be completed before job j may be started'. A FABP with precedence constraints is a family of alternative components, the components corresponding to sets of jobs such that there are no precedence constraints between jobs in different sets. Continuation of a component corresponds to continuation of (or processing) one of its constituent jobs. Arrivals may be allowed provided each of them is assigned to the component which is being continued when it arrives, and that the probability of an arrival between successive decision times depends only on the state of the component being continued and the continuation control. In particular, these conditions are satisfied if the arriving jobs form a Poisson process.

For a FABP with precedence constraints, as for a SFABP, the role of the policy followed is to splice together portions of the reward streams for the individual bandit processes. However, the introduction of precedence constraints means that forwards induction and DAI policies are not always optimal. To see this, note that for a particular component M the availability of more than one bandit process for continuation means there is more than one order in which information relevant to the future rewards available from M may be accumulated. The first stage of a forwards induction policy is chosen so as to maximise the expected reward rate. For this purpose it is important to obtain information about future rewards, as well as to schedule large rewards to occur early. This objective may be irrelevant in achieving an optimal policy; for example, if M is the only component, so that the information gained can not be used to indicate a switch to a more promising component. This observation quickly leads to examples of FABP's for which forwards induction and DAI policies are not optimal.

What we can do is to use the idea of splicing together portions of reward streams to obtain a condition under which a DAI theorem for FABP's holds. Let γ be a forwards induction policy for a FABP F with precedence and arrival constraints. The reward stream from any particular bandit process may belong entirely to the first stage of γ, or be entirely excluded from the first stage, or be divisible into two successive portions, only the first of which belongs to the first stage of γ. Under any other policy g, unless this simply rearranges within the same time period the reward stream portions forming the first stage of γ, these occur later, on average, than they do under γ, and those reward stream portions which do not belong to this first stage occur earlier, again on average. Since the first stage of a forwards induction policy is chosen to maximise the expected reward rate, this rearrangement tends to schedule less profitable reward stream portions before more profitable portions to a greater extent than under γ, and so to produce a lower expected total discounted reward. This argument fails only if under g the reward stream portions forming the first stage of γ occur in a sequence with the property that there is at least one reward stream portion for which we do not know whether it belongs to the first stage of γ until after it has occurred under g. The reason why the argument fails is that for a sequence with the stated property the definition of the first stage of γ does not rule out the possibility of a rearrangement of the first stage, with interruptions by other reward stream portions, which yields a higher expected reward rate than the first stage does under γ.

This possibility can be ruled out by requiring that under any policy g the reward stream portions belonging to the first stage of γ occur in a (possibly interrupted) sequence which forms the initial portion of the reward stream sequence generated by the superprocess (F,h) under some freezing rule f, where h is appropriately chosen. This requirement will be termed *Condition B*. If it holds, the maximum expected reward rate under any policy g of the reward stream portions forming the first stage of γ is no greater than can be achieved if any interruptions under g are removed by replacing f by a suitable stopping time. By definition of a forwards induction policy, this expected reward rate is, in turn, no greater than that generated by the first stage of γ. Thus Condition B closes the gap in the argument outlined in the previous paragraph for the optimality of forwards induction policies. DAI policies are also optimal, for reasons similar to those mentioned in connection with Theorem 1, and the following result holds.

Theorem 3 (The DAI Theorem for a FABP). For a family of alternative bandit processes satisfying Condition B, and which may include precedence and arrival constraints, a policy is optimal if and only if it is a DAI policy, and forwards induction policies are optimal.

One case for which Condition B is satisfied is when for each job at most one immediate predecessor is defined by the set of precedence constraints. In terms of the directed graph in which the vertices represent jobs, and the edges precedence constraints, this means that the in-degree of each vertex is at most one; a graph of this form is known as an *out-tree*. Initially, then, there is just one job in each component which may be started, so the difficulty of more than one order in which information relevant to future rewards may be accumulated does not arise. This situation is preserved at later decision times if components are redefined in terms only of uncompleted jobs, and hence Condition B may be shown to hold. Poisson arrivals of further jobs, subject to precedence constraints only between jobs which arrive at the same time, may also be allowed.

Index results in this context have been obtained by several authors under a variety of particular assumptions. Glazebrook [10] reported on precedence constraints forming an out-tree. Nash [28], Simonovits [31], Harrison [21], Meilijson and Yechiali [27], and Gittins and Nash [9], consider the optimisation of priority systems for queues with Poisson arrivals. Whittle [34] allows the arrivals to be general bandit processes. Meilijson and Weiss [26] consider a more general arrival structure which implicitly includes an out-tree of precedence constraints. Numerous earlier papers are referenced by these authors.

Condition B also holds for an arbitrary set of precedence constraints if the only decision times after time zero occur on completion of a job, a situation which will be termed the *non-pre-emptive* case. Under these circumstances the first stage of a forwards induction policy consists of a non-random sequence of the jobs which belong to it. Thus there is no question of not knowing which reward stream portions belong to it, and Condition B follows.

The optimality of forwards induction and DAI policies has been demonstrated for this situation, under increasingly general conditions, by Sidney [30], Kadane and Simon [22], and Glazebrook and Gittins [19], all without the benefit of Theorem 3; these authors also suggest algorithms for the explicit determination of such policies. More efficient algorithms are available when the precedence constraints are what is known as *series parallel*; these have been developed by Lawler [24] and others under conditions which are implied by Theorem 3. Glazebrook [14], [15] and [16] establishes conditions under which an optimal policy does not interrupt the processing time of any job, so that Theorem 3 holds for general precedence constraints, and the aforementioned algorithms lead to optimal policies, and investigates [17] the extent to which such policies are sub-optimal if an optimal policy does lead to interruptions.

3. APPLICATIONS

3.1. Single Machine Scheduling

Here we revert to the more conventional terminology of Problem 1, where a typical job yields a single reward $Ve^{-\beta t}$ on completion at time t. Note that the more general jobs mentioned in connection with precedence constraints reduce to this case if it can be shown that it is never optimal to switch from an uncompleted job. Other assumptions are also as for Problem 1, except that the total processing time τ required for completion has an arbitrary distribution function F.

A set of jobs to be processed thus forms a SFABP in the limit as the intervals between decision times tends to zero. Optimal policies may be obtained (Nash and Gittins [27]) by taking the same limit in the expression for the dynamic allocation index. The state of a job is the amount of processing time x which it has so far received, if it has not been completed, and C if it has been completed. For a job in state x the expected reward rate must be maximised by processing for a further time $\sigma = \min[t,\tau] - x$, for some non-random, and possibly infinite, $t > x$.

Thus our expression for the limit of the DAI when every time is a decision time is

$$\nu(x) = \sup_{\{\sigma>0\}} \nu_\sigma(x) = \sup_{\{\sigma>0\}} [R_\sigma(x)/W_\sigma(x)],$$

where $R_\sigma(x) = E$ (total reward up to process time $\sigma + x |$ no completion up to process time x)

$$= V[1 - F(x)]^{-1} \int_x^t e^{-\beta(s-x)} dF(s),$$

and $W_\sigma(x) = E$ (discounted time from x to $\sigma + x |$ no completion up to process time x)

$$= [1-F(x)]^{-1} \{ \int_x^t \beta^{-1} [1-e^{-\beta(s-x)}] dF(s) + \beta^{-1} [1-e^{-\beta(t-x)}] \}$$

$$= [1-F(x)]^{-1} \int_x^t e^{-\beta(s-x)} [1-F(s)] ds.$$

Finally, then,

$$\nu(x) = \sup_{\{t>0\}} \frac{V\int_x^t e^{-\beta s} dF(s)}{\int_x^t e^{-\beta s}[1-F(s)]ds}, \text{ or } \sup_{\{t>0\}} \frac{V\int_x^t f(s)e^{-\beta s} ds}{\int_x^t [1-F(s)]e^{-\beta s} ds} \qquad (2)$$

if F has a density f.

The quantity $\rho(s) = f(s)[1-F(s)]^{-1}$ is the completion (hazard) rate for the job. If $\rho(s)$ is increasing it follows from (1) that

$$\nu(x) = \frac{V\int_x^\infty f(s)e^{-\beta s}ds}{\int_x^\infty [1-F(s)]e^{-\beta s}ds}, \tag{3}$$

and from Theorem 1 that, if the completion rate is increasing for every job in a SFABP, switches between jobs occur under an optimal policy only at the completion times of the various jobs. Thus the jobs should be processed in decreasing order of $\nu(0)$ which, for job i, may be expressed in the form (1). This result holds for arbitrary process time distributions if switching is prohibited except at completion times. If $\rho(s)$ is decreasing it follows from (2) that $\nu(x) = V\rho(x)$.

An optimal policy for this discounted rewards problem is one which maximises the expectation of $\Sigma V_i e^{-\beta t_i}$, where t_i is the completion time of job i. This quantity may be written as

$$E\Sigma V_i e^{-\beta t_i} = E\Sigma V_i - \beta E\Sigma V_i t_i + o(\beta).$$

In this expression the only quantities affected by the policy followed are the completion times t_i. Thus in the limit as $\beta \to 0$, an optimal policy minimises $E\Sigma V_i t_i$, i.e. it is an optimal policy for the minimal expected weighted flow-time problem, for which V_i is the weight, or cost per unit time of delay, for job i. This policy is the DAI policy obtained by setting $\beta = 0$ in the expression (2) for the index $\nu(x)$.

When $\beta = 0$ the expression (3) for $\nu(x)$ may be written as

$$\nu(x) = V/E\,(\text{remaining processing time received} \mid \text{no completion up to time } x), \tag{4}$$

and it is not difficult to show that (4) holds if and only if the denominator of the right hand side is decreasing, so that an optimal policy processes a set of jobs in decreasing order of $\nu(0) = V/E\tau$, a well-known result (e.g. see Simonovits [31]).

Expressions similar to, but more complicated than, (2) may be derived for the DAI's of the respective components of a FABP subject to precedence and arrival constraints. These define optimal policies by Theorem 3, provided Condition B holds. The monotonicity conditions leading to the simplified expressions (3) and (4) are again important when there are constraints; in all cases expressions for DAI's are simplified, and for general precedence constraints these conditions allow us to restrict attention to the non-pre-emptive case, for which Theorem 3 holds.

Consider, for example, a component C consisting of two jobs, 1 and 2, and suppose job 2 may not be started until job 1 has been completed. If $\rho_1(s)$ and $\rho_2(s)$ are both increasing the processing of neither job will be interrupted under an optimal policy. For discounted rewards the initial DAI is

$$\nu(C) = \max\left\{\frac{V_1\int_0^\infty e^{-\beta s}dF_1(s)}{\int_0^\infty e^{-\beta s}P_1(s)ds},\right.$$

$$\left.\frac{\int_0^\infty [V_1+V_2\int_0^\infty e^{-\beta t}dF_2(t)]e^{-\beta s}dF_1(s)}{\int_0^\infty e^{-\beta s}P_1(s)ds+\int_0^\infty\int_0^\infty e^{-\beta t}P_2(t)dte^{-\beta s}dF_1(s)}\right\}. \tag{5}$$

The first expression inside the curly bracket corresponds to the possibility that the first stage of a forwards induction policy for C may complete job 1 only, and the second to the possibility that it also completes job 2. This result also holds if pre-emption is prohibited. Putting $\beta = 0$ in (5) we get, for the case of weighted flow times,

$$\nu(C) = \max\{V_1/E\tau_1,\ (V_1+V_2)/(E\tau_1+E\tau_2)\}.$$

This result holds if and only if the expected remaining processing time for both jobs satisfies the monoticity condition for (4).

Further details may be found in the papers cited at the end of Section 2. In general the resulting expressions for the DAI of a component are complicated. Nash [28], however, showed that for Poisson arrivals, and no precedence constraints, optimal policies are defined by priority indices which are independent of the arrival rates, and the DAI's for the case of no arrivals therefore define optimal policies in all cases. Gittins and Glazebrook [8] showed that it is possible to incorporate learning, using Bayes theorem, about the unknown parameters of a stochastic scheduling model of this type.

3.2. Search

In Problem 2 let p_i now refer only to the initial probability of the object being in box i. A deterministic policy is simply a listing of the boxes in the order that they are to be searched, up to that point in the sequence when the object is found. For a particular policy, let $t(i,j)$ be the total cost incurred if the object is found at the j'th search of box i. The probability of this happening is $p_i(1-q_i)^{j-1}q_i$, and so the expected total cost is

$$\sum_{i=1}^{n} p_i \sum_{j=1}^{\infty} (1-q_i)^{j-1} q_i t(i,j).$$

Now
$$\sum_{i=1}^{n} p_i \sum_{j=1}^{\infty} (1-q_i)^{j-1} q_i e^{-\beta t(i,j)}$$

$$= 1- \beta \sum_{i=1}^{n} p_i \sum_{j=1}^{\infty} (1-q_i)^{j-1} q_i t(i,j)+o(\beta) . \tag{6}$$

We may therefore minimise the expected total cost of discovery by finding a policy which maximises (6), and then letting β tend to zero.

Each of Problems 2 to 5 fits naturally into the framework of a SFABP, though in each case there is some form of linkage between the objects which most closely resemble bandit processes which must be removed by some form of transformation before Theorem 1 can be applied. The difficulty in Problem 2 is that the posterior probability of the object being in each box changes when any one of them is searched, so these probabilities may not be used as states of bandit processes. The trick in this case is to note that (6) is the total discounted reward from the deterministic SFABP for which decision times for bandit process i occur at fixed intervals c_i, and $p_i(1-q_i)^{j-1}q_i$ is the reward when bandit process i is continued for the j'th time. Thus, by Theorem 1, (6) is maximised by a DAI policy, the DAI for bandit process i when it has been continued j times being

$$\nu_i(j) = \max_{\{k>j\}} \frac{\beta p_i \sum_{m=j+1}^{k} (1-q_i)^{m-1} q_i \exp[-\beta(m-j)c_i]}{1-\exp[-\beta(m-j)c_i]}$$

$$= \frac{\beta p_i(1-q_i)^j q_i}{1-\exp(-\beta c_i)} . \tag{7}$$

Letting β tend to 0 gives $\nu_i(j)=p_i(1-q_i)^j q_i/c_i$, and the DAI policy using this index is therefore optimal for the original search problem. Note that the posterior probability that the object is in box i after $j(i)$ unsuccessful searches $(i=1, 2, .., n)$ is proportional to $p_i(1-q_i)^{j(i)}$, so this confirms the well-known result (e.g. Black [3]) quoted in Section 1.

Theorems 1 and 3 rapidly lead to some generalisations of this result. First, the above argument may be carried through assuming $c(i,j)$ to be the cost of the j'th search of box i. The left hand equality in (7) now holds with $(m-j)c_i$ replaced by

$$\sum_{r=j+1}^{m} c(i,r).$$

The probability of discovering the object on the j'th search of box i, if it is in box i, and conditional on no prior discovery, may also be allowed to depend on j, so that the factor $(1-q_i)^{m-1}q_i$ in (7) becomes

$$\prod_{j=1}^{m-1} [1-q(i,j)]q(i,m).$$

The right hand equality in (7) depends on the expected reward rate from bandit process i being decreasing. In the more general case this may not be so, though it is, for example, if $c(i,j)$ is increasing in j and $q(i,j) = q_i$, independent of j, in which case

$$\nu_i(j) = \frac{\beta p_i(1-q_i)^j q_i}{1-\exp[-\beta c(i,j+1)]} .$$

For the solution of the search problem we again let β tend to 0.

Further generalisation may be achieved in two different ways. In the first place we can (as observed by Kelly [22]) note what happens if we allow the SFABP leading to the DAI policy solution to be stochastic instead of deterministic. This simply means allowing the sequence $\{[q(i,j),c(i,j)]: j=1,2,\ldots\}$ to be a stochastic process, both for the search problem and for bandit process i. Again, an expression for the DAI for bandit process i may be written down, and we let β tend to 0. This expression simplifies, as it does for any bandit process, if the DAI is almost surely a monotone function of processing time.

Alternatively, we may regard the deterministic bandit process i (=1,2,..,n) as consisting of a sequence of jobs $\{(i,j) : j=1,2,\ldots\}$, none of which may be pre-empted once started. The m'th job has a process time $c(i,m)$ and yields the reward

$$p_i \prod_{j=1}^{m-1} [1-q(i,j)]q(i,m) .$$

Thus the entire SFABP may be regarded as made up of jobs, with the obvious precedence constraints holding between jobs forming part of the same original bandit process. In other words we have a FABP, and since pre-emption is not allowed Condition B holds, and so we can use Theorem 3. The generalisation arises because these properties are preserved if an arbitrary set of additional precedence constraints between jobs is introduced, provided, as always, that these do not cause any precedence constraint circuits to occur. Given that our boxes must in practice be located in different places this seems a worthwhile feature to be able to add, as was done by Kadane and Simon [22]. For precedence constraints forming an out-tree Theorem 3 still holds if the jobs within each component form independent stochastic processes.

There is an extensive literature on search theory. Useful reviews are given by Stone [32], Gal [5], and Lehnerdt [25].

3.3. Bi-terminal Bandit Processes

Problems 3 and 4 may both be modelled as SFABP's. For Problem 3 we need terminal rewards which ensure that research stops as soon as the chemist has succeeded in synthesising the compound by one of the routes. For Problem 4 we need terminal rewards which ensure that testing continues on a formulation until it has passed every test, or until the first test which it fails, after which it is subjected to no further tests.

Consider a bandit process D for which, after time zero, decision times occur at intervals in processing time of $t_1, t_2, t_3, \ldots$, and costs $c_1, c_2, c_3, \ldots$ (> 0) occur at those times until D moves into one of the terminal states S and T. Once D reaches state S or state T it never leaves this state and behaves like a standard bandit process with parameter λ or μ respectively. Provided D has not already reached a terminal state, the probabilities of transition to state S or state T at the i'th decision time are p_i and q_i. A bandit process with these properties will be termed a *bi-terminal* bandit process.

Problem 3 may be modelled by a simple family F of alternative bi-terminal bandit processes $\{D_j : j = 1,2,..,n\}$ for which $\lambda_j = 0$, $\mu_j = -\beta M$, $M >> 0$, $q_{ji} \doteq 0$ $(i = 1,2,..,m_j)$, and $q_{ji} = 1(i = m_j + 1)$, $(j = 1,2,..,n)$, in the limit as the discount parameter β tends to zero. As soon as any of the D_j reaches state S, no further costs are incurred if this bandit process is continued, and any other policy does lead to further costs. Thus state S corresponds to success in synthesising the compound, continuing a bandit process which is in state S corresponds to doing no more research, and any optimal policy must have this property, as required. If D_j is in state T this corresponds to every stage in route j being completed without succeeding in synthesising the compound. The large cost incurred by continuing D_j in state T ensures that this does not happen unless all the other bandit processes have also reached state T, so that any optimal policy for $\mathcal{F}$ will result in synthesising the compound if this is possible. If it is not possible, then at some time t, which is the same for any policy with the properties just described, the last of the bandit processes reaches state T, and a terminal cost $\exp(-\beta t)M$ is incurred by continuing one of the D_j from time t to infinity. Since this terminal cost, and the probability of its occurrence, are the same for any policy with the required properties, it follows that any optimal policy for $\mathcal{F}$ minimises the expected cost of synthesising the compound if it can be synthesised at all.

The same model serves for Problem 4. State S for bandit process i must now be interpreted as the formulation having failed one of the stages in the test sequence i, and state T as having passed every stage in sequence i.

Optimal policies for $\mathcal{F}$ are DAI policies. The index takes the forms given in Section 1 under the monotonicity conditions there stated, which, in terms of $\mathcal{F}$, are the same for Problems 3 and 4.

As an example of a less restricted bi-terminal bandit process, suppose D is such that $\lambda = \beta V > 0$, $\mu << 0$, $q_i = 0 (i = 1,2,..,m)$, $q_{im+1} = 1$, and that the intervals t_i between decision times are independent random variables. The stopping time σ marking the end of the first stage of a forwards induction policy for D starting at process time zero must be of the form

$$\sigma = \begin{cases} \sum_{i=1}^{N} t_i & \text{if D does not reach state S by then,} \\ \infty & \text{if D reaches state S not later than process time } \sum_{i=1}^{N} t_i, \end{cases}$$

where $1 \leq N \leq m$. Writing $a_i = E\exp(-\beta t_i)$, $b_i = \prod_{j\leq i} a_j$, $r_1 = 1$, and $r_i = \prod_{j=1}^{i-1} (1 - p_j)$ for $i > 1$, it follows that the initial DAI value is

$$\nu(0) = \max_{1\leq N\leq m} \{ \sum_{i\leq N} r_i b_i (-c_i + p_i V)/(1 - r_N b_N)\}.$$

If $a_i(-c_i + p_i V)/[1 - (1 - p_i)a_i]$ is decreasing with i, the maximum occurs for N = 1 and

$$\nu(0) = a_1(-c_1 + p_1 V)/[1 - (1-p_1)a_1].$$

If this ratio increases with i, the maximum occurs for N = m. This might be a suitable model for one of a number of possible routes to some long term research goal, success in the route being worth V. Models of this kind for research projects have been discussed by Glazebrook [11], [12]. A more extensive coverage of the practicalities of using such models is given by Bergman and Gittins [2] together with Bergman's original solution of Problem 4.

As in Section 3.2, further generalisation falling within the conditions of Theorem 1 may be achieved by allowing the sequence $\{(c_i, p_i, q_i): i = 1,2,\ldots\}$ to be a stochastic process. However, since bi-terminal bandit processes are not deterministic, there are no obvious parallels to the generalisation in Section 3.2 using Theorem 3.

3.4. Stochastic Discounting

The formal equivalence between discounting a future reward, and receiving the same reward with a probability less than one, is well known. Kelly [23] has pointed out that it provides a means whereby Problem 5 may be modelled as a SFABP.

A deterministic policy is a fixed schedule of the mines to be worked on each day until the machine breaks down. Under a particular policy let R_i be the amount of gold mined on day i if there is no breakdown on day i, conditional on no breakdown during the previous i - 1 days. Write $T_i = -\log P_i$. Thus the expected total reward may be written as

$$\sum_{i=1}^{\infty} R_i \exp[-\sum_{j\le i} T_j].$$

This is also the total reward from a deterministic SFABP $\mathcal{F}$ for which R_i is the reward at the i'th decision time (including time zero), T_i is the interval between the i'th and (i + 1)'th decision times, and the discount parameter is 1. There is an obvious one-one correspondence between gold-mines and the bandit processes in $\mathcal{F}$.

For a particular gold-mine D, with r_i the gold mined and p_i the breakdown probability on day i, and $t_i = -\log p_i$, the initial DAI value is

$$\nu(0) = \max_{N\ge 1} \frac{\sum_{i\le N} r_i \exp\{-\sum_{j\le i} t_j\}}{1-\exp\{-\sum_{j\le N} t_j\}} = \max_{N\ge 1} \frac{\sum_{i\le N} r_j \prod_{j\le i} p_j}{1-\prod_{j\le N} p_j}.$$

If $r_i p_i (1 - p_i)^{-1}$ is decreasing in i the maximum occurs for N = 1 and $\nu(0) = r_1 p_1 (1 - p_1)^{-1}$. If $r_i p_i (1 - p_i)^{-1}$ is increasing in i the maximum occurs for $N = \infty$.

Since F is a deterministic SFABP, these results may be extended to establish the optimality of forwards induction and DAI policies under more general conditions in two different ways, just as in Section 3.2. Firstly, Theorem 1 still applies if $\{(r_i, p_i) : i = 1,2,\ldots\}$ is a stochastic process. Secondly, precedence constraints of the form 'the j'th day's mining on mine i must precede the k'th day's mining on mine m' may be introduced. Theorem 3 holds if the precedence constraints form an out-tree, or if the random pairs (r_i, p_i) are independent for different mines and/or different days. Partial results along these lines have been obtained by other methods. Hall [19] allows the (r_i, p_i) to form a stochastic process under monotonicity conditions. Kadane and Simon [21] consider precedence constraints for the deterministic case.

Stochastic discounting also provides an alternative method of solution for Problems 3 and 4. For 'the machine breaks down' simply substitute 'the compound is synthesised' or 'the formulation fails the test', as appropriate. This means that arbitrary precedence constraints may be introduced, as just mentioned, though the use of stochastic discounting seems to preclude incorporation of ordinary discounting.

3.5. Stoppable Bandit Processes

In Problem 3 the success or otherwise of a route is clear cut. Either the compound has been synthesised or not. More generally some degree of partial attainment of a research objective is possible. As in Problem 3, there may be more than one approach, or route, to the attainment of the objective.

This situation may be modelled by a family F of alternative components each of which is a *stoppable* bandit process: a bandit process, that is to say, with an additional control, control 2 say, representing the possibility of permanently stopping work on the route, and realising, in whatever manner is appropriate, the investment which it represents. Such a representation may be achieved by stipulating that so long as control 2 is applied the component (which is not strictly a bandit process) behaves like a standard bandit process with a parameter $\mu(x)$ which depends on the state x of the component, x remaining unchanged unless control 1 is applied. If every component has this form a stationary policy for F must apply control 2 permanently to the first component which receives this control. Thus if control 2 is applied to component M first in state x and at time t, a total discounted reward of $\exp(-\beta t)\beta^{-1} \mu(x)$ accrues from that time onwards, and this represents the reward resulting from stopping work on all the approaches aimed at the objective in question, and exploiting the position reached on the approach represented by component M. This model for the alternative routes problem was suggested by Glazebrook [13], who notes elsewhere [18] that it provides a neat application of Theorem 2 if $\mu(x(t_i))$ is almost surely increasing in i, where t_i is the i'th decision time for component M when control 1 is applied.

To see this, suppose component M belongs to a family G of two alternative components, the other of which is a standard bandit process Λ with parameter λ. We must show that for those values of μ for which an optimal policy for G continues M when it is in state x, the control which such a policy applies to M is independent of λ. This can be done by showing that the control applied must be optimal for the semi-Markov decision process M formed by removing the freeze control 0 from the control set for M, leaving control 1, the usual continuation control for a bandit process, and control 2.

Suppose then that for G in state x (i.e. when M is in state x) it is optimal to continue M. Note first that if control 2 is optimal for M or G in state x, then it must be optimal to apply control 2 permanently, since this causes no change of state. Suppose control 1 is optimal for M in state x. This is equivalent to saying that at least as high an expected total reward may be achieved by a policy which starts with control 1 as by applying control 2 permanently. If this is true for M it must also be true for G, since every policy for M is also a policy for G (i.e. one which never continues Λ). Thus control 1 is also optimal for G in state x.

Now suppose control 1 is optimal for G in state x. If $\mu(x) \geq \lambda$, then, since $\mu(x(t_i))$ increases in i, the inclusion of Λ makes no difference to an optimal policy, so control 1 is also optimal for M in state x. Suppose then that $\mu(x) < \lambda$. Let g be an optimal stationary policy for G which starts with control 1, and τ the time at which g first continues Λ. (Note that g must then continue Λ permanently, and that g may never select Λ, in which case $\tau = \infty$.) Since g is optimal it must be at least as good as the policy which always continues Λ. It must therefore achieve at least as high an expected reward up to time τ. In symbols

$$R_\tau((G,g)) \geq \lambda\beta^{-1}E(1 - e^{-\beta\tau}).$$

Thus

$$R_\tau((M,g)) = R_\tau((G,g)) > \mu(x)\beta^{-1}E(1 - e^{-\beta\tau}),$$

i.e. we can achieve a higher expected reward up to time τ by applying g to G (or equivalently to M) than by using control 2. Since $\mu(x(t_i))$ increases with i, the policy which coincides with g up to time τ and then switches permanently to control 2 must therefore achieve a higher expected total reward than one which uses control 2 from the outset. Thus control 1 is also optimal for M in state x if $\mu(x) > \lambda$.

We have shown, then, that if it is optimal to continue component M when G is in state x, then it is also optimal to apply control 1 if and only if it is optimal to apply control 1 to the semi-Markov decision process M in state x. Thus the control which an optimal policy for G applies to M is independent of λ. This is Condition A, and it therefore follows from Theorem 2 that forwards induction and DAI policies are optimal for a family of alternative stoppable bandit processes, for each of which $\mu(x(t_i))$ is increasing in i.

APPENDIX. PROOF OF THEOREM 2

Let F be a family of alternative components M_j (j = 1,2,..,n). Let γ denote an arbitrary policy for F, and σ a stopping time for the superprocess (F,γ). Let $R_{\gamma\sigma}(M_j)$ be the expected total discounted reward from M_j up to time σ when policy γ is applied to F, and $W_{\gamma\sigma}(M_j)$ the expected total discounted time, within the same period, for which M_j is continued. Thus

$$R_{\gamma\sigma}(F) = \sum_{j=1}^{n} R_{\gamma\sigma}(M_j) \text{ and } W_{\gamma\sigma}(F) = \sum_{j=1}^{n} W_{\gamma\sigma}(M_j).$$

The obvious extension of Theorem 1 does not hold in full generality. The following lemma is, however, easily obtained, where now γ is a forwards induction policy, and σ the stopping time which defines the first stage of such a policy.

Lemma 1. (i) $\nu(F) = \max_j \nu(M_j)$. (ii) $\nu(F) = \nu(M_j)$ for all j such that $W_{\gamma\sigma}(M_j) > 0$. (iii) A forwards induction policy for F is a DAI policy.

The proof of the first two parts of the lemma is virtually identical to that of Lemma 3.5 of [7], with M_j in place of D_j. The essential point is that the maximum expected rate attainable from F is the maximum expected rate attainable by any of the components of F. The last part of the lemma follows quickly from the first two parts.

Let $\{M,\Lambda\}$ denote a family of two alternative components, where Λ is a standard bandit process with parameter λ. In general an optimal deterministic stationary Markov policy for $\{M,\Lambda\}$ is defined by a policy g for M, which excludes control 0, and a stopping time τ for the superprocess (M,g), after which Λ is continued permanently. Both g and τ may depend on λ.

Condition A. The component M will be said to satisfy Condition A if the policy g for M, which together with a stopping time for the superprocess (M,g) defines an optimal deterministic stationary Markov policy for $\{M,\Lambda\}$, does not depend on λ, and if the only optimal deterministic stationary Markov policies for $\{M,\Lambda\}$, for any λ, are of this form, for some g with the stated property. A simple family $\mathcal{F}$ of alternative components M_j (j = 1,2,..,n) satisfies Condition A if the same is true for each of its components. Note that if M satisfies Condition A the policy g must be optimal for M when regarded as an isolated semi-Markov decision process. This follows by comparing the maximum expected total rewards from the SFABP $\{(M,g),\Lambda\}$ and from M, and letting λ tend to $-\infty$. Note too that if $\mathcal{F}$ satisfies Condition A, and a DAI policy selects component M_i for continuation, the control which it applies to M_i must be one given by a policy g_i for M_i of the type defined by Condition A.

Theorem 2 (The General DAI Theorem). (i) For a simple family $\mathcal{F}$ of alternative components satisfying Condition A a policy is optimal if and only if it is a DAI policy,and forwards induction policies are optimal. (ii) The classes of DAI policies and optimal policies coincide for the simple family of alternative components formed by merging an arbitrary SFABP with a fixed component M if and only if M satisfies Condition A.

Proof. We first show that DAI policies are the only optimal policies. From a standard result in dynamic programming it follows that it is sufficient to prove that no other deterministic policy is optimal in the class S of those policies which coincide with DAI policies from t_1, the first decision time after time zero, onwards. Since portions of optimal policies may be interchanged without destroying optimality we may restrict attention to stationary deterministic Markov DAI policies for which the continuation control within each component M_j is selected by a stationary deterministic Markov policy g_j of the type defined by Condition A. This means that from t_1 onwards the system reduces to the SFABP formed by the superprocesses (M_j,g_j) (j = 1,2,..,n). Call this the SFABP $\mathcal{A}$.

Suppose then that $h \in S$, and that at time zero h applies control $u(\neq 0)$ to component M_k, and thus control 0 to every other component. Let the initial state of M_k be y_k. Policy h may be regarded as stationary and Markov if state y_k is replaced by two states y_k^1 and y_k^2, each with identical properties to y_k, except that return to the initial state y_k^1 is impossible. Let M_k^* denote the modified version of M_k defined by this new terminology, and ug_k the policy for M_k^* which initially applies control u, and thereafter applies policy g_k, treating y_k^2 as equivalent to y_k. Thus applying policy h to $\mathcal{F}$ is equivalent to applying some policy h' to the SFABP $\mathcal{B}$ formed from $\mathcal{A}$ by replacing (M_k,g_k) by (M_k^*,ug_k).

For both $\mathcal{A}$ and $\mathcal{B}$ optimal policies are DAI policies (Theorem 1), and a DAI policy for $\mathcal{A}$ is equivalent to a DAI policy for $\mathcal{F}$. To show then that h is not optimal in S unless it is a DAI policy it suffices to show that the total expected return $R(\mathcal{A})$ from $\mathcal{A}$ under a DAI policy is greater than the expected total return $R(\mathcal{B})$ from $\mathcal{B}$ under a DAI policy, unless there is a DAI policy for $\mathcal{F}$ which starts by applying control u to M_k, i.e. unless $\nu((M_k^*, ug_k)) = \nu(\mathcal{F})$.

There are two possible ways in which this condition may fail to be fulfilled: (i) there may be a control $v(\neq u)$ such that $\nu(M_k) = \nu((M_k^*, vg_k)) = \nu(\mathcal{F})$, (ii) there may be no such control, so that $\nu(M_k) < \nu(\mathcal{F})$. In either case policy h does not define a DAI policy for $\mathcal{B}$, and consequently any DAI policy δ for $\mathcal{B}$ defines a policy for $\mathcal{F}$ which yields a higher expected total reward. In case (ii) there may be a zero probability that policy δ ever selects M_k^* for continuation, and consequently policy δ is indistinguishable from the obvious analogous policy for $\mathcal{A}$, thus yielding an expected reward no greater than a DAI policy for $\mathcal{A}$, which is therefore better than policy h, as required. Alternatively, there is a non-zero probability that under policy δ control v is first applied to M_k^* at some finite time T. Again policy δ is indistinguishable from the analogous policy for $\mathcal{A}$ up to time T. To show that for this variant of case (ii) policy h yields a lower expected reward than a DAI policy for $\mathcal{A}$, it is therefore sufficient to show that the expected reward yielded by δ from time T onwards, conditional on the state of $\mathcal{F}$ at T, is less than the expected reward from this point under a DAI policy for $\mathcal{A}$. This will follow immediately if we can show that $R(\mathcal{A}) > R(\mathcal{B})$ in case (i), to which we now turn.

To make the necessary comparison between the SFABP's $\mathcal{A}$ and $\mathcal{B}$ in case (i) it is useful to have a notation which clearly exhibits their common component. To this end write $\mathcal{A} = \{A, \mathcal{C}\}$ and $\mathcal{B} = \{B, \mathcal{C}\}$, where $A = (M_k, g_k)$, $B = (M_k^*, ug_k)$, and $\mathcal{C}$ denotes the component formed by the alternative superprocesses (M_j, g_j) $(j = 1,2,\ldots n; j \neq k)$. Regarding $\mathcal{C}$ as a SFABP: let R(t) be the expected total reward from $\mathcal{C}$ up to time t, and x(t) the state at time t, under a forwards induction policy γ; let $t_i (i = 1,2,\ldots)$ denote the end of the i'th stage of such a policy, with $t_0 = 0$; and write $\mu_i = \nu(\mathcal{C}, x(t_i))$ $(i = 0,1,\ldots)$. Let $R_A(t)$ and $R_B(t)$ be the expected total reward up to time t from A and B, respectively, when control 1 is applied at all times, so that t is also the process time. Let $\alpha(t)$ and $\beta(t)$ denote the states of A and B, respectively, at process time t, and let

$$\tau_A(\mu) = \inf\{t : \nu(A, \alpha(t)) < \mu, \text{ t is a decision time for A}\},$$

$$\tau_B(\mu) = \inf\{t : \nu(B, \beta(t)) < \mu, \text{ t is a decision time for B}\}.$$

Consider now the DAI policy γ_A for A which selects superprocess A for continuation whenever $\nu(A) \geq \nu((M_j,g_j))$ for all $j \neq k$, and chooses between the superprocesses in C according to γ (which is a DAI policy by Theorem 1). The total return from superprocess A under policy γ_A is

$$R_A(\tau_A(\mu_0)) + \sum_{i=1}^{\infty} e^{-\beta t_i} [R_A(\tau_A(\mu_i)) - R_A(\tau_A(\mu_{i-1}))] , \tag{8}$$

the factors $e^{-\beta t_i}$ arising from the periods during which one of the superprocesses in C is continued, thereby delaying the sequence of rewards from A, and causing them to be further discounted. Similarly, the total return from C is

$$e^{-\beta\tau_A(\mu_0)} R(t_1) + \sum_{i=1}^{\infty} e^{-\beta\tau_A(\mu_i)} [R(t_{i+1}) - R(t_i)] . \tag{9}$$

Note that either of the increasing sequences t_i and $\tau_A(\mu_i)$ may attain the value infinity for a finite value of i. Writing $I_1 = \min\{i : t_i = \infty\}$, $I_2 = \min\{i : \tau_A(\mu_i) = \infty\}$, we make the obvious conventions: $e^{-\beta t_i} = 0$ and $R(t_{i+1}) = R(t_i)$ for $i \geq I_1$; $e^{-\beta t_A(\mu_i)} = 0$ and $R(\tau_A(\mu_{i+1})) = R(\tau_A(\mu_i))$ for $i \geq I_2$. With these conventions the expressions (8) and (9) are valid in all cases.

The total return from the SFABP A is obtained by adding the expressions (8) and (9), giving

$$\sum_{i=0}^{\infty} \{R_A(\tau_A(\mu_i))(e^{-\beta t_i} - e^{-\beta t_{i+1}}) + e^{-\beta\tau_A(\mu_i)} [R(t_{i+1}) - R(t_i)]\}. \tag{10}$$

The total return from B is given by the expression derived from (10) by replacing A by B throughout, which we refer to as the B-analogue of (10).

The sequence of rewards from superprocess A, and the process times at which they occur, are independent of the similar quantities for the superprocess formed by applying policy γ to C. Moreover μ_i is a function of $x(t_i)$, so that applying the expectation operator conditional on t_i and $x(t_i)$ to the i'th term in the summation in (10), and denoting this operator by E_i, gives

$$E_i R_A(\tau_A(\mu_i)) E_i(e^{-\beta t_i} - e^{-\beta t_{i+1}}) + E_i e^{-\beta\tau_A(\mu_i)} E_i[R(t_{i+1}) - R(t_i)]. \tag{11}$$

From the definitions of t_i, t_{i+1} and μ_i it also follows that

$$E_i[R(t_{i+1}) - R(t_i)] = e^{-\beta t_i}\mu_i E_i \int_0^{t_{i+1}-t_i} e^{-\beta s} ds = \beta^{-1}\mu_i E_i(e^{-\beta t_i} - e^{-\beta t_{i+1}}),$$

so that (11) may be rewritten in the form

$$E_i[R_A(\tau_A(\mu_i)) + e^{-\beta\tau_A(\mu_i)}\beta^{-1}\mu_i]E_i(e^{-\beta t_i} - e^{-\beta t_{i+1}}). \qquad (12)$$

Applying E_i to the corresponding term in the expression for the total return from $\mathcal{B}$ gives us the B-analogue of (12).

Now by definition of A and g_k, and using Condition A, it follows that, for all $\lambda < \nu(M_k)$, $E[R_A(\tau_A(\lambda)) + e^{-\beta\tau_A(\lambda)}\beta^{-1}\lambda]$ is the maximum expected total reward from the simple family of alternative components $\{\mathcal{C}, \Lambda\}$. Moreover, using in addition the definition of B,

$$E[R_A(\tau_A(\lambda)) + e^{-\beta\tau_A(\lambda)}\beta^{-1}\lambda] \geq E[R_B(\tau_B(\lambda)) + e^{-\beta\tau_B(\lambda)}\beta^{-1}\lambda] \qquad (13)$$

for $\lambda < \nu(M_k)$, with equality if and only if there is a deterministic stationary Markov policy g_k' for M_k which starts by applying control u and is optimal in the sense of Condition A. However, the existence of such a g_k' implies, for case (i) and using the second note following the statement of Condition A, that $\nu(\mathcal{F}) = \nu(M_k) = \nu((M_k, g_k')) = \nu((M_k^*, ug_k))$. Since we are assuming that $\nu((M_k^*, ug_k)) < \nu(\mathcal{F})$ it follows that (13) must hold with strict inequality.

We therefore have

$$E_i[R_A(\tau_A(\mu_i)) + e^{-\beta\tau_A(\mu_i)}\beta^{-1}\mu_i] > E_i[R_B(\tau_B(\mu_i)) + e^{-\beta\tau_A(\mu_i)}\beta^{-1}\mu_i] \qquad (14)$$

for every i for which μ_i is defined. Now taking the expectation of (10), subtracting the expectation of its B-analogue, and using (11), (12), their B-analogues, and (14), it follows that $R(\mathcal{A}) > R(\mathcal{B})$. Thus only DAI policies are optimal in S, and the classes of DAI policies and optimal policies therefore coincide. This conclusion, together with part (iii) of Lemma 1, completes the proof of the first part of the theorem.

The 'if' statement of the second part of the theorem is simply a special case of the first part. The proof of 'only if' is immediate if we take our arbitrary SFABP to be just a single standard bandit process with parameter λ, and then consider how the class of optimal policies changes with λ.

REFERENCES

1. R.E. BELLMAN (1957) *Dynamic Programming*, Princeton University Press, Princeton.
2. S.W. BERGMAN, J.C. GITTINS (1982) *The Uses of Statistical Methods in the Planning of New-Product Chemical Research*, to appear.
3. W.L. BLACK (1965) Discrete sequential search. *Inform. and Control* 9,159-162.
4. Y.S. CHOW, H. ROBBINS, S. SIEGMUND (1971) *Great Expectations, the Theory of Optimal Stopping*, Houghton Mifflin, New York.
5. S. GAL (1980) *Search Games*, Academic Press, New York.
6. J.C. GITTINS (1979) Bandit processes and dynamic allocation indices. *J. Roy. Statist. Soc. Ser. B* 14,148-177.
7. J.C. GITTINS (1980) *Sequenctial Resource Allocation: Progress Report*, obtainable from the author.
8. J.C. GITTINS, K.D. GLAZEBROOK (1977) On Bayesian models in stochastic scheduling. *J. Appl. Probab.* 14,556-565.
9. J.C. GITTINS, P. NASH (1977) Scheduling, queues, and dynamic allocation indices. *Proc. EMS, Prague, 1974*, Czechoslovak Academy of Sciences, Prague, 191-202.
10. K.D. GLAZEBROOK (1976) Stochastic scheduling with order constraints. *Internat. J. Systems Sci.* 7,657-666.
11. K.D. GLAZEBROOK (1976) A profitability index for alternative research projects. *Omega* 4,79-83.
12. K.D. GLAZEBROOK (1978) Some ranking formulea for alternative research projects. *Omega* 6,193-194.
13. K.D. GLAZEBROOK (1979) Stoppable families of alternative bandit processes. *J. Appl. Probab.* 16,843-854.
14. K.D. GLAZEBROOK (1980) On single-machine sequencing with order constraints. *Naval Res. Logist. Quart.* 27,123-130.
15. K.D. GLAZEBROOK (1981) On non-preemptive strategies in stochastic scheduling, *Naval Res. Logist. Quart.* 28,289-300.
16. K.D. GLAZEBROOK (1980) On non-preemptive strategies for stochastic scheduling problems in continuous time. Unpublished manuscript.
17. K.D. GLAZEBROOK (1980) Methods for the evaluation of permutations as strategies in stochastic scheduling problems. *Management Sci.*, to appear.
18. K.D. GLAZEBROOK (1981) On a sufficient condition for superprocesses due to Whittle. *J. Appl. Probab.*, to appear.
19. K.D. GLAZEBROOK, J.C. GITTINS (1981) On single-machine scheduling with precedence relations and linear or discounted costs. *Oper. Res.* 29,161-173.
20. G.J. HALL (1978) Nonstationary stochastic gold-mining: a time-sequential tactical allocation problem. *Naval Res. Logist. Quart.* 25,81-93.
21. J.M. HARRISON (1975) Dynamic scheduling of a multiclass queue: discount optimality. *Oper. Res.* 23,270-282.
22. J.B. KADANE, H.A. SIMON (1977) Optimal strategies for a class

of constrained sequential problems. *Ann. Statist.* 5,237-255.
23. F.P. KELLY (1979) Contribution to the discussion of [5]. *J. Roy. Statist. Soc. Ser. B* 41,167-168.
24. E.L. LAWLER (1978) Sequencing problems with series parallel precedence constraints. *Proc. Summer School in Combinatorial Optimization, Urbino, 1978*, to appear.
25. M. LEHNERDT (1980) *On the Structure of Discrete Sequential Search Problems and of their Solutions*, Doctoral dissertation, University of Hamburg.
26. I. MEILIJSON, G. WEISS (1977) Multiple feedback at a single server station. *Stochastic Process. Appl.* 5,195-205.
27. I. MEILIJSON, U. YECHIALI (1977) On optimal right-of-way policiesat a single-server station when insertion of idle times is permitted. *Stochastic Process. Appl.* 6,25-32.
28. P. NASH (1973) *Optimal Allocation of Resources between Research Projects*, Ph.D. thesis, Cambridge University.
29. P. NASH, J.C. GITTINS (1977) A Hamiltonian approach to optimal stochastic resource allocation. *Adv. in Appl. Probab.* 9,55-68.
30. J.B. SIDNEY (1975) Decomposition algorithms for single machine sequencing with precedence relations and deferral costs. *Oper. Res.* 23,283-298.
31. A. SIMONOVITS (1973) Direct comparison of different priority queueing disciplines. *Studia Sci. Math. Hungar.* 8,225-243.
32. L.D. STONE (1975) *Theory of Optimal Search*, Academic Press, New York.
33. P. WHITTLE (1980) Multi-armed bandits and the Gittins index. *J. Roy. Statist. Soc. Ser. B* 42,143-149.
34. P. WHITTLE (1981) Arm acquiring bandits. *Ann. Probab.* 9,284-292.

MULTISERVER STOCHASTIC SCHEDULING

Gideon Weiss

Tel-Aviv University

ABSTRACT

m Parallel machines are available for the processing of n jobs. The jobs require random amounts of processing. When processing times are exponentially distributed, SEPT (shortest expected processing time first) minimizes the flowtime, LEPT (longest expected processing time first) minimizes the makespan and maximizes the time to first machine idleness. For m = 2, various other problems can be optimized by different rules. Optimality of preemptive SEPT and LEPT also holds when processing times are drawn from a common MHR (monotone hazard rate) distribution.

1. INTRODUCTION

Unlike the wide range of solved problems in deterministic scheduling, and the adequate solution given to many stochastic single server scheduling problems by the forward induction principle, results for stochastic scheduling of more than one machine are scant and the area offers a wide range of interesting unsolved problems.

In this survey we concentrate mainly on one group of problems for which fairly conclusive results exist, problems of scheduling stochastic jobs on parallel machines. m Machines are available to perform n jobs, requiring processing times $X_1,\dots,X_n$. The processing times are assumed to be random variables. A schedule is an assignment of jobs to machines at any moment, and it influences the (random) completion times $C_1,\dots,C_n$ of the jobs, or their ordered values $T_1 < \dots < T_n$. Various cost functions are of interest here, the classic ones being minimize ΣC_i (flowtime), minimize $T_n = C_{max}$ (makespan) and maximize T_{n-m+1} (time to first idleness,

M. A. H. Dempster et al. (eds.), Deterministic and Stochastic Scheduling, 157–179.

the nylon stocking problem). One seeks schedules which optimize such cost functions in the sense of giving optimal expected values of the (random) objective. General simple solutions cannot be expected to exist, and results to date are under simplifying assumptions on $X_1,\ldots,X_n$.

A special role is played by the family of exponential distributions, X being an exponential random variable, with rate λ, if

$$P(X > x) = e^{-\lambda x}, \quad x > 0, \tag{1.1}$$

and $E(X) = 1/\lambda$. It is characterized by the memoryless property

$$P(X > x+t \mid X > t) = P(X > x). \tag{1.2}$$

This property is shared by the discrete geometric distribution, X being geometric, with parameter q (equivalent to $e^{-\lambda}$), if

$$P(X = x) = (1-q)q^{x-1}, \quad x = 1,2,\ldots, \tag{1.3}$$

and $E(X) = 1/(1-q)$. If $X_1,\ldots,X_n$ are exponential, then because of the memoryless property, optimal schedules are usually non-preemptive.

Two scheduling rules which are intuitively appealing are SEPT (shortest expected processing time first) and LEPT (longest expected processing time first), and indeed in the cases studied, SEPT optimizes flowtime, while LEPT optimizes makespan and time to first idleness.

The survey is in three parts. In Section 2 we deal with two machines and exponential jobs. Here we cover minimization as well as maximization of C_{max}, and also some machine shop problems, using direct proofs. In Section 3 we give a more general formulation for m parallel machines with exponential jobs, and prove optimality of SEPT and LEPT for various cost functions, using Markov decision processes. Section 4 presents results for nonexponential jobs - here processing times are assumed to be drawn from a common distribution, with a monotone hazard rate, but different jobs start after being partly processed for different amounts of time. Again SEPT and LEPT, applied preemptively, turn out to be optimal, sometimes even in the sense of stochastic minimization or maximization. We shall denote stochastic ordering by $U >_{ST} V$ when $P(U > x) \geq P(V > x)$, for all x.

2. TWO MACHINE PROBLEMS, EXPONENTIAL JOBS

In this section we consider some problems involving two machines only, with n jobs requiring random processing times which are exponentially distributed, with known rates. Generalizations to more machines and nonexponential jobs are discussed in Sections 3 and 4. The emphasis is on those problems which can be solved by direct methods, thus providing more insight than the general methods of Sections 3 and 4.

2.1. Two parallel machines, minimize expected C_{max}

Two identical parallel machines are available to perform n jobs requiring random amounts of processing $X_1,\dots,X_n$, where X_i is exponentially distributed with rate μ_i (and expectation $1/\mu_i$). The aim is to minimize the expected completion time of the last job (makespan, C_{max}). This problem was first solved by Bruno and Downey [5], the presentation here follows Pinedo and Weiss [22].

A permutation $\lambda_1,\dots,\lambda_n$ of $\mu_1,\dots,\mu_n$ defines a schedule as follows: we assume that at time $t = 0$ one machine is occupied by a job requiring processing for time X_0 (nonnegative, random fixed or zero), and put a job with rate λ_1 on the other machine; thereafter, at the i-th job completion, we put a job with rate λ_{i+1} on the machine that was freed. Let the random completion times be $T_0 \le T_1 \le \dots \le T_n$. We wish to minimize the expected value of $C_{max} = T_n$, the last job completion. Since $2T_n = (X_0+X_1+\dots+X_n)+(T_n-T_{n-1})$, and $X_0+X_1+\dots+X_n$ is independent of the schedule, this is equivalent to minimizing the expected value of $D(X_0;\lambda_1,\dots,\lambda_n) = T_n-T_{n-1}$.

$D(X_0;\lambda_1,\dots,\lambda_n)$ is the duration in which one machine is idle while the other machine is completing the service of the last remaining job. Because of the randomness of the job processing times the remaining job may be any of the jobs $1,\dots,n$, or job 0 (unless $X_0 = 0$), with probabilities $p_1,\dots,p_n$, p_0, where $p_1,\dots,p_n$ depend on the schedule; p_0 and $D(X_0;\lambda_1,\dots,\lambda_n)$ conditional on the remaining job being job 0 are the same for all schedules. To minimize $D(X_0;\dots)$ we should try and minimize the probability that the last remaining job is a long job, and this can be achieved by scheduling long jobs as early as possible. Indeed we have:

THEOREM 2.1. *The* LEPT *(longest expected processing time first) rule, which defines the schedule* $\lambda_1 \le \lambda_2 \le \dots \le \lambda_n$, *minimizes expected* C_{max} *for the two parallel machine problem.*

We can prove the slightly stronger result that LEPT stochastically minimizes $D(X_0;\dots)$, by the following exchange argument.

LEMMA 2.1. *If* $\mu_1 \le \mu_i$, $i = 2,\dots,n$, *then*

$$D(X_0;\mu_1,\mu_2,\dots,\mu_n) <_{ST} D(X_0;\mu_2,\mu_1,\mu_3,\dots,\mu_n). \tag{2.1}$$

Proof of Lemma 2.1. Let p_j (q_j), $j = 0,1,\dots,n$, be the probabilities associated with $\mu_1,\mu_2,\dots,\mu_n$ $(\mu_2,\mu_1,\mu_3,\dots,\mu_n)$; the lemma follows by showing that

$$p_0 = q_0,\ p_1 \le q_1,\ p_j \ge q_j,\quad j = 2,\dots,n. \tag{2.2}$$

For $n = 2$, we have, by conditioning on the first completion and on $X_0 = x$ that

$$p_0 = q_0 = P(X_1+X_2 \le x) = 1 - \frac{\mu_2}{\mu_2-\mu_1} e^{-\mu_1 x} + \frac{\mu_1}{\mu_2-\mu_1} e^{-\mu_2 x},$$

$$p_1 = e^{-\mu_1 x} \frac{\mu_2}{\mu_1+\mu_2}, \quad q_1 = 1-q_0-q_1, \tag{2.3}$$

$$p_2 = 1-p_0-p_1, \quad q_2 = e^{-\mu_2 x} \frac{\mu_1}{\mu_1+\mu_2},$$

so that

$$q_1-p_1 = \frac{2\mu_1\mu_2}{\mu_2^2-\mu_1^2}(e^{-\mu_1 x}-e^{-\mu_2 x}) \ge 0, \tag{2.4}$$

and the proof is completed by induction on n. □

Proof of Theorem 2.1. We note that

$$D(X_0;\lambda_1,\dots,\lambda_{k-1},\lambda_k,\lambda_{k+1},\dots,\lambda_n) \\ = D(D(X_0;\lambda_1,\dots,\lambda_{k-1});\lambda_k,\lambda_{k+1},\dots,\lambda_n), \tag{2.5}$$

and therefore, by applying Lemma 2.1, any non-LEPT schedule can be improved by a finite number of pairwise exchanges of jobs, involving the longest remaining job and its preceding job, until an LEPT schedule is reached. □

We have shown that among all permutations of $\mu_1,\dots,\mu_n$ the one with $\lambda_1 \le \dots \le \lambda_n$ has smallest expected C_{max}. A simple inductive argument shows:

LEMMA 2.2. *The schedule which is optimal at* $t = 0$ *remains optimal if we allow rescheduling at job completions or if we allow any finite number of preemptions, and minimizes at any moment the remaining expected value of the objective function.*

We remark here that when the job durations are deterministic, preemptive application of the LEPT rule will lead to an optimal schedule; however, if no preemptions are allowed, LEPT is suboptimal and the problem is NP-complete.

The problem of minimizing makespan on two parallel machines has the following reliability interpretation: n spares with life lengths $X_1,\dots,X_n$, exponentially distributed with rates $\mu_1,\dots,\mu_n$, are available to keep a two-component series system in operation. We seek a schedule $\lambda_1,\dots,\lambda_n$ for replacing failed components by spares, so as to maximixe the system expected life time. Here the system lifetime is T_{n-1}, and its expectation is maximized (simultaneously with minimal expected T_n) by LEPT.

2.2. Two parallel machines, maximize expected C_{max}

With the same setup as in Section 2.1 we now wish to find a schedule $\lambda_1,\ldots,\lambda_n$ which will maximize expected completion time of the last job. This problem was partly solved by Pinedo [17] which led to a solution by Weiss [32].

The motivation to this seemingly artificial problem is its reliability equivalent, where C_{max} is the lifetime of a two-component parallel system with n spares. In the reliability context, one may also consider schedules which "take risks" by not replacing a failed component for a certain amount of time. However, Pinedo [17] has shown:

LEMMA 2.3. *For a two-component parallel system with* n *spares with exponential lifetimes, an optimal schedule with a finite number of decision moments does not take risks.*

Proof. The problem can be formulated as a Markov decision process for which, by general theory, a stationary optimal policy exists. However, if a stationary policy ever starts to take a risk, it will not replace the failed component until the system fails, which is strictly suboptimal. □

If we define $d(X_0;\mu_1,\ldots,\mu_n) = E\{D(X_0;\mu_1,\ldots,\mu_n)\}$, we have the following recursion formula to calculate $d(0;\mu_1,\ldots,\mu_n)$:

$$d(0;y) = \frac{1}{y}, \tag{2.6}$$

$$d(0;y,\mu_2,\ldots,\mu_n) = \frac{\mu_2}{y+\mu_2}\{d(0;y,\mu_3,\ldots,\mu_n)-d(0;\mu_2,\ldots,\mu_n)\} + d(0;\mu_2,\ldots,\mu_n).$$

We define:

ALGORITHM 2.1.
(i) Set $\lambda_n = \min_{1\le i\le n}\{\lambda_i\}$, and calculate $d(0;\mu_i)$, $i = 1,\ldots,n$.
(ii) If $\lambda_{k+1},\ldots,\lambda_n$ have already been defined, and A is the set of remaining rates, calculate $d(0;\mu,\lambda_{k+1},\ldots,\lambda_n)$, $\mu \in A$ from (2.6).
(iii) Set

$$\begin{aligned}\lambda_k &= \{\lambda \in A \mid d(0;\lambda,\lambda_{k+1},\ldots,\lambda_n)\} \\ &= \max_A\{d(0;\mu,\lambda_{k+1},\ldots,\lambda_n)\}.\end{aligned} \tag{2.7}$$

If k = 1, stop; else set k to k-1 and A to $A-\{\lambda_k\}$ and return to (ii).

It turns out from the proof of the following theorem that the maximum over A in (iii) is either $\lambda_k = \min\{\mu \in A\}$ or $\lambda_k = \max\{\mu \in A\}$.

THEOREM 2.2. *The schedule* $\lambda_1,\ldots,\lambda_n$ *obtained by Algorithm* 2.1 *maximizes the expected value of* C_{max}.

We need the following exchange argument.

LEMMA 2.3. *If* $d(0;\mu_2,\mu_3,\ldots,\mu_n) \geq d(0;\mu_1,\mu_3,\ldots,\mu_n)$ *then* $d(X_0;\mu_1,\mu_2,\mu_3,\ldots,\mu_n) \geq d(X_0;\mu_2,\mu_1,\mu_3,\ldots,\mu_n)$.

Proof of Lemma 2.3. Similar to the argument of Lemma 2.1; for $X_0 = x$ we have

$$d(x;\mu_1,\mu_2,\ldots,\mu_n) - d(x;\mu_2,\mu_1,\ldots,\mu_n) = \tag{2.8}$$

$$\frac{2\mu_1\mu_2}{\mu_2^2-\mu_1^2}(e^{-\mu_1 x}-e^{-\mu_2 x})\{d(0;\mu_2,\mu_3,\ldots,\mu_n)-d(0;\mu_1,\mu_3,\ldots,\mu_n)\} \geq 0. \quad \square$$

The function $d(0;y,\lambda_{k+1},\ldots,\lambda_n)$ is described schematically in Figure 2.1, for $k = n,n-1,n-2$. For $k = 1,\ldots,n-3$ the function behaves similarly to $d(0;y,\lambda_{n-1},\lambda_n)$, as stated in the next lemma, cited without proof.

LEMMA 2.4. *For* $k = 2,\ldots,n$ *let* $\lambda_k,\ldots,\lambda_n$ *be obtained by Algorithm* 2.1, *with* $\mu_1,\ldots,\mu_{k-1}$ *the remaining rates. Then:*

(i) *The equation* $d(0;y,\lambda_k,\ldots,\lambda_n) = d(0;\lambda_k,\ldots,\lambda_n)$ *has two roots,* $\lambda_k' \leq \lambda_k''$, *one of which is* λ_k. *For* $k = n$, $\lambda_n'' = \infty$, *else* $\lambda_k'' < \infty$.

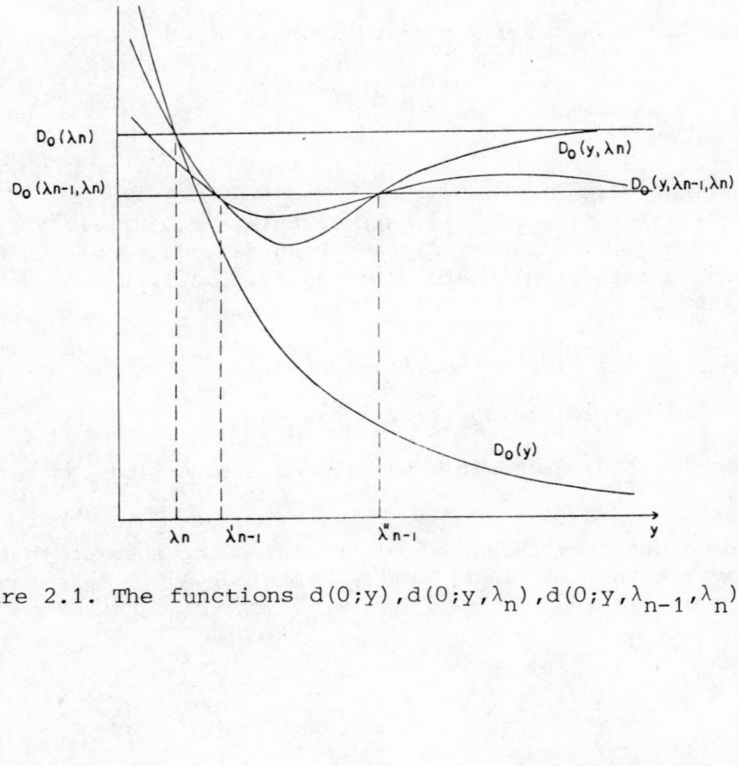

Figure 2.1. The functions $d(0;y), d(0;y,\lambda_n), d(0;y,\lambda_{n-1},\lambda_n)$

(ii) $d(0;y,\lambda_k,\ldots,\lambda_n)$ *is less than (greater than)* $d(0;\lambda_k,\ldots,\lambda_n)$ if $\lambda_k' < y < \lambda_k''$ ($y < \lambda_k'$ *or* $y > \lambda_k''$).

(iii) $d(0;y,\lambda_k,\ldots,\lambda_n)$ *is unimodal in the interval* (λ_k',λ_k''), *and convex in* (λ_k',λ_k^0) *where* λ_k^0 *is the minimum of the function.*

(iv) $\lambda_k' \le \mu_1,\ldots,\mu_{k-1} \le \lambda_k''$.

Proof of Theorem 2.2. Proceed by induction on n where n = 1 is void. Let $\mu_1,\ldots,\mu_n$ be any schedule, let $\lambda_2,\ldots,\lambda_n$ be the schedule given by Algorithm 2.1 for $\mu_2,\ldots,\mu_n$, with λ_i',λ_i'' defined for $i = 2,\ldots,n$ as in Lemma 2.4. By the induction and Lemma 2.3, $d(X_0;\mu_1,\mu_2,\ldots,\mu_n) \le d(X_0;\mu_1,\lambda_2,\ldots,\lambda_n)$. If $\mu_1,\lambda_2,\ldots,\lambda_n$ is not yet the schedule given by Algorithm 2.1, then by Lemma 2.4 $\mu_1 \notin (\lambda_2',\lambda_2'')$, and so $d(X_0;\mu_1,\lambda_2,\lambda_3,\ldots,\lambda_n) \le d(X_0;\lambda_2,\mu_1,\lambda_3,\ldots,\lambda_n)$. Repeat the argument a finite number of times to obtain the schedule given by the algorithm. □

We remark that Lemma 2.2 holds for this problem.

2.3. Two machine flowshop, minimize C_{max}

Each of n jobs, 1,...,n, needs to be processed first on machine A, then on machine B, and job j requires time A_j (B_j) on machine A (machine B). For the deterministic case, Johnson [15] shows that C_{max}, the time of completion of all jobs, is minimized if jobs are arranged in the following (transitive) order on both machines:

$$\text{job } j \text{ precedes job } k \iff \min\{A_j,B_k\} \le \min\{A_k,B_j\}. \qquad (2.9)$$

We assume that A_j (B_j) are exponentially distributed, with rates a_j (b_j). Now $E(\min\{A_j,B_k\}) = 1/(a_j+b_k)$, and taking expectations on both sides of Johnson's rule one obtains the rule:

$$\text{job } j \text{ precedes job } k \iff a_j-b_j \ge a_k-b_k. \qquad (2.10)$$

Bagga [1] and Cunningham and Dutta [10] prove that the order (2.10) minimizes the expected value of C_{max}. We give here a modified proof.

THEOREM 2.3. *Ordering the jobs on both machines by descending* a_j-b_j *minimizes expected* C_{max}.

Proof. Johnson [15] shows that for any given order of jobs on machine A, C_{max} is minimized by using the same order on Machine B, for any realization; we need therefore consider only permutation schedules. The theorem is established by showing that the "remainder" $C_{max} - \Sigma_{j=1}^n A_j$ is stochastically minimized.

The following two lemmas show that the remainder decreases stochastically with every pairwise exchange of two consecutively scheduled jobs which are not ordered according to (2.10), thus completing the proof. □

We consider now jobs 1 and 2 only, and assume that at time $t = 0$ machine B is occupied by a "remainder" of duration z. Let $D(z;j,k) = C_{max}-A_1-A_2$ when job j precedes job k, $j,k = 1,2$. Then:

LEMMA 2.5. *If* $z_1 \leq_{ST} z_2$ *then* $D(z_1;j,k) \leq_{ST} D(z_2;j,k)$.

Proof. Let $f_i(z)$ be the probability density of z_i, $\bar{F}_i(z) = P(z_i > z)$, and $g_z(w) = P\{D(z_i;j,k) > w | z_i = z\}$, $i = 1,2$. Clearly $g_z(w)$ increases with z, and by assumption $\bar{F}_2(z) > \bar{F}_1(z)$. So:

$$P\{D(z_2;j,k) > w\} - P\{D(z_1;j,k) > w\}$$
$$= \int_0^\infty g_z(w)\{f_2(z)-f_1(z)\}dz \tag{2.11}$$
$$= \int_0^\infty \frac{d}{dz} g_z(w)\{\bar{F}_2(z)-\bar{F}_1(z)\}dz > 0$$

which completes the proof. □

LEMMA 2.6. *If* $a_1-b_1 \geq a_2-b_2$ *then for any fixed* z, $D(z;1,2) <_{ST} D(z;2,1)$.

Proof. We have to compare $P\{D(z;1,2) > w\}$ with $P\{D(z;2,1) > w\}$. When $z > A_1+A_2$ these are equal, and the probability that $z > A_1+A_2$ is independent of the schedule. We thus need only look at $P\{D(z;j,k) > w|A_1+A_2 > z\}$. It is easy to see that

$$P\{D(z;1,2) > w|A_1+A_2 > z\}$$
$$= \frac{b_1}{a_2+b_1} e^{-b_2 w} + \frac{a_2}{a_2+b_1}\left(\frac{b_2}{b_2-b_1} e^{-b_1 w} - \frac{b_1}{b_2-b_1} e^{-b_2 w}\right) \tag{2.12}$$

with a similar expression for $P\{D(z;2,1) > w|A_1+A_2 > z\}$, and so

$$P\{D(z;2,1) > w|A_1+A_2 > z\} - P\{D(z;1,2) > w|A_1+A_2 > z\}$$
$$= \frac{b_1 b_2}{(a_1+b_2)(a_2+b_1)} \frac{e^{-b_1 w} - e^{-b_2 w}}{b_2-b_1}(a_1+b_2-b_1-a_2) \geq 0. \ \square \tag{2.13}$$

Cunningham and Dutta [10] also provide a recursive scheme to calculate expected C_{max} for any permutation schedule. We note that Lemma 2.2 holds for this problem. Also, the ordering by ascending a_j-b_j will maximize expected C_{max}. The question whether the optimization in this section holds also in the stochastic sense remains open.

Examination of the deterministic case indicates that results for the two machine flowshop cannot be directly generalized for the m machine flowshop. There is an equivalence between flowshops and queues with servers in tandem - the reader is directed to

Tembe and Wolff [26] and to Pinedo [18].

Recently, Pinedo [20] and Pinedo and Ross [21] have found optimal policies to minimize makespan for the 2-machine job shop and for special cases of the 2-machine open shop.

3. m UNIFORM MACHINES IN PARALLEL, EXPONENTIAL JOBS

3.1. The scheduling problems

In this section we consider m machines which are available to process N jobs. We assume again that jobs require $X_1,\dots,X_N$ random amounts of processing, exponential with rates $\lambda_1,\dots,\lambda_N$. The machines are arranged in parallel so that every machine can process any job, where at any moment operating machines and processed jobs are in one to one correspondence. The machines are not identical, but work at speeds $s_1,\dots,s_m$, where s_i indicates that machine i provides s_i units of processing per unit of time. Thus job j will require time T_{ij} to be processed completely on machine i where T_{ij} is exponential with rate $\lambda_j s_i$, but more typically, job j will be processed for a total time T_{ij}, $i = 1,\dots,m$, on machine i so that $s_1T_{1j}+s_2T_{2j}+\dots+s_mT_{mj} = X_j$ is exponential with rate λ_j.

The SEPT and LEPT rules can be generalized for this situation. At time t let $U = \{j_1,\dots,j_k\}$ be the set of uncompleted jobs, and let $s_1 \geq s_2 \geq \dots \geq s_m$. The generalized preemptive SEPT (LEPT) rule at time t says rearrange $j_1,\dots,j_k$ so that $\lambda_{j_1} \geq \dots \geq \lambda_{j_k}$ ($\lambda_{j_1} \leq \dots \leq \lambda_{j_k}$) and assign job j_i to processor i, $i = 1,\dots,\min\{k,m\}$. These rules cause rescheduling only at job completions, and involve preemptions in the form of moving jobs to faster processors. When $s_1 = \dots = s_m$ the rules are equivalent to non-preemptive rules.

For any schedule let $C_1,C_2,\dots,C_N$ be the completion times of jobs $1,2,\dots,N$ and let $T_1 \leq T_2 \leq \dots \leq T_N$ be the completion times in their order of occurrence. Three problems are of interest.

PROBLEM 1. Minimize expected flowtime - the flowtime is the total waiting time to completion of all jobs, that is $\Sigma_{j=1}^N C_j$ or $\Sigma_{j=1}^N T_j$.

PROBLEM 2. Minimize expected makespan - the makespan is the time taken to complete all jobs, that is $C_{max} = \max\{C_i\} = T_N$.

PROBLEM 3. Maximize the expected time to first machine idleness - here we assume that all m machines will be occupied as long as jobs are still available. The time to first machine idleness is T_{N-m+1}. This problem has a reliability interpretation, since T_{N-m+1} is the lifetime of an m component series system operated with N spares. It is also known as the Nylon Stocking Problem (Cox [8]).

Problem 1 is solved by using the SEPT rule, Problems 2 and 3 (which are equivalent for $m = 2$) are solved by the LEPT rule.

Bruno and Downey [5] solved Problems 1 and 2 for $m = 2$ identical machines and these results were extended to general m identical machines by Bruno, Downey and Frederickson [6]. Van der Heyden [27] presents a similar proof for Problem 2. The approach in these papers is via Markov decision processes. Weiss and Pinedo [33] used the approach of [6] to solve Problems 1, 2 and 3 for m uniform parallel machines, and the following is a description of their proof. A different approach to the proof for Problem 1 is presented in Glazebrook [13].

3.2. Formulation for general cost rate and optimality criterion

The scheduling problems of 3.1 can be formulated as semi-Markov decision processes (see Ross [23], Chapter 7). At time $t = 0$ we have jobs $\{1,\dots,N\}$ and processors $\{1,\dots,m\}$. We order the processors to have $s_1 \geq s_2 \geq \dots s_m \geq 0$, with $s_1 > 0$, and we assume throughout that $m \geq N$; this is no loss of generality since one can always add machines with speeds $s = 0$. Jobs $\{1,\dots,N\}$ are then assigned to machines and processed until a completion occurs, and a new assignment of machines is chosen. We consider $t = 0$ and times at which jobs are completed as decision moments. The state U at a decision moment is the set of uncompleted jobs, $U \subseteq \{1,\dots,n\}$. The set of actions available at U, J(U), is defined as

$$J(U) = \{f \mid f \text{ is a 1-1 function from } U \text{ into } \{1,\dots,m\}\}. \quad (3.1)$$

The action f assigns job j in U to machine f(j). For state U and action $f \in J(U)$, the next decision moment occurs time T later, where T is an exponential random variable with rate $\Lambda_f(U) = \Sigma_{j\in U} \lambda_j s_{f(j)}$. The transition probabilities to state U' are

$$P(U' = U-\{k\}) = \lambda_k s_{f(k)}/\Lambda_f(U), \quad k \in U. \quad (3.2)$$

For the cost structure, we assume that the cost incurred between the two decision moments is g(U)T, where g is a set function, the rate of cost function, and we assume $g(U) \geq 0$, $g(\emptyset) = 0$. Problems 1, 2 and 3 can each be described by a rate of cost function of that type.

We wish to find a policy which minimizes the expected cost over time $[0,\infty)$. We need to consider mainly stationary policies - π being stationary if it takes action $f \in J(U)$ whenever the process is in state U. Starting at t with state U and using π from t onwards we denote the expected cost over $[t,\infty)$ by $G_\pi(U)$; $G_{f'|\pi}(U)$ denotes the expected cost if at t we take action $f' \in J(U)$, and subsequently we use π. Because the action space is finite and the state $\emptyset$ is absorbing for policies with finite expected cost, we have (Strauch [25], Theorems 9.1 and 6.5):

THEOREM 3.1.
(i) *There exists a stationary optimal policy* π.
(ii) *A stationary policy* π *is optimal iff for all* U

$$G_\pi(U) = \min_{f' \in J(U)} \{G_{f'|\pi}(U)\}, \tag{3.3}$$

and π *is unique iff the minimum* (3.3) *is unique for all* U.

Let P_0 be the class of all policies which take decisions at the completion times $0 \le T_1 \le T_2 \ldots \le T_N$. We can also define the class P_Δ which takes decisions at $0 < \Delta < 2\Delta \ldots \le T_1 < T_1+\Delta \ldots \le T_N$, that is at completion times or Δ intervals, whichever occur first. This defines a new semi-Markov decision process, for which Theorem 3.1 holds. Obviously $P_\Delta \supseteq P_0$, but both classes have the same stationary policies and hence the same optimal stationary policies. Thus to show that π is optimal we need only check (3.3) when f' is used for a period of min{Δ,next completion}.

We call an action $f \in J(U)$ fast if for $|U| = n$, f is 1-1 from U to $\{1,\ldots,n\}$; a policy π is fast if it uses only fast actions. Fast policies use at any moment the fastest processors available, in particular they exclude inserted idle time. We call a policy a priority policy if for some particular renumbering of the jobs as $(1,\ldots,N)$, for every state $U = \{j_1,j_2,\ldots,j_n\}$, $j_1 \le j_2 < \ldots < j_n$, π uses action $f(j) = j$, $j = 1,\ldots,n$. Examples of priority policies are SEPT, LEPT, as well as the "cμ" rule.

Let π be a priority policy, and let U be an arbitrary fixed subset of the jobs $\{1,\ldots,N\}$. To simplify notation renumber the jobs in U by their priorities so that $U = \{1,\ldots,n\}$, and π takes the action $f(j) = j$, $j = 1,\ldots,n$ at U. Let

$$G = G_\pi(U),\ G_k = G_\pi(U)-\{k\}),\ G_{k\ell} = G_\pi(U-\{k,\ell\}),$$

$$\Lambda = \Lambda_f(U),\ \Lambda_k = \Lambda_f(U-\{k\}) = \Sigma_{j=1}^{k-1}\lambda_j s_j + \Sigma_{j=k+1}^{n}\lambda_j s_{j-1},$$

$$g = g(U),\ g_k = g(U-\{k\}),\ g_{k\ell} = g(U-\{k,\ell\}).$$

The following theorem gives an optimality condition for a priority policy π.

THEOREM 3.2. *Let* π *be a priority policy. Assume* $s_1 > s_2 > \ldots > s_m > 0$, $m > N$.
(i) π *is a unique optimal policy if for every arbitrary fixed* U

$$G-G_k > 0,\quad k = 1,\ldots,n, \tag{3.4}$$

$$\lambda_k(G-G_k)-\lambda_\ell(G-G_\ell) > 0,\quad k < \ell. \tag{3.5}$$

(ii) *Weak inequalities in* (3.4),(3.5) *are necessary conditions for optimality of* π.

Proof. For an arbitrary action $f' \in J(U)$,

$$G_{f'|\pi} = \Delta g + \Sigma_{j=1}^{n} \Delta\lambda_j s_{f'(j)} G_j + (1-\Sigma_{j=1}^{n} \Delta\lambda_j s_{f'(j)}) G + o(\Delta). \quad (3.6)$$

One can define a sequence of actions $f' = f^1, f^2, \ldots, f^r = f$ such that by (3.6) $G_{f^i|\pi} > G_{f^{i+1}|\pi}$. In particular condition (3.4) ensures that all optimal policies are fast. For details see [33]. □

We note that conditions (3.4) and (3.5) involve only expected costs when the priority policy π is used.

3.3. Optimality of SEPT and LEPT policies

To check the conditions (3.4) and (3.5) one can use:

$$G = (g + \Sigma_{j=1}^{n} \lambda_j s_j G_j)/\Lambda, \quad (3.7)$$

$$G_k = (g_k + \Sigma_{j=1}^{k-1} \lambda_j s_j G_{jk} + \Sigma_{j=k+1}^{n} \lambda_j s_{j-1} G_{jk})/\Lambda_k. \quad (3.8)$$

We now quote conditions on the cost rate function g under which SEPT or LEPT are optimal.

LEMMA 3.1. *Let* g *satisfy* $g(\emptyset) = 0$, *and for* $U \supset V$, $g(U) > g(V)$. *Then condition* (3.4) *holds, and there exist fast optimal policies.*

THEOREM 3.3. *Let the jobs be ordered so that* $\lambda_1 \geq \lambda_2 \ldots \geq \lambda_N$. *The* SEPT *priority rule defined by this ordering is optimal if* g *satisfies for every* $U \subset \{1,\ldots,N\}$

$$g(\emptyset) = 0, \quad g(U) \geq 0,$$

$$g(U-\{k\}) \geq g(U-\{\ell\}), \quad k < \ell,\ k,\ell \in U, \quad (3.9)$$

$$g(U)-g(U-\{k\})-g(U-\{\ell\})+g(U-\{k,\ell\}) \geq 0, \quad k,\ell \in U.$$

THEOREM 3.4. *Let the jobs be ordered so that* $\lambda_1 \leq \lambda_2 \ldots \leq \lambda_N$. *The* LEPT *priority rule defined by that ordering is optimal if* g *is of the form* $g(U) = h_{|U|}$, *for every* $U \subset \{1,\ldots,N\}$, *where* $|U|$ *is the size of* U, *and*

$$0 = h_0 \leq h_1 \leq \ldots \leq h_n,$$

$$\frac{h_1-h_0}{s_1} \geq \frac{h_2-h_1}{s_2} \geq \ldots \geq \frac{h_n-h_{n-1}}{s_n} \quad (3.10)$$

(where we define 0/0 *as* 0*)*.

The proof of Lemma 3.1 and Theorems 3.3 and 3.4 consists of verifying conditions (3.4),(3.5) by induction on $|U|$, using (3.7),

(3.8). In the proof of Theorem 3.4, one has to show that, under the conditions of the theorem,

$$G-G_n \geq \frac{h_n - h_{n-1}}{\lambda_n s_n}. \tag{3.11}$$

To prove (3.11) we condition on the completion times of jobs $1,\ldots,n-1$ being $t_0 = 0 < C_{i_1} = t_1 < C_{i_2} = t_2 < \ldots < C_{i_{n-1}} = t_{n-1}$, to get

$$G-G_n =$$

$$\Sigma_{k=0}^{n-2} \frac{h_{n-k} - h_{n-k-1}}{\lambda_n s_{n-k}} P(t_k < C_n \leq t_{k+1}) + \frac{h_1}{\lambda_n s_1} P(C_n > t_{n-1}) \tag{3.12}$$

and (3.11) follows by (3.10).

Theorem 3.3 shows directly that SEPT minimizes the flowtime, by using $g(U) = |U|$, while Theorem 3.4 shows that LEPT minimizes makespan by using $g(U) = 1$ for $U \neq \emptyset$, $g(\emptyset) = 0$. To see that LEPT maximizes T_{N-m+1} among fast policies, we proceed as follows: Let $M_1,\ldots,M_m$ be the total times during which machines $1,\ldots,m$ are occupied. Then for all fast policies, $M_1 = T_N$, $M_2 = T_{N-1}$, $\ldots$, $M_m = T_{N-m+1}$. By Theorem 3.4, one can see that LEPT minimizes

$$W_r = E(\Sigma_{i=1}^r s_i T_{N-i+1}) = E(\Sigma_{i=1}^r s_i M_i), \quad r = 1,\ldots,m \tag{3.13}$$

among all fast policies. But, since W_r is the expected total work done by machines $1,\ldots,r$, we have

$$\bar{W}_r = E(\Sigma_{i=r+1}^m s_i M_i) = \Sigma_{j=1}^N \frac{1}{\lambda_j} - W_r, \quad r = 1,\ldots,m-1 \tag{3.14}$$

and so $\bar{W}_r$ is maximized by LEPT among all fast policies, for $r = 1,\ldots,m-1$, and in particular $T_{N-m+1} = \bar{W}_{m-1}/s_m$ is maximized. Various other cost criteria can also be seen to be optimized by the SEPT or LEPT rules, using Theorems 3.3 and 3.4.

4. PARALLEL IDENTICAL MACHINES AND NONEXPONENTIAL JOBS

In this section we extend some of the results of Section 3 to nonexponential stochastic jobs. This work is due to Weber [29,30]; similar results were obtained by Gittins [12] who extended the use of a method of Glazebrook [13]. We are again concerned with preemptive scheduling of jobs on parallel identical machines. Weber shows that under certain assumptions on the distributions of the job lengths, preemptive SEPT is optimal for flowtime while preemptive LEPT is optimal for makespan and for time to first idleness. For some of these problems, under additional assumptions, the objective is optimized stochastically. Time varying numbers of machines and arrivals of new jobs are also considered. In Section 4.1 we discuss order relations between random variables, to motivate the results. In Section 4.2 we give the results of Weber, and in Section 4.3 we outline some of the proofs.

4.1. Order relations between random variables

We discuss general random processing times, $X > 0$, and denote by $F(x) = P(X \leq x)$ the distribution of X, by $\bar{F}(x) = 1-F(x)$ the survival function, and by $f(x)$ the density of X, defined as $f(x) = dF(x)/dx$ if X is continuous, and as $f(x) = P(X = x)$ if X is discrete. An important role is played by:

DEFINITION 4.1. The hazard rate of a random variable X is defined for all x such that $\bar{F}(x) > 0$, as

$$h(x) = \begin{cases} f(x)/\bar{F}(x), & X \text{ continuous}, \\ f(x+1)/\bar{F}(x), & X \text{ discrete}. \end{cases} \tag{4.1}$$

The hazard rate in the discrete case is the probability that a job be completed by processing for a period x+1, conditional on it not being completed by processing for a period x. It is related to $\bar{F}(x)$ by

$$\bar{F}(x) = \begin{cases} \exp(-\int_0^x h(t)\,dt), & X \text{ continuous}, \\ \prod_{t=0}^{x-1}(1-h(t)), & X \text{ discrete}. \end{cases} \tag{4.2}$$

For X exponential with rate λ, $h(x) = \lambda$ for all $x > 0$. For X geometric, with $f(x) = p(1-p)^{x-1}$, $h(x) = p$ for $x = 0,1,\ldots$.

DEFINITION 4.2. (i) A random variable X is called MHR (monotone hazard rate) if $h(x)$ is monotone, IHR (increasing hazard rate) if $h(x)$ is nondecreasing, DHR (decreasing hazard rate) if $h(x)$ is nonincreasing. (ii) A random variable X is called MLR (monotone likelihood ratio), if $\log f(t)$ is concave or convex, ILR (increasing likelihood ratio) in the convex case, DLR (decreasing likelihood ratio) in the concave case. This is equivalent in the discrete time case to:

$$\text{ILR (DLR)} \iff \frac{h(x+1)(1-h(x))}{h(x)} \text{ nonincreasing (decreasing)}. \tag{4.3}$$

There are various ways of comparing two random variables X and Y in order to say that X is greater than Y. The simplest comparison, which always exists (for $X,Y > 0$), is to compare their expectations, $E(X) > E(Y)$. Two more informative order relations are (we index $F,\bar{F}$, etc. of X,Y by 1,2 respectively):

DEFINITION 4.3. (i) X is stochastically greater than Y if

$$X >_{ST} Y \iff \bar{F}_1(x) \geq \bar{F}_2(x), \quad \text{all } x > 0. \tag{4.4}$$

(ii) X is greater than Y in likelihood ratio if

$$X >_{LR} Y \iff f_1(x)/f_2(x) \text{ is nondecreasing in } x. \tag{4.5}$$

LEMMA 4.1. *For any* X,Y,

$$X >_{LR} Y \Rightarrow X >_{ST} Y \Rightarrow E(X) \geq E(Y).$$

To describe a system with m machines and n jobs at time t, when a job that has total processing time X has already received processing for a period s, we need to define remaining processing time.

DEFINITION 4.4. For a job with processing time $X > 0$, we let ${}_sX$, the remaining processing time after s, be the random variable with distribution $P({}_sX \leq x) = (F(x+s)-F(s))/\bar{F}(s)$ and hazard rate $h(s+x)$.

LEMMA 4.2. *For any* $X > 0$,

$$\begin{aligned} X \textit{ is } \text{IHR (DHR)} &\iff {}_sX <_{ST} {}_tX \textit{ for all } s > t \ (s < t), \\ X \textit{ is } \text{ILR (DLR)} &\iff {}_sX <_{LR} {}_tX \textit{ for all } s > t \ (s < t). \end{aligned} \tag{4.6}$$

COROLLARY 4.1. ILR ⇒ IHR, DLR ⇒ DHR.

Proofs and further readings on the above topics can be found in Barlow and Proschan [2], Lehman [16], and Brown and Solomon [3].

We return now to the scheduling problems. It is easily seen that SEPT and LEPT do not remain optimal when one allows general random processing times. The following is a typical counterexample (Weber [29]).

Counterexample. There are two jobs, with discrete processing times, X_1, X_2. X_1 is geometric, while $X_2 = 2$ with probability .8 and X_2 is distributed like X_1+2 with probability .2. Thus $h_1(x) = 1/3$, $x = 0,1,\ldots$, $h_2(0) = 0$, $h_2(1) = .8$, $h_2(x) = 1/3$, $x = 2,3,\ldots$. There are two machines, one available at $t = 0$, the other occupied until $t = 1$. We want to minimize the makespan. Since $E(X_1) = 3 > E(X_2) = 2.6$, LEPT would schedule job 1 alone until $t = 1$, with expected makespan 4.382. But if we schedule job 2 alone until $t = 1$, the expected makespan is 4.36.

Why does LEPT fail here? To minimize makespan, we would like to process longer jobs first, and $E(X_1) > E(X_2)$ indicates that job 1 is longer than job 2. However, if we are only concerned with scheduling until $t = 1$, we have $h_1(0) = 1/3$, $h_2(0) = 0$, so that job 2 is less likely to finish and appears longer. The optimal schedule has to balance the overall length of the job with its current hazard rate, which is not done by LEPT or SEPT. Would LEPT be optimal if $X_1 >_{ST} X_2$? The answer is still no, as a slight change in the counterexample shows: just assume the second machine becomes available at $t = 3$, and replace X_2 by $X_2' = X_2+2$. Now $X_2' >_{ST} X_1$, and the optimal schedule is to process job 2 alone until $t = 3$. But this policy is LEPT only until $t = 2$. Here, although $X_2' >_{ST} X_1$, the remainder of job 2 after $t = 1$ is already not stochastically comparable with X_1. □

These counterexamples lead us to consider jobs which are comparable in the stochastic ordering sense at any time, no matter how much processing they have already received. This is a very strong requirement, as the following lemma shows.

LEMMA 4.3. *Let* $X, Y > 0$ *be random variables with continuous hazard rates* h_1, h_2. *If for every* s, t, *either* ${}_sX >_{ST} {}_tY$ *or* ${}_sX <_{ST} {}_tY$, *then one of* (i)-(iii) *holds.*

(i) *For all* s,t, $h_1(s)-h_2(t)$ *has constant sign.*

(ii) *There exists* $t \geq 0$ *such that for all* $x > 0$, $h_2(x) = h_1(x+t)$ *and* $h_2(x)$ *is monotone.*

(iii) *Same as* (ii) *with* 1,2 *exchanged.*

Proof. By definition

$$X >_{ST} Y \Longleftrightarrow \bar{F}_X(x) \geq \bar{F}_Y(x) \tag{4.7}$$

$$\Longleftrightarrow \int_0^x h_1 \leq \int_0^x h_2 \text{ for all } x > 0.$$

Since the hazard rates of ${}_tX, {}_sY$ are $h_1(x+t), h_2(x+s)$, the assumption that ${}_tX$ and ${}_sY$ are comparable is equivalent by (4.7) to

$$\int_t^{t+x} h_1 - \int_s^{s+x} h_2 \text{ has constant sign for all } x > 0. \tag{4.8}$$

By the continuity of h_1 and h_2 this implies that the functions $h_1(t+x)$ and $h_2(s+x)$ cannot cross each other, for any s,t. Thus, for every u, $h_1(x)-h_2(x+u)$ has constant sign for all x. If $h_1(x)-h_2(x+u)$ has the same sign for all u,x, we are in case (i). If there exist u,v such that $h_1(x)-h_2(x+u) \geq 0$, $h_1(x)-h_2(x+v) \leq 0$ for all x, then there exists u_0, between u and v, for which $h_1(x) = h_2(x+u_0)$, and we are in case (ii) or (iii). If $u_0 > 0$, we define $h(x) = h_1(x)$, and have for all u, $h(x)-h(x+u)$ has constant sign for all x. But then $h(x)-h(x-u)$ has opposite sign to $h(x)-h(x+u)$ for all x, and so h has no local minima, and therefore h is monotone. □

We see from the lemma that if for all s,t ${}_sX >_{ST} {}_tY$ or ${}_sX <_{ST} {}_tY$, then $E({}_sX) > E({}_tY)$ will imply ${}_sX >_{ST} {}_tY$ and $h_1(s) < h_2(t)$, so LEPT and SEPT will be taking into account both expected processing times and current hazard rates. Indeed in this case LEPT and SEPT are optimal.

To obtain even stronger results one can require that ${}_sX, {}_tY$ be comparable in the likelihood ratio sense. Similar to Lemma 4.3, one has:

COROLLARY 4.2. *If* $X, Y > 0$ *have continuous hazard rates* h_1, h_2, *and for every* s,t, *either* ${}_sX >_{LR} {}_tY$ *or* ${}_sX <_{LR} {}_tY$, *then one of the cases* (i)-(iii) *of Lemma* 4.3 *holds. In addition, in case* (ii), $Y = {}_tX$ *is* MLR, *in case* (iii) $X = {}_tY$ *is* MLR.

We make one final comment on the application of preemptive LEPT and SEPT to jobs whose processing times are continuous random variables. Assume X,Y are as in case (iii) of Lemma 4.3, with ${}_sX$ distributed like ${}_tY$, and both having IHR, and we start processing job ${}_sX$. If we use SEPT all is well, we would continue processing ${}_sX$ until it is completed before processing ${}_tY$. But if we want to use LEPT, we would have to switch to ${}_tY$ immediately, and then back to ${}_sX$ etc. Thus we can only talk of preemptive LEPT in discrete time, say in time units of length Δ, where $\Delta > 0$ may be as small as we want but fixed. For this reason all the results in Sections 4.2, 4.3 are in terms of discrete processing times.

4.2. Optimality of SEPT and LEPT policies

We consider now jobs 1,...,n with processing times $X_1, \ldots, X_n$ which are discrete random variables, and which form one of the following two families:

- *an* MHR *family of random processing times*. $X_1, \ldots, X_n$ form an MHR family if

(i) there exist random variables $Z_1, \ldots, Z_k$ with hazard rates $h_1, \ldots, h_k$ such that for all s,t, $h_i(s) \geq h_{i+1}(t)$, $i = 1, \ldots, k-1$;
(ii) for every j there exist i and $s > 0$ such that X_j is distributed as ${}_sZ_i$;
(iii) $Z_1, \ldots, Z_k$ are all of them either IHR or DHR.

- *an* MLR *family of random processing times*. $X_1, \ldots, X_n$ form an MLR family if (i),(ii) hold and
(iii) $Z_1, \ldots, Z_k$ are all of them either ILR or DLR.

We note that all variables in an MHR family are comparable in the stochastic order sense at all "ages", and those of an MLR family which belong to the same Z_i are in addition comparable in the likelihood ratio sense. In addition each variable is comparable to itself at any two ages.

The following results extend those of Sections 2,3 in three additional directions.

Variable number of machines. We assume that the number of machines available for processing jobs during (t-1,t), t = 0,1,..., is m(t), rather than a fixed m. Some of the results to be described hold for any machine availability function m(t), others hold only for m(t) nondecreasing. The results hold for any fixed m(t) (nondecreasing) and hence for a random stream of machine arrivals and departures (arrivals).

Job arrivals or release dates. Some of the results hold also when some of the jobs only become available at $t > 0$. These results again hold for any fixed release dates and hence for any random stream of arrivals.

Stochastic optimization. In Sections 2,3, we showed that LEPT or SEPT optimize the expected value of the objective function among all policies. These results can in some of the cases be extended to stochastic optimization, *i.e.* the objective function under LEPT (or SEPT) is stochastically smaller (or greater) than it would be under any other policy.

The problems covered in this section are as in Section 3, minimizing flowtime, minimizing makespan, and maximizing time to first idleness. The results proved by Weber [29] are summarized in Table 4.1.

problem	model	optimal strategy	optimization in	server number m(t)	arrivals
maximize time to first idleness	MLR	LEPT	distribution	arbitrary	yes
	MHR		distribution	non-decreasing	no
minimize makespan	MLR	LEPT	distribution	arbitrary	yes
	MHR		expectation	non-decreasing	no
minimize flowtime	MLR	SEPT	distribution	non-decreasing	no
	MHR		expectation	non-decreasing	no

Table 4.1. Job scheduling results

These results are conclusive for MHR and MLR models, in the sense that there exist counterexamples to all extensions.

4.3. Outline of some proofs

The results of Weber [29] summarized in Table 4.1 are quite laborious to prove. No single general proof exists, each of the three problems has to be considered separately, once for MHR and once for MLR, though the proofs are very similar.

We shall discuss here only the minimal makespan problem, for the MLR and the MHR case. We shall assume that the MLR (MHR) family of jobs $1,\ldots,n$ as defined in Section 4.2, all belongs to a single Z, in other words there exists a hazard rate function $h(x)$ and $s_1,\ldots,s_n$ such that $X_1,\ldots,X_n$ have hazard rates given by $h(s_1+x),\ldots,h(s_n+x)$. For the sake of definiteness we assume $h(x)$ is strictly increasing, so we are in the IHR or ILR case.

The state of the system at each moment t is given by the state of each of the jobs at t. Job j at time t can still not be present (in case of arrivals), denoted by (.), or it may have been completed, denoted by $(*)$, or it may have received an amount s of processing and not been completed yet, in which case its remaining processing time will have hazard rate $h(s_j+s+x)$, $x = 0,1,\ldots,$ and it will be uniquely characterized by its "age" s_j+s. Thus the state of the system at time t can be described by $x = (x_1,x_2,\ldots,x_n)$, where x_j is $(.)$, $(*)$, or $s+s_j$, and we call it the "age" of job j at time t. We remind here that $h(x_j)$ is the probability that job j will then be completed if processed in $(t,t+1]$.

In what follows we shall need to use differencing of a function of the state, $g(x_1,\ldots,x_n)$, with respect to the age of job i. We let

$$\begin{aligned}&\Delta_i g(x_1,\ldots,x_n)\\&= g(x_1,\ldots,x_i+1,\ldots,x_n) - g(x_1,\ldots,x_i,\ldots,x_n).\end{aligned} \tag{4.9}$$

THEOREM 4.1. *Under the above assumptions, for an* MLR *model,* LEPT *minimizes the makespan stochastically, for any* m(t)*, and for any release dates.*

Proof. We want to show that for any policy π,

$$\begin{aligned}&P(C_{max} \le s^*|\text{using LEPT})\\&\ge P(C_{max} \le s^*|\text{using }\pi), \quad \text{all } s^*.\end{aligned} \tag{4.10}$$

Fix s^*, and let

$$\begin{aligned}P^\pi(x,s) &= P(C_{max} \le s^*|\text{state at s is x, and we use }\pi),\\P(x,s) &= P^{LEPT}(x,s),\end{aligned} \tag{4.11}$$

then we have to show that

$$P(x,s) \ge P^\pi(x,s), \quad s = 0,1,\ldots,s^*, \tag{4.12}$$

which is done by backward induction, with $s = s^*$ trivial. Simple dynamic programming arguments show that the induction step in (4.12) is equivalent to showing that if $x_i < x_j$ (and therefore $h(x_i) < h(x_j)$ and i should precede j by LEPT), then

$$P_i(x,s)-P_j(x,s) \ge 0, \quad s = 0,1,\ldots,s^*, \tag{4.13}$$

where

$$\begin{aligned}P_i(x,s) &= h(x_i)P(x_1,\ldots,*,\ldots,x_n,s)\\&\quad + (1-h(x_i))P(x_1,\ldots,x_i+1,\ldots,x_n,s).\end{aligned} \tag{4.14}$$

Note that $P_i(x,s)$ is the probability that $C_{max} \le s^*$ when we are in state x at s-1, we work on i alone until s, and we then use LEPT. The condition (4.13) is the exact analogue of condition (3.5) of Theorem 3.2.

We show (4.13) by induction on $x_j - x_i$. For $x_j = x_i$ it is simply 0, and the induction is completed if we show that

$$\Delta_i\{P_i(x,s) - P_j(x,s)/h(x_i)\} \le 0, \quad s = 0,\dots,s^*. \tag{4.15}$$

We again use backward induction on s, $s = s^*$ being trivial. So we assume (4.13),(4.15) hold for $s = t,\dots,s^*$, and we need to check (4.15) for $s = t-1$.

We need to analyze $P_i(x,t-1)$ first. Let $m(t-1)$ be the number of machines available at t-1. Let $A = A(x,t-1)$ be the set of $m(t-1)$ jobs which would be processed at $(t-1,t)$ by LEPT, when the state is x at t-1. Let $X = (X_1,\dots,X_n)$ denote the random state at time t, reached by that processing. Then $X_\ell = x_\ell+1$ or $X_\ell = *$ for $\ell \in A$, $X_\ell = x_\ell$ for $\ell \notin A$, $\ell \ne (.)$, and $X_\ell = s_\ell$ if job ℓ arrives at t. Let k be the job such that $k \notin A$ but if one job in A were dropped before t-1, LEPT would choose k instead, in other words, x_k is minimal among jobs not in A. Then

$$P_i(x,t-1) = E_X[P_{i(X)}(X,t)], \tag{4.16}$$

where

$$i(X) = \begin{cases} i, & X_i \le x_k, \quad i \in A, \\ k, & X_i > x_k \text{ or } X_i = *, \quad i \in A, \\ i, & \quad i \notin A. \end{cases} \tag{4.17}$$

Note also that if $i \notin A$, $X_i = x_i$.

We return to checking (4.15) for $s = t-1$. There are two cases.

Case (i). $x_i \ge x_k$, so that $i \notin A(x,t-1)$, $j \notin A(x,t-1)$, and $i \notin A(x_1,\dots,x_i+1,\dots,x_n,t-1)$. Then by (4.16)

$$\begin{aligned} &\Delta_i\{P_i(x,t-1) - P_j(x,t-1)/h(x_i)\} \\ &= E_X[\Delta_i\{P_i(X,t) - P_j(X,t)/h(x_i)\}], \end{aligned} \tag{4.18}$$

which is ≤ 0 by induction, since $X_i = x_i$, $X_j = x_j$, and the time is t.

Case (ii). $x_i < x_k$, so that $i \in A(x,t-1)$ and $i \in A(x_1,\dots,x_i+1,\dots,x_n,t-1)$. Then by (4.16)

$$\begin{aligned} P_i(x,t-1) - P_j(x,t-1)/h(x_i) &= E_X[P_k(X,t) - P_{j(X)}(X,t) \mid X_i = *] \\ &+ \frac{h(x_i+1)(1-h(x_i))}{h(x_i)} E_X[P_i(X,t) - P_{j(X)}(X,t)/h(x_i+1) \mid X_i = x_i+1], \end{aligned} \tag{4.19}$$

and by differencing one obtains

$$\Delta_i\{P_i(x,t-1)-P_j(x,t-1)/h(x_i)\} = \tag{4.20}$$

$$[\Delta_i(\frac{h(x_i+1)(1-h(x_i))}{h(x_i)})]E_X[P_i(X,t)-P_{j(X)}(X,t)/h(x_i+1)|X_i=x_i+1]$$

$$+\frac{h(x_i+2)(1-h(x_i+1))}{h(x_i+1)}E_X[\Delta_i\{P_i(X,t)-P_{j(X)}(X,t)/h(x_i+1)\}|X_i=x_i+1].$$

In (4.20), $X_i = x_i+1$, while $j(X)$ is either k or j, with $X_{j(X)}$ being either x_k, or x_j+1 (for $j \in A$), or x_j where $x_j \geq x_k$. But $x_i+1 \leq x_k$, $x_i+1 < x_j+1$. By (4.13) for t, the 1st expectation is ≥ 0. By (4.15) for t, the 2nd expectation is ≤ 0. The term multiplying the 1st expectation is ≤ 0, by the ILR property of h (see (4.3)). So the expression is ≤ 0. □

THEOREM 4.2. *Under the above assumptions, for an* MHR *model,* LEPT *minimizes the expected makespan for nondecreasing* $m(t)$ *and no arrivals.*

Proof. Denote by $C^\pi(x,s,\infty)$ the expected makespan given that we are in state x at time s, and we use policy π. Let s^* be some fixed time, $s^* \geq s$, and assume now that we employ policy π from s onwards until s^*, when the state is $y_1,\ldots,y_n$ and we then freeze the ages of the jobs at the hazard rates $h(y_1),\ldots,h(y_n)$, so that all jobs still not completed at s^* will need geometrical processing times. We then proceed with π for the frozen jobs. Let $C^\pi(x,s,s^*)$ be the expected makespan for this frozen case.

Clearly

$$C^\pi(x,s,s^*) \to C^\pi(x,s,\infty) \quad \text{as } s^* \to \infty. \tag{4.21}$$

So we need to show that

$$C^\pi(x,s,s^*)-C^{LEPT}(x,s,s^*) \geq 0, \quad \text{all } s \leq s^*. \tag{4.22}$$

This is proved by backward induction on s, starting at $s = s^*$. For $s = s^*$, (4.22) is equivalent to saying that LEPT minimizes makespan for exponential jobs, which is true. The induction step is similar to the proof of Theorem 4.1, though it requires that there are no arrivals and that $m(t)$ is nondecreasing. We omit the details. □

REFERENCES

1. P.C. BAGGA (1970) n-Job, 2-machine sequencing problem with stochastic service times. *Opsearch* 7,184-197.
2. R.E. BARLOW, F. PROSCHAN (1975) *Statistical Theory of Reliability and Life Testing: Probability Models*, Holt, Rinehart and Winston, New York.
3. M. BROWN, H. SOLOMON (1973) Optimal issuing policies under stochastic field lives. *J. Appl. Probab.* 10,761-768.
4. J. BRUNO (1976) Sequencing tasks with exponential service times on parallel machines. Technical Report, Department of Computer Science, Pennsylvania State University.
5. J. BRUNO, P. DOWNEY (1977) Sequencing tasks with exponential service times on two machines. Technical Report, Department of Electrical Engineering and Computer Science, University of California, Santa Barbara.
6. J. BRUNO, P. DOWNEY, G.N. FREDERICKSON (1981) Sequencing tasks with exponential service times to minimize the expected flowtime or makespan. *J. Assoc. Comput. Mach.* 28,100-113.
7. V.YA. BURDYUK (1969) The stochastic problem of two machines. *Cybernetics* 5,651-661 (translated from Russian).
8. D.R. COX (1959) A renewal problem with bulk ordering of components. *J. Roy. Statist. Soc. Ser. B* 21,180-189.
9. T.B. CRABILL, D. GROSS, M.J. MAGAZINE (1977) A classified bibliography of research on optimal design and control of queues. *Oper. Res.* 25,219-232.
10. A.A. CUNNINGHAM, S.K. DUTTA (1973) Scheduling jobs with exponentially distributed processing times on two machines of a flow shop. *Naval Res. Logist. Quart.* 16,69-81.
11. G.N. FREDERICKSON (1978) Sequencing tasks with exponential service times to minimize the expected flow time or makespan. Technical Report, Department of Computer Science, Pennsylvania State University.
12. J.C. GITTINS (1981) Multiserver scheduling of jobs with increasing completion rates. *J. Appl. Probab.* 18,321-324.
13. K.D. GLAZEBROOK (1979) Scheduling tasks with exponential service times on parallel processors. *J. Appl. Probab.* 16,685-689.
14. K.D. GLAZEBROOK, P. NASH (1976) On multiserver stochastic scheduling. *J. Roy. Statist. Soc. Ser. B* 38,67-72.
15. S.M. JOHNSON (1954) Optimal two- and three-stage production schedules with setup times included. *Naval Res. Logist. Quart.* 1,61-68.
16. E.L. LEHMAN (1959) *Testing Statistical Hypotheses*, Wiley, New York.
17. M. PINEDO (1980) Scheduling spares with exponential lifetimes in a two component parallel system. *J. Appl. Probab.* 17, 1025-1032.
18. M. PINEDO (1980) Minimizing the makespan in a stochastic flowshop. *Oper. Res.*, to appear.

19. M. PINEDO (1980) Minimizing makespan with bimodal processing time distributions. *Management Sci.*, to appear.
20. M. PINEDO (1980) A note on the two machine job shop with exponential processing times. *Naval Res. Logist. Quart.*, to appear.
21. M. PINEDO, S.M. ROSS (1981) Minimizing expected makespan in stochastic open shops. *J. Appl. Probab.*, to appear.
22. M. PINEDO, G. WEISS (1979) Scheduling of stochastic tasks on two parallel processors. *Naval Res. Logist. Quart.* 26,527-535.
23. S.M. ROSS (1970) *Applied Probability Models with Optimization Applications*, Holden Day, San Francisco.
24. M.H. ROTHKOPF (1966) Scheduling with random service times. *Management Sci.* 12,707-713.
25. R. STRAUCH (1966) Negative dynamic programming. *Ann. Math. Statist.* 37,871-890.
26. S.V. TEMBE, R.W. WOLFF (1974) The optimal order of service in tandem queues. *Oper. Res.* 22,824-832.
27. L. VAN DER HEYDEN (1979) A note on scheduling jobs with exponential processing times on identical processors so as to minimize makespan. *Math. Oper. Res.*, to appear.
28. R.R. WEBER (1978) On the optimal assignment of customers to parallel servers. *J. Appl. Probab.* 15,406-413.
29. R.R. WEBER (1979) Optimal organization of multiserver systems. Ph.D. thesis, University of Cambridge.
30. R.R. WEBER, P. NASH (1979) An optimal strategy in multiserver stochastic scheduling. *J. Roy. Statist. Soc. Ser. B* 40,322-327.
31. G. WEISS (1977) A 2-machine n job scheduling problem. Technical Report, Department of Statistics, Tel-Aviv University.
32. G. WEISS (1981) Scheduling spares with exponential lifetimes in a two-component parallel system. Technical Report, Department of Statistics, Tel-Aviv University.
33. G. WEISS, M. PINEDO (1980) Scheduling tasks with exponential service times on non identical processors to minimize various cost functions. *J. Appl. Probab.* 17,187-202.

STOCHASTIC SHOP SCHEDULING: A SURVEY

Michael Pinedo[1], Linus Schrage[2]

[1]Georgia Institute of Technology
[2]University of Chicago

ABSTRACT

In this paper a survey is made of some of the recent results in stochastic shop scheduling. The models dealt with include:
(i) Open shops.
(ii) Flow shops with infinite intermediate storage (permutation flow shops).
(iii) Flow shops with zero intermediate storage and blocking.
(iv) Job shops.
Two objective functions are considered: Minimization of the expected completion time of the last job, the so-called makespan, and minimization of the sum of the expected completion times of all jobs, the so-called flow time. The decision-maker is not allowed to preempt. The shop models with two machines and exponentially distributed processing times usually turn out to have a very nice structure. Shop models with more than two machines are considerably harder.

1. INTRODUCTION AND SUMMARY

In this paper an attempt is made to survey the recent results in stochastic shop scheduling. Four shop models are considered; a short description of these follows.
(i) *Open Shops*. We have n jobs and m machines. A job requires an execution on each machine. The order in which a job passes through the machines is immaterial.
(ii) *Flow Shops with Infinite Intermediate Storage*. We have n jobs and m machines. The order of processing on the different machines is the same for all jobs; also the sequence in which the

M. A. H. Dempster et al. (eds.), Deterministic and Stochastic Scheduling, 181–196.

jobs go through the first machine has to be the same as the sequence in which the jobs go through any subsequent machine, i.e. one job may not pass another while waiting for a machine. A flow shop with these restrictions is often referred to as a *permutation* flow shop.

(iii) *Flow Shops with Zero Intermediate Storage and Blocking.* This shop model is similar to the previous one. The only difference is that now there is no storage space in between two successive machines. This may cause the following to happen: Job j after finishing its processing on machine i cannot leave machine i when the preceding job (job j-1) still is being processed on the next machine (machine i+1); this prevents job j+1 from starting its processing on machine i. This phenomenon is called blocking.

(iv) *Job Shops.* We have n jobs and m machines. Each job has its own machine order specified.

Throughout this paper will be assumed that the decision-maker is not allowed to preempt, i.e. interrupt the processing of a job on a machine. For results where the decision-maker *is* allowed to preempt, the reader should consult the references. In this paper two objectives will be considered, namely (i) minimization of the expected completion time of the last job (the so-called makespan) and (ii) minimization of the sum of the expected completion times of all jobs (the so-called flow time).

This survey is organized as follows: In Section 2 we give a short description of the most important results in *deterministic* shop scheduling (without proofs). The purpose of this section is to enable the reader to compare the results for the stochastic versions of the different models, presented in Section 3, with their deterministic counterparts. For the stochastic models in Section 3 we will not present any rigorous proofs either. However, we will provide for each model heuristic arguments that may make the results seem more intuitive. In Section 4 we discuss the similarities and differences between the deterministic and the stochastic results.

The notation used in this paper is the one developed by Graham et al.(5). For example, p_{ij} represents the processing time of job j on machine i. When this processing time is a random variable it will be denoted by $\underset{\sim}{p}_{ij}$. A second example: $O2|\underset{\sim}{p}_{ij} \sim \exp(1)|E(C_{max})$ represents a two machine open shop where the processing times of each job on the two machines are random variables, exponentially distributed with rate one and where the objective to be minimized is the expected makespan ($E(C_{max})$).

2. DETERMINISTIC SHOP MODELS

This section consists of three subsections: In the first subsection we deal with open shops, in the second one with flow shops and in the last one with job shops.

2.1 Deterministic Open Shops

Consider the two machine case where the makespan has to be minimized. In (5) this problem is referred to as $O2||C_{max}$. Gonzalez and Sahni (4) developed an algorithm that finds an optimal sequence in O(n) time. We present here a much simpler method that appears to be new.

Theorem 2.1.1. Let $p_{hk} = \max(p_{ij}, i=1,2, j=1,...,n)$. The following schedule minimizes the makespan: If h=2(1), job k has to be started at t=0 on machine 1(2); after finishing this processing on machine 1(2) job k's processing on machine 2(1) has to be postponed as long as possible. All other jobs may be processed in an arbitrary way on machines 1 and 2. Job k may only be started on machine 2(1) either when no other job remains to be processed on machine 2(1) or when only one other job needs processing on machine 2(1) but this job is just then being processed on machine 1(2).

The reader should have little difficulty in proving this theorem. Gonzalez and Sahni (4) showed that the open shop problem with more than two machines is NP-complete.

2.2. Deterministic Flow Shops

Consider first the two machine flow shop with infinite intermediate storage between the machines. We are interested in minimizing the makespan. This problem is usually referred to as $F2||C_{max}$. Johnson (7) developed the well-known rule for obtaining the optimal sequence in this problem.

Theorem 2.2.1. The sequence, that puts the jobs with $p_{1j} \leq p_{2j}$ first, in order of nondecreasing p_{1j} and puts the remaining jobs afterwards, in order of nonincreasing p_{2j}, is optimal.

When there are more than two machines in series, the problem is NP-complete (see Garey et al. (3)). Research in this area is still going on, focussing mainly on enumerative methods. One special case, however, is easy: Consider the case where the processing times of job i, i=1,...,n at all m machines is p_i. In practice such a situation would occur in a communication channel where messages do not change in length when they pass from one station to the next. For this special case Avi-Itzhak (1) established the following theorem.

Theorem 2.2.2. When $p_{1j}=p_{2j}=\ldots=p_{mj}$ for $j=1,\ldots,n$, any sequence is optimal.

Reddi and Ramamoorthy (14) considered the flow shop with zero intermediate storage and blocking. This problem is not covered in the survey paper of Graham et al.(5). We will refer to this flow shop problem as $F|blocking|C_{max}$. Reddi and Ramamoorthy (14) found that $F2|blocking|C_{max}$ can be formulated as a Travelling Salesman Problem with a special structure, a structure that enables one to use an $O(n^2)$ algorithm.

2.3. Deterministic Job Shops

Consider the two machine case $J2||C_{max}$. In this two machine model one set of jobs has to be processed first on machine 1 and after that on machine 2. This set of jobs will be referred to as set A. A second set of jobs has to be processed first on machine 2 and after that on machine 1. This set of jobs will be referred to as set B. Jackson (6), using Johnson's algorithm for $F2||C_{max}$, obtained an optimal schedule for $J2||C_{max}$.

Theorem 2.3.1. The following schedule is optimal: All the jobs of set A (B) are to be processed on machine 1 (2) before any job of set B (A) is to be processed on machine 1 (2). The jobs of set A (B) are to be processed on machine 1 (2) in the following order: The jobs with $p_{1j} \leq p_{2j}$ ($p_{2j} \leq p_{1j}$) first in order of nondecreasing p_{1j} (p_{2j}) and the remaining jobs afterwards in order of nonincreasing p_{1j} (p_{2j}). The order in which the jobs of set A (B) are processed on machine 2 (1) does not affect the makespan.

Job shops with more than two machines are NP-complete, even when all processing times are equal to one. But a considerable amount of effort has been dedicated to the research in enumerative methods (see McMahon and Florian (8)).

3. STOCHASTIC SHOP MODELS

This section consists of four subsections: In the first subsection we deal with stochastic open shops. In the second one with stochastic flow shops with infinite intermediate storage between the machines. In the third one we consider stochastic flow shops with zero intermediate storage and blocking. In the last subsection stochastic job shops are considered.

3.1. Stochastic Open Shops

In this subsection we assume that there are two machines available to process n jobs. Each job has to undergo operations on both machines, the order in which this happens being immaterial. Every time a machine finishes an operation the decision-maker has to decide which job will be processed next on the machine just freed. A policy prescribes the decision-maker which actions to take at the various decision moments; such an action at a decision moment depends on the state of the system at that moment. Observe that a policy only has to instruct the decision-maker what to do as long as there are still jobs which have not yet undergone processing on either machine. This is true for the following reason: When machine 1 (2) becomes free, the decision-maker otherwise only can choose from jobs which have to be processed only on machine 1 (2) and the sequence in which these jobs will be processed on machine 1 (2) does not affect the makespan. Clearly

$$C_{max} \geq \max \left(\sum_{j=1}^{n} \underset{\sim}{p}_{1j}, \sum_{j=1}^{n} \underset{\sim}{p}_{2j} \right).$$

When one machine is kept idle for some time in between the operations of two jobs, the makespan may be strictly larger than the R.H.S. of the above expression. We may distinguish between two types of idle periods, see Figure 1.

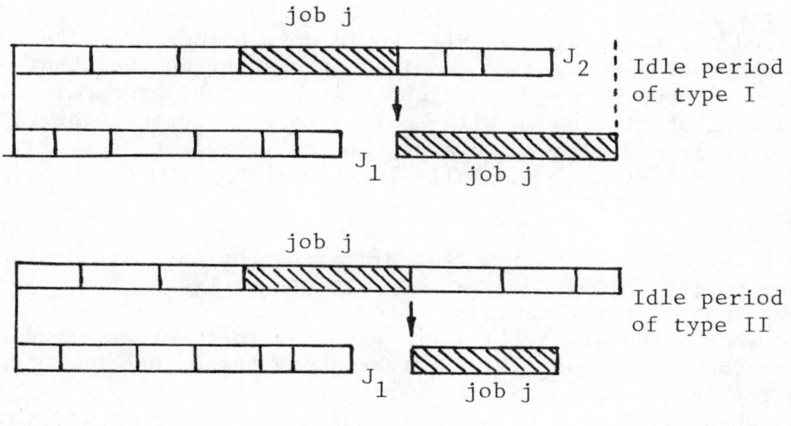

Figure 1

In an idle period of type II, a machine is kept idle for some time, say J_1, then processes its last job, say job j, and finishes processing job j while the other machine is still busy processing other jobs. It is clear that, although a machine has been kept idle for some time, $C_{max} = \max\left(\Sigma p_{1j}, \Sigma p_{2j}\right)$. In an idle period of type I, a machine is kept idle for some time, say J_1, then processes the last job, say job j, and finishes processing this job some time, say J_2, after the other machine has finished all its jobs. Now

$$C_{max} = \max\left(\sum_{j=1}^{n} p_{1j}, \sum_{j=1}^{n} p_{2j}\right) + \min(J_1, J_2).$$

It can be verified easily that only one job can cause an idle period and an idle period has to be either of type I or type II. As the first term on the R.H.S. of the above expression does *not* depend on the policy, it suffices to find a policy that minimizes $E(\min(J_1,J_2))$.

We will consider now a special case of the two machine open shop model, namely the model $O2|p_{ij} \sim \exp(\mu_j)|E(C_{max})$, where the operations of job j on the two machines are independent and exponentially distributed, both with rate μ_j. From the explanation above it appears intuitive that, in order to minimize the probability that an idle period of type I occurs, jobs that have not received any processing at all should have higher priority than the jobs that already have received processing on the other machine. Moreover among the jobs that have not yet received any processing at all, those with smaller expected processing times should be processed more towards the end. In fact, for $O2|p_{ij} \sim \exp(\mu_j)|E(C_{max})$ the following theorem has been shown in Pinedo and Ross (13).

Theorem 3.1.1. The policy that minimizes the expected makespan is the policy which, whenever any one of the two machines is freed, instructs the decision-maker:
(i) When there are still jobs which have not yet received processing on either machine, to start among these jobs the one with the largest expected processing time and
(ii) when all jobs have been processed at least once to start any one of the jobs still to be processed on the machine just freed.

Consider now the model $O2|p_{ij} \sim G_i|E(C_{max})$. In this model we have n identical jobs and two machines. These two machines have different speeds. The distribution of the processing time of a job on machine i, i=1,2, is G_i. We assume that G_i, i=1,2, is New Better than Used (NBU), i.e.

$$\bar{G}_i(x+y)/\bar{G}_i(x) \le \bar{G}_i(y), \qquad x \ge 0,\ y \ge 0.$$

Under this assumption the following theorem can be proven (see Pinedo and Ross (13)):

Theorem 3.1.2. The makespan is stochastically minimized if the decision-maker starts, whenever a machine is freed, when possible with a job which has not yet been processed on either machine.

The proof of this theorem is a proof by induction, on which we shall not elaborate here. However, the special case $O2|p_{ij} \sim \exp(1)|E(C_{max})$ can be analyzed further. For this model the following closed form expression for $E(C_{max})$ under the optimal policy can be obtained:

$$E(C_{max}) = 2n - \sum_{k=n}^{2n-1} k\binom{k-1}{n-1}\left(\frac{1}{2}\right)^k + \left(\frac{1}{2}\right)^n.$$

This expression is obtained by calculating the probability of each job causing an idle period of type I.

Up to now the only objective under consideration has been minimization of expected makespan. Our second objective is minimization of expected flow time, i.e. $E(\Sigma C_j)$. Consider the model $O2|p_{ij} \sim \exp(\mu_i)|E(\Sigma C_j)$: Again we have n identical jobs on two machines with different speeds. For the case where the processing times on machine i, i=1,2, are exponentially distributed with rate μ_i, we have the following theorem (see Pinedo (9)):

Theorem 3.1.3. The flow time is stochastically minimized if the decision-maker starts, whenever a machine is freed, when possible with a job that already has been processed on the other machine.

3.2. Stochastic Flow Shops with Unlimited Intermediate Storage Between Machines

In this subsection we consider m machines and n jobs. The n jobs are to be processed on the m machines with the order of processing on the different machines being the same for all jobs. Each job has to be processed first on machine 1, after that on machine 2, etc. At t=0 the jobs have to be set up in a sequence, in which they have to traverse the system. We want to determine the job sequence that minimizes either $E(C_{max})$ or $E(\Sigma C_j)$.

We will consider first the case m=2. In Figure 2 is depicted a realization of the process. Intuitively we may expect that in order to minimize the expected makespan, jobs with shorter expected processing times on machine 1 and larger expected processing times on machine 2 should be scheduled more towards the beginning of the sequence, while jobs with larger expected processing times on machine 1 and shorter expected processing times

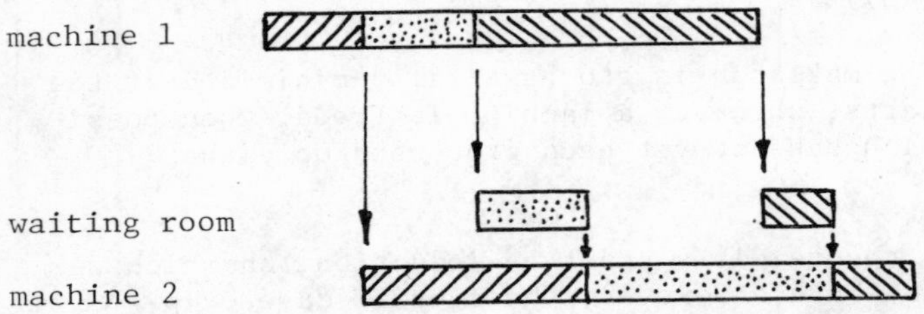

Figure 2

on machine 2 should be scheduled more towards the end of the sequence. In the deterministic version of this problem the optimal sequence is determined by Johnson's rule. Bagga (2) considered $F2|p_{ij} \sim \exp(\mu_{ij})|E(C_{max})$, i.e. the case where the processing times are exponentially distributed, and proved, through an adjacent pairwise switch argument, the following theorem.

Theorem 3.2.1. Sequencing the jobs in decreasing order of $\mu_{1j} - \mu_{2j}$ minimizes the expected makespan.

This theorem implies that when the processing time of job j on machine 1 (2) is zero, i.e. $\mu_{1j} = \infty$ $(\mu_{2j}=\infty)$, it has to go first (last). If there is a number of jobs with zero processing times on machine 1, these jobs have to precede all the others in the sequence. The sequence in which these jobs go through machine 2 does not affect the makespan. A similar remark can be made if there is more than one job with zero processing time on machine 2.

One special case of the flow shop model is of particular importance, namely the case where the processing times of a job on the different machines are independent draws from the same distribution, i.e. $F|p_{ij} \sim G_j|E(C_{max})$ and $F|p_{ij} \sim G_j|E(\Sigma C_j)$. Of this case one can easily find examples in real life: Consider a communication channel, where messages do not lose their identity when they pass from one station to the next. From Theorem 3.2.1 follows that for the model $F2|p_{ij} \sim \exp(\mu_j)|E(C_{max})$ any sequence will be optimal. Weber (15) considered $F|p_{ij} \sim \exp(\mu_j)|E(C_{max})$, the case for an arbitrary number of machines. With regard to this model he showed the following theorem.

Theorem 3.2.2. The distribution of the makespan does not depend on the sequence in which the jobs traverse the system.

In Theorem 2.2.2 was stated that in the case where the G_j, $j=1,\ldots,n$, are deterministic, not necessarily identical, the makespan does not depend on the job sequence either. This property which holds for the exponential and deterministic distributions does not hold for arbitrary distributions. One can easily find counterexamples. In (10) Pinedo considered other examples of $F|\underset{\sim}{p}_{ij} \sim G_j|E(C_{max})$. Before discussing the results presented in (10) we need two definitions:

Definition 1. A sequence of jobs $j_1, j_2, \ldots, j_n$ is a SEPT-LEPT sequence if there exists a k such that

$$E\left(\underset{\sim}{p}_{ij_1}\right) \leq E\left(\underset{\sim}{p}_{ij_2}\right) \leq \ldots \leq E\left(\underset{\sim}{p}_{ij_k}\right)$$

and

$$E\left(\underset{\sim}{p}_{ij_k}\right) \geq E\left(\underset{\sim}{p}_{ij_{k-1}}\right) \geq \ldots \geq E\left(\underset{\sim}{p}_{ij_n}\right)$$

(observe that both the SEPT and the LEPT sequences are SEPT-LEPT sequences).

Definition 2. Distribution G_k and G_ℓ are said to be nonoverlappingly ordered if $P\left(p_{ik} \geq p_{i\ell}\right)$ is either zero or one. This implies that the probability density functions do not overlap.

Based on these two definitions we can present the following theorem concerning $F|\underset{\sim}{p}_{ij} \sim G_j|E(C_{max})$.

Theorem 3.2.3. For n jobs with nonoverlapping processing time distributions, any SEPT-LEPT sequence minimizes the expected makespan.

Note that this theorem does *not* state that SEPT-LEPT sequences are the only sequences that minimize $E(C_{max})$. However it is important to observe that $E(C_{max})$ *does* depend on the sequence and that there *are* sequences which do not minimize $E(C_{max})$. The next theorem, also concerning $F|p_{ij} \sim G_j|E(C_{max})$, gives us some idea of how the variance in the processing time distributions affect the job sequences that minimize $E(C_{max})$.

Theorem 3.2.4. Let n-2 jobs have deterministic processing times, not necessarily identical, and let 2 jobs have nondeterministic processing time distributions. Then, any sequence that schedules either one of the stochastic jobs first in the sequence and the other one last minimizes the makespan stochastically.

Based on Theorems 3.2.3 and 3.2.4 and some computational results the following rule of thumb for $F|\underset{\sim}{p}_{ij} \sim G_j|E(C_{max})$ was stated in (10): Schedule jobs with smaller expected processing times and larger variances in the processing times more towards the beginning and towards the end of the sequence and schedule jobs with larger expected processing times and smaller variances more towards the middle of the sequence. This implies that the optimal sequences have a unimodal form, both as a function of the expectations of the processing time distributions and as a function of the variances of the processing time distributions. Because of this form these optimal sequences may also be referred to as "bowl"-sequences.

Observe that for the problem $F|p_{ij} \sim G_j|E(\Sigma C_j)$ when the processing time distributions of the jobs are nonoverlapping the SEPT sequence is the only optimal sequence.

Instead of different jobs on identical machines we will consider now the case of identical jobs on different machines, i.e. the processing times of the jobs on a machine are independent draws from the same distribution. The objective now is to find the optimal *machine* sequence (the machine sequence that minimizes the expected makespan) instead of job sequence. This model, which also can be viewed as a tandem queueing model where n customers are waiting at time t=0, will be referred to as $F|p_{ij} \sim G_i|E(C_{max})$. It can be shown easily that interchanging machines and jobs, i.e. transforming $F|p_{ij} \sim G_j|E(C_{max})$ into $F|p_{ij} \sim G_i|E(C_{max})$, results in a problem with exactly the same structure. So for $F|p_{ij} \sim G_i|E(C_{max})$ Theorems 3.2.3 and 3.2.4, after replacing the words "jobs" for "machines", also hold. One should observe now that the optimal machine sequences stated in these theorems not only minimize $E(C_{max})$, but minimize $E(C_j)$ for all j=1,...,n. So these machine sequences also minimize $E(\Sigma C_j)$ (for $F|p_{ij} \sim G_j|E(\Sigma C_j)$ with nonoverlapping processing time distributions SEPT was the only job sequence that minimized $E(\Sigma C_j)$).

3.3. Stochastic Flow Shops with Zero Intermediate Storage Between Machines

The model in this subsection is rather different from the model in the preceding subsection as now there is no intermediate storage space between the machines. This may have the following consequences: When job j has finished its processing on machine i but cannot be further processed because job j-1 is still being processed on machine i+1, job j will be held on machine i. However, as long as machine i is holding job j, job j+1 may not start its processing on machine i, i.e. job j+1 may not leave machine i-1. This phenomenon is called blocking. These models will therefore be referred to as $F|blocking|E(C_{max})$ and $F|blocking|E(\Sigma C_j)$.

Again, we consider first the case m=2. It is clear that whenever a job starts on machine 1, the preceding job starts on machine 2. Let the total time during $[0,C_{max}]$ that only one machine functions be denoted by I, which is equivalent to the total time during $[0,C_{max}]$ that one machine is not busy processing a job. A machine is idle when either a machine is empty or when a job in the first machine is being blocked by a job in the second. During the time period that job j occupies machine 1, there will be some time that only one machine is processing a job: In case $p_{1j} > p_{2j-1}$ machine 1 will keep on processing job j the moment job j-1 leaves machine 2. When $p_{1j} < p_{2j-1}$ machine 2 will still be processing job j-1 after job j has finished on machine 1. Minimizing $E(C_{max})$ is equivalent to minmizing E(I) which is equivalent to maximizing the total expected time that both machines are busy processing jobs. Based on this analysis the following result was shown in (10).

Theorem 3.3.1. Minimizing $E(C_{max})$ in $F2|\text{blocking}|E(C_{max})$ is equivalent to *maximizing* the total distance in the following *deterministic* Travelling Salesman Problem. Consider a travelling salesman who starts out from city 0 and has to visit cities 1,2,...,n and return to city 0, while maximizing the total distance travelled, where the distance between cities k and ℓ is defined as follows:

$$d_{k0} = 0,$$

$$d_{0\ell} = 0,$$

$$d_{k\ell} = E\left(\min(p_{1k}, p_{2\ell})\right). \qquad k\neq 0,\ \ell\neq 0.$$

Observe that when the processing times on the two machines are exponentially distributed the distance matrix of the TSP has a very nice structure.

Consider now the model $F2|\text{blocking}, p_{ij} \sim G_j|E(C_{max})$ where the processing times of a job on the different machines are independent draws from the same distribution. We will say that G_j is stochastically larger than G_k, $G_j >_{st} G_k$, when $P(p_{ij}>t) > P(p_{ik}>t)$ for all t. Again, by minimizing E(I) the following theorem can be shown, see (10).

Theorem 3.3.2. When $G_1 >_{st} G_2 \geq_{st} \cdots \geq_{st} G_n$ and n is even job sequences n,n-2,n-4,...,4,2,1,3,5,...,n-3,n-1 and n-1,n-3,..., 5,3,1,2,4,...,n-4,n-2,n minimize $E(C_{max})$. When n is odd job sequences n,n-2,n-4,...,3,1,2,4,...,n-3,n-1 and n-1,n-3,..., 4,2,1,3,5,...,n-4,n-2,n minimize $E(C_{max})$.

Note that the sequences stated in this theorem are SEPT-LEPT sequences and therefore "bowl"-sequences. This theorem gives us some indication of how the optimal job sequence is influenced by the expected processing times.

Now we will discuss the influence of the variance in the processing times given that the expected values of the processing

times of all jobs are equal, say μ. Consider the following special case: Let the probability density functions of the processing times be symmetric around the mean μ. This implies that the random variables have an upper bound 2μ. We will say that the processing time of job j is more variable than the processing time of job k, $G_j >_v G_k$, when $G_j(t) \geq G_k(t)$ for $0 \leq t \leq \mu$ and (because of symmetry) $G_j(t) \leq G_k(t)$ for $\mu \leq t \leq 2\mu$. Distributions which satisfy these symmetry conditions are:
(i) The Normal Distribution, truncated at 0 and at 2μ;
(ii) The Uniform Distribution.
The probability density functions of these distributions are depicted in Figure 3.

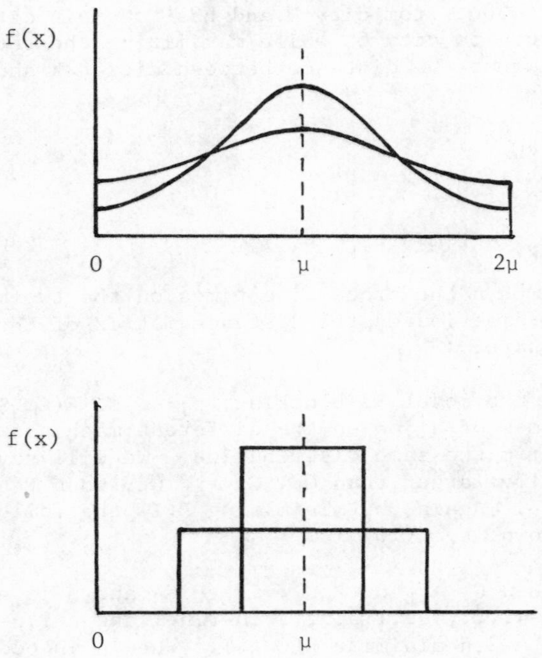

Figure 3

Theorem 3.3.3. When $G_1 \leq_v G_2 \leq_v \ldots \leq_v G_n$ and when n is even job sequences n,n-2,n-4,...,4,2,1,3,5,...,n-3,n-1 and n-1,n-3, ...,5,3,1,2,4,...,n-4,n-2,n minimize $E(C_{max})$. When n is odd job sequences n,n-2,n-4,...,3,1,2,4,...,n-3,n-1 and n-1,n-3,..., 4,2,1,3,...,n-4,n-2,n minimize $E(C_{max})$.

So from Theorems 3.3.2 and 3.3.3 we observe that here too "bowl"-sequences are optimal and therefore the rule of thumb stated in subsection 3.2 would be valid here, too.

In (10) Pinedo also considered $F|blocking, p_{ij} \sim G_j|E(C_{max})$. For this problem a theorem very much like Theorem 3.2.3 could be proven.

Theorem 3.3.4. For n jobs with nonoverlapping processing time distributions a job sequence minimizes $E(C_{max})$ *if and only if* it is SEPT-LEPT.

So this theorem, too, emphasizes the importance of "bowl"-sequences. In (10) the author was unable to present a theorem similar to Theorem 3.2.4. However, the following conjecture was stated.

Conjecture. Let n-2 jobs have identical deterministic processing times, say with unit processing times and let two jobs have nondeterministic processing times with symmetric probability density functions and mean one. Then, any sequence which schedules either one of the stochastic jobs first and the other one last minimizes $E(C_{max})$.

Observe that for $F|blocking, p_{ij} \sim G_j|E(\Sigma C_j)$, when the processing times of the jobs are nonoverlappingly distributed, the SEPT sequence is the only optimal sequence.

Consider now again the case of n identical jobs and m different machines. This implies that the processing times of the jobs on the different machines are independent draws from the same distribution. Again, we would like to know the optimal order in which to set up the machines in order to minimize $E(C_{max})$. This model will be referred to as $F|blocking, p_{ij} \sim G_i|E(C_{max})$. In subsection 3.2 it was mentioned that, in the case of infinite intermediate storage, interchanging jobs and machines results in a model with exactly the same structure. With blocking, however, interchanging machines and jobs does change the structure of the model significantly. In (11) Pinedo showed the following result with regard to $F|blocking, p_{ij} \sim G_i|E(C_{max})$.

Theorem 3.3.5. The expected makespan of n jobs in a system with m-2 identical deterministic machines with unit processing times and 2 nonidentical stochastic machines, both with mean one and symmetric density functions, is minimized if one of the stochastic machines is set up at the beginning of the sequence and the other at the end of the sequence.

This theorem appears to be the perfect dual of the conjecture stated before. The next theorem, however, will illustrate the difference between $F|\text{blocking}, \tilde{p}_{ij} \sim G_j|E(C_{max})$ and $F|\text{blocking}, \tilde{p}_{ij} \sim G_i|E(C_{max})$.

Theorem 3.3.6. The expected makespan of n jobs in a system with m-2 identical machines with distributions G_1 and two identical machines with distribution G_2, where G_2 is nonoverlappingly larger than G_1, is minimized, when one of the two slow machines is set up at the beginning of the sequence and the other one at the end of the sequence.

This theorem is quite different from Theorem 3.3.4. Based on these last two theorems, some other minor results and extensive simulation work, it appeared that the optimal machine sequences are not "bowl"-sequences but so-called "sawtooth"-sequences. These may be described as follows: Suppose we have m-1 identical machines with distributions G_1 and m identical machines with distribution G_2 (for a total of 2m-1 machines), where G_2 is nonoverlappingly larger than G_1, then we conjecture that the optimal machine sequence puts a slow machine at the beginning of the sequence, followed by a fast machine in the second place, a slow machine in the third place, etc. This sequence has the shape of a "sawtooth". Suppose now we have m-1 identical deterministic machines with unit processing times and m identical machines with mean one and symmetric density function, then we conjecture that the optimal sequence puts a stochastic machine at the beginning of the sequence, followed by a deterministic machine in the second place, a stochastic machine in the third place, etc. This sequence, too, has the shape of a "sawtooth".

3.4. Stochastic Job Shops

Very little work has been done on stochastic job shops. The main reason is that these models are even harder than the stochastic flow shops with infinite intermediate storage between machines; these flow shops are just very special cases of job shops. The only result known up to now concerns $J2|\tilde{p}_{ij} \sim \exp(\mu_{ij})|E(C_{max})$. In this two machine model one set of jobs has to be processed first on machine 1 and after that on machine 2. This set of jobs will be referred to as set A. A second set of jobs has to be processed first on machine 2 and after that on machine 1. This set of jobs will be referred to as set B. In (12) Pinedo showed that Bagga's Theorem for $F2|\tilde{p}_{ij} \sim \exp(\mu_{ij})|E(C_{max})$ can be generalized into a theorem for $J2|\tilde{p}_{ij} \sim \exp(\mu_{ij})|E(C_{max})$. However, for $J2|\tilde{p}_{ij} \sim \exp(\mu_{ij})|E(C_{max})$ we cannot speak anymore of an optimal sequence, we have to speak of an optimal policy which instructs the decision-maker in any state what action to take.

Theorem 3.4.1. The optimal policy instructs the decision-maker, whenever machine 1 (2) is freed, to start processing of the remaining jobs of set A (B) that did not yet undergo processing on machine 1 (2) the one with the highest value of $\mu_{1j} - \mu_{2j}$ $(\mu_{2j}-\mu_{1j})$. If no jobs of set A (B) remain that did not yet undergo processing on machine 1 (2), the decision-maker may start any one of the jobs of set B (A) that already have finished their processing on machine 2 (1).

4. CONCLUSIONS

It is clear that the stochastic shop scheduling problems are not all that easy. The two machine models with exponential processing times are in general tractable, just like the two machine models with deterministic processing times. For the deterministic versions of $F2||C_{max}$ and $J2||C_{max}$ the algorithms are O(n log n); for the versions of $F2||E(C_{max})$ and $J2||E(C_{max})$ with exponential processing times the algorithms are also O(n log n). For the deterministic version of $O2||C_{max}$ the algorithm is O(n). The version of $O2||E(C_{max})$ with exponentially distributed processing times is harder. For the special case $O2|p_{ij} \sim \exp(\mu_j)|E(C_{max})$ the algorithm is already O(n log n). We have not yet been able to determine the complexity of the more general case $O2|p_{ij} \sim \exp(\mu_{ij})|E(C_{max})$. Models with three machines or more appear to be very hard. Only results of a qualitative nature were obtained (e.g. "bowl"-sequences, "sawtooth"-sequences). This jump in complexity when we go from two machines to three occurs also when the processing times are deterministic; non-preemptive shop scheduling with three or more machines are consistently NP-complete.

REFERENCES

(1) Avi-Itzhak, B., "A Sequence of Service Stations with Arbitrary Input and Regular Service Times": 1965, Man. Sci. 11, pp. 565-573.

(2) Bagga, P.C., "N-Job, 2 Machine Sequencing Problem with Stochastic Service Times": 1970, Opsearch 7, pp. 184-199.

(3) Garey, M.R., Johnson, D.S. and Sethi, R., "The Complexity of Flowshop and Jobshop Scheduling": 1976, Math. Operations Res. 1, pp. 117-129.

(4) Gonzalez, T., and Sahni, S., "Open Shop Scheduling to Minimize Finish Time": 1976, J. Assoc. Comput. Mach. 23 pp. 665-679.

(5) Graham, R.L., Lawler, E.L., Lenstra, J.K. and Rinnooy Kan, A.H.G., "Optimization and Approximation in Deterministic Sequencing and Scheduling: A Survey": 1979, Ann. Discr. Math. 5, pp. 287-326.

(6) Jackson, J.R., "An Extension of Johnson's Results on Job Lot Scheduling": 1956, Naval Res. Logist. Quart. 3, pp. 201-203.
(7) Johnson, S.M., "Optimal Two and Three Stage Production Schedules with Setup Times Included": 1954, Naval Res. Logist. Quart. 1, pp. 61-68.
(8) McMahon, G.B. and Florian, M., "On Scheduling with Ready Times and Due Dates to Minimize Maximum Lateness": 1975, Operations Res. 23, pp. 475-482.
(9) Pinedo, M.L., "Minimizing the Expected Flow Time in a Stochastic Open Shop with and without Preemptions": 1981, Technical Report, Georgia Institute of Technology.
(10) Pinedo, M.L. "Minimizing the Makespan in a Stochastic Flow Shop": 1981, to appear, Operations Research.
(11) Pinedo, M.L., "On the Optimal Order of Stations in Tandem Queues": 1981, to appear, Proceedings of the Applied Probability-Computer Science: The Interface, Conference at Boca Raton, Florida.
(12) Pinedo, M.L. "A Note on the Two Machine Job Shop with Exponential Processing Times": 1981, to appear, Naval Research Logistics Quarterly.
(13) Pinedo, M.L. and Ross, S.M., "Minimizing the Expected Makespan in a Stochastic Open Shop": 1981, submitted to Journal of Applied Probability.
(14) Reddi, S.S. and Ramamoorthy, C.V., "On the Flowshop Sequencing Problem with No Wait in Process": 1972, Operational Res. Quart. 23, pp. 323-330.
(15) Weber, R.R., "The Interchangeability of Tandem $\cdot|M|1$ Queues in Series": 1979, J. of Applied Prob. 16, pp. 690-695.
(16) Wismer, D.A., "Solution of Flowshop-Scheduling Problem with No Intermediate Queues": 1972, Operations Res. 20, pp. 689-697.

MULTI-SERVER QUEUES

Sheldon M. Ross

University of California, Berkeley

We will survey a variety of multiserver models in which the arrival stream is a Poisson process. In particular, we will consider the Erlang loss model in which arrivals finding all servers busy are lost. In this system, we assume a general service distribution. We will also consider finite and infinite capacity versions of this model. Another model of this type is the shared processor system in which service is shared by all customers.

Another model to be considered is the G/M/k in which arrivals are in accordance with a renewal process and the service distribution is exponential. We will analyze this model by means of the embedded Markov chain approach.

0. INTRODUCTION

We will consider some multiserver queueing models. In Section 1, we deal with the Erlang loss model which supposes Poisson arrivals and a general service distribution G. By use of a "reversed process" argument (see [2]) we will indicate a proof of the well-known result that the distribution of number of busy servers depends on G only through its mean. In Section 2 we then analyze a shared-processor model in which the servers are able to combine forces. Again making use of the reverse process, we obtain the limiting distribution for this model. In Section 3 we review the embedded Markov chain approach for the G/M/k model; and in the final section we present the model M/G/k.

M. A. H. Dempster et al. (eds.), Deterministic and Stochastic Scheduling, 197–209.

1. THE ERLANG LOSS SYSTEM

One of the most basic types of queueing system are the loss systems in which an arrival that finds all servers busy is presumed lost to the system. The simplest such system is the M/M/k loss system in which customers arrive according to a Poisson process having rate λ, enter the system if at least one of the k servers is free, and then spend an exponential amount of time with rate μ being served. The balance equations for the stationary probabilities are

State	Rate leave = rate enter
0	$\lambda P_0 = \mu P_1$
$i,\ 0 < i < k$	$(\lambda + i\mu)P_i = (i+1)\mu P_{i+1} + \lambda P_{i-1}$
k	$k\mu P_k = \lambda P_{k-1}$.

Using the equation $\sum_0^k P_i = 1$, the above equations can be solved to give

$$P_i = \frac{(\lambda/\mu)^i/i!}{\sum_{j=0}^{k} (\lambda/\mu)^j/j!}, \quad i = 0,1, \ldots, k .$$

Since $E[S] = 1/\mu$, where $E[S]$ is the mean service time, the above can be written as

$$P_i = \frac{(\lambda E[S])^i/i!}{\sum_{j=0}^{k} (\lambda E[S])^j/j!}, \quad i = 0,1, \ldots, k .$$

The above was originally obtained by Erlang who then conjectured that it was valid for an arbitrary service distribution. We shall present a proof of this result, known as the Erlang loss formula when the service distribution G is continuous and has density g.

Theorem 1. The limiting distribution of the number of customers in the Erlang loss system is given by

$$P\{n \text{ in system}\} = \frac{(\lambda E[S])^n/n!}{\sum_{i=0}^{k} (\lambda E[S])^i/i!}, \quad n = 0,1, \ldots, k$$

and given that there are n in the system, the ages (or the residual times) of these n are independent and identically distributed according to the equilibrium distribution of G.

Proof. We can analyze the above system as a Markov process by letting the state at any time be the ordered ages of the customers in service at that time. That is, the state will be $\underline{x} = (x_1, x_2, \ldots, x_n)$, $x_1 \leq x_2 \leq \ldots \leq x_n$, if there are n customers in service, the most recent one having arrived x_1 time units ago, the next most recent arrival being x_2 time units ago, and so on. The process of successive states will be a Markov process in the sense that the conditional distribution of any future state given the present and all the past states will depend only on the present state. In addition, let us denote by $\lambda(t) = g(t)/\bar{G}(t)$ the hazard rate function of the service distribution.

We will attempt to use the reverse process to obtain the limiting probability density $p(x_1, x_2, \ldots, x_n)$, $1 \leq n \leq k$, $x_1 \leq x_2 \leq \ldots \leq x_n$, and $P(\phi)$ the limiting probability that the system is empty. Now since the age of a customer in service increases linearly from 0 upon its arrival to its service time upon its departure, it is clear that if we look backwards, we will be following the excess or additional service time of a customer. As there will never be more than k in the system, we make the following conjecture.

Conjecture. In steady state, the reverse process is also a k server loss system with service distribution G in which arrivals occur according to a Poisson process with rate λ. The state at any time represents the ordered residual service times of customers presently in service. In addition, the limiting probability density is

$$p(x_1, \ldots, x_n) = \frac{\lambda^n \prod_{i=1}^{n} \bar{G}(x_i)}{\sum_{i=0}^{k} (\lambda E[S])^i/i!}, \quad x_1 \leq x_2 \leq \ldots \leq x_n$$

and

$$P(\phi) = \left[\sum_{i=0}^{k} (\lambda E[S])^i/i! \right]^{-1} .$$

To verify the conjecture, for any state $\underline{x} = (x_1, \ldots, x_i, \ldots, x_n)$, let $e_i(\underline{x}) = (x_1, \ldots, x_{i-1}, x_{i+1}, \ldots, x_n)$. Now in the original process when the state is $\underline{x}$, it will instantaneously go to $e_i(\underline{x})$ with a probability density equal to $\lambda(x_i)$ since the person whose time in service is x_i would have to instantaneously complete its service. Similarly in the reversed process if the state is $e_i(\underline{x})$, then it will instantaneously go to $\underline{x}$ if a customer having service time x_i instantaneously arrives. So we see that

in forward: $\underline{x} \to e_i(\underline{x})$ with probability intensity $\lambda(x_i)$,
in reverse: $e_i(\underline{x}) \to \underline{x}$ with (joint) probability intensity $\lambda g(x_i)$.

Hence if $p(\underline{x})$ represents the limiting density, then we would need that

$$p(\underline{x})\lambda(x_i) = p(e_i(\underline{x}))\lambda g(x_i)$$

or, since $\lambda(x_i) = g(x_i)/\bar{G}(x_i)$,

$$p(\underline{x}) = p(e_i(\underline{x}))\lambda\bar{G}(x_i)$$

which is easily seen to be satisfied by the conjectured $p(\underline{x})$

To complete our proof of the conjecture, we must consider transitions of the forward process from $\underline{x}$ to $(0,x) = (0,x_1,x_2, \ldots, x_n)$ when $n < k$. Now

in forward: $x \to (0,x)$ with instantaneous intensity λ,
in reverse: $(0,\underline{x}) \to \underline{x}$ with probability 1.

Hence we must verify that

$$p(\underline{x})\lambda = p(0,\underline{x})$$

which easily follows since $\bar{G}(0) = 1$.

Hence we see that the conjecture is true and so, upon integration, we obtain

$P\{n$ in the system$\}$

$$= P(\phi)\lambda^n \int\int_{x_1 \le x_2 \ldots \le x_n}\cdots\int \prod_{i=1}^{n} \bar{G}(x_i)dx_1\, dx_2 \ldots dx_n$$

$$= P(\phi)\,\frac{\lambda^n}{n!}\int\int_{x_1,x_2,\ldots,x_n}\cdots\int \prod_{i=1}^{n} \bar{G}(x_i)dx_1\, dx_2 \ldots dx_n$$

$$= P(\phi)(\lambda E[S])^n/n!\ ,\ n = 1,2, \ldots, k$$

where $E[S] = \int \bar{G}(x)dx$ is the mean service time. Also, we see that the conditional distribution of the ordered ages given that there are n in the system is

$$p\{\underline{x} \mid n \text{ in the system}\} = p(\underline{x})/P\{n \text{ in the system}\}$$

$$= n! \prod_{i=1}^{n} (\bar{G}(x_i)/E[S]) \ .$$

As $\bar{G}(x)/E[S]$ is just the density of G_e, the equilibrium distribution of G, this completes the proof.

In addition, by looking at the reversed process, we also have the following corollary.

Corollary 1. In the Erlang loss model, the departure process (including both customers completing service and those that are lost) is a Poisson process at rate λ.

Proof. The above follows since in the reversed process arrivals of all customers (including those that are lost) constitutes a Poisson process.

2. THE SHARED PROCESSOR SYSTEM

Suppose that customers arrive in accordance with a Poisson process having rate λ. Each customer requires a random amount of work, distributed according to G. The server can process work at a rate of one unit of work per unit time, and divides his time equally among all of the customers presently in the system. That is, whenever there are n customers in the system, each will receive service work at a rate of $1/n$ per unit time.

Let $\lambda(t)$ denote the failure rate function of the service distribution, and suppose that $\lambda E[S] < 1$ where $E[S]$ is the mean of G.

To analyze the above, let the state at any time be the ordered vector of the amounts of work already performed on customers still in the system. That is, the state is $\underline{x} = (x_1, x_2, \ldots, x_n)$, $x_1 \leq x_2 \leq \cdots \leq x_n$ if there are n customers in the system and $x_1, \ldots, x_n$ is the amount of work performed on these n customers. Let $p(\underline{x})$ and $P(\phi)$ denote the limiting probability density and the limiting probability that the system is empty. We make the following conjecture regarding the reverse process.

Conjecture. The reverse process is a system of the same

type, with customers arriving at a Poisson rate λ, having workloads distributed according to G and with the state representing the ordered residual workloads of customers presently in the system.

To verify the above conjecture and at the same time obtain the limiting distribution let $e_i(\underline{x}) = (x_1, \ldots, x_{i-1}, x_{i+1}, \ldots, x_n)$ when $\underline{x} = (x_1, \ldots, x_n)$, $x_1 \leq x_2 \leq \ldots \leq x_n$. Note that

in forward: $\underline{x} \to e_i(\underline{x})$ with probability intensity $\frac{\lambda(x_i)}{n}$,
in reverse: $e_i(x) \to \underline{x}$ with (joint) probability intensity $\lambda G'(x_i)$.

The above follows as in the previous section with the exception that if there are n in the system then a customer who already had the amount of work x_i performed on it will instantaneously complete service with probability $\lambda(x_i)/n$.

Hence, if $p(x)$ is the limiting density then we need that

$$p(\underline{x}) \frac{\lambda(x_i)}{n} = p(e_i(\underline{x}))\lambda G'(x_i)$$

or, equivalently,

$$\begin{aligned} p(\underline{x}) &= n\bar{G}(x_i)p(e_i(\underline{x}))\lambda \\ &= n\bar{G}(x_i)(n-1)\bar{G}(x_j)p(e_j(\underline{x}))\lambda^2 \quad, i \neq i \\ &\;\vdots \\ &= n!\lambda^n P(\phi) \prod_{i=1}^{n} \bar{G}(x_i) \quad . \end{aligned} \tag{1}$$

Integrating over all vectors $\underline{x}$ yields

$$P\{n \text{ in system}\} = (\lambda E[S])^n P(\phi) .$$

Using

$$P(\phi) + \sum_{n=1}^{\infty} P\{n \text{ in the system}\} = 1$$

gives

$$P\{n \text{ in the system}\} = (\lambda E[S])^n (1 - \lambda E[S]) \quad, n \geq 0 .$$

Also, the conditional distribution of the ordered amounts of work already performed, given n in the system is, from (1)

$$p(\underline{x} \mid n) = p(\underline{x})/P\{n \text{ in system}\}$$

$$= n! \prod_{i=1}^{n} (\bar{G}(x_i)/E[S]) \ .$$

That is, given n customers in the system the unordered amount of work already performed are distributed independently according to G_e, the equilibrium distribution of G.

All of the above is based on the assumption that the conjecture is valid. To complete the proof of its validity, we must verify that

$$p(\underline{x})\lambda = p(0,\underline{x}) \frac{1}{n+1} \ .$$

The above being the relevant equation since the reverse process when in state $(\varepsilon,\underline{x})$ will go to state $\underline{x}$ in time $(n+1)\varepsilon$. As the above is easily verified, we have thus shown

Theorem 2. For the Processor Sharing Model, the number of customers in the system has the distribution

$$P\{n \text{ in system}\} = (\lambda E[S])^n(1 - \lambda E[S]) \ , \ n \geq 0 \ .$$

Given n in the system, the completed (or residual) workloads are independent and have distribution G_e. The departure process is a Poisson process with rate λ.

If we let L denote the average number in the system, and W, the average time a customer spends in the system then

$$L = \sum_{n=0}^{\infty} n(\lambda E[S])^n(1 - \lambda E[S])$$

$$= \frac{\lambda E[S]}{1 - \lambda E[S]} \ .$$

We can obtain W from the well-known formula $L = \lambda W$ and so

$$W = L/\lambda = \frac{E[S]}{1 - \lambda E[S]} \ .$$

Another interesting computation in this model is that of the conditional mean time an arrival spends in the system given its workload is y. To compute this quantity, fix y and say that a customer is "special" if its workload is between y and

$y + \varepsilon$. By $L = \lambda W$, we thus have that

Average Number of Special Customers in the System
=
Average Arrival Rate of Special Customer x Average Time a Special Customer Spends in the System.

To determine the average number of special customers in the system, let us first determine the density of the total workload of an arbitrary customer presently in the system. Suppose such a customer has already received the amount of work x . Then the conditional density of its workload is

$$f(w \mid \text{has received } x) = g(w)/\bar{G}(x) \ , \ x \le w \ .$$

But, from Theorem 2, the amount of work an arbitrary customer in the system has already received has the distribution G_e . Hence the density of the total workload of someone present in the system is

$$\begin{aligned} f(w) &= \int_0^w \frac{g(w)}{\bar{G}(x)} \, dG_e(x) \\ &= \int_0^w \frac{g(w)}{E[S]} \, dx \ , \text{ since } dG_e(x) = \bar{G}(x)/E[S] \\ &= wg(w)/E[S] \ . \end{aligned}$$

Hence the average number of special customers in the system is

$$\begin{aligned} &E[\text{number in system having workload between } y \text{ and } y + \varepsilon] \\ &= Lf(y)\varepsilon + o(\varepsilon) \\ &= Lyg(y)\varepsilon/E[S] + o(\varepsilon) \ . \end{aligned}$$

In addition, the average arrival rate of customers whose workload is between y and $y + \varepsilon$ is

$$\text{Average arrival rate} = \lambda g(y)\varepsilon + o(\varepsilon) \ .$$

Hence we see that

$$\begin{aligned} &E[\text{time in system} \mid \text{workload in } (y \ , y + \varepsilon)] \\ &= \frac{Lyg(y)\varepsilon}{E[S]\lambda g(y)\varepsilon} + \frac{o(\varepsilon)}{\varepsilon} \ . \end{aligned}$$

Letting $\varepsilon \to 0$, we obtain

$$E[\text{time in system} \mid \text{workload is } y] = \frac{y}{\lambda E[S]} L$$

$$= \frac{y}{1 - \lambda E[S]} .$$

Thus the average time in the system of a customer needing y units of work also depends on the service distribution only through its mean.

3. THE G/M/k QUEUE

In this model we suppose that there are k servers, each of whom serves at an exponential rate μ. We allow the time between successive arrivals to have an arbitrary distribution G. In order to ensure that a steady-state (or limiting) distribution exists, we assume $1/\mu_G < k\mu$ where μ_G is the mean of G.

To analyze this model, we will use an embedded Markov chain approach. Define X_n as the number in the system as seen by the nth arrival. Then it is easy to see that $\{X_n, n \geq 0\}$ is a Markov chain.

To derive the transition probabilities of the Markov chain, it helps to note the relationship

$$X_{n+1} = X_n + 1 - Y_n, \quad n \geq 0$$

where Y_n denotes the number of departures during the interarrival time between the nth and $(n + 1)$st arrival. The transition probabilities can be calculated as

<u>Case (i)</u>: $j > i + 1$. In this case, $P_{ij} = 0$.

<u>Case (ii)</u>: $j \leq i + 1 \leq k$. In this case,

$$P_{ij} = P\{i + 1 - j \text{ of } i + 1\} \text{ services are completed in an interarrival time}$$

$$= \int_0^\infty P\{i + 1 - j \text{ of } i + 1 \text{ are completed} \mid \text{interarrival time is } t\} dG(t)$$

$$= \int_0^\infty \binom{i+1}{j} (1 - e^{-\mu t})^{i+1-j} (e^{-\mu t})^j \, dG(t) .$$

Case (iii): $i + 1 \geq j \geq k$. To evaluate P_{ij}, in this case, we first note that when all servers are busy the departure process is a Poisson process with rate $k\mu$. Hence,

$$P_{ij} = \int_0^\infty P\{i + 1 - j \text{ departures in time } t\} dG(t)$$

$$= \int_0^\infty e^{-k\mu t} \frac{(k\mu t)^{i+1-j}}{(i + 1 - j)!} dG(t) .$$

Case (iv): $i + 1 \geq k > j$. Conditioning first on the interarrival time and then on the time until there are only k in the system (call this latter random variable T_k) yields

$$P_{ij} = \int_0^\infty P\{i + 1 - j \text{ departures in time } t\} dG(t)$$

$$= \int_0^\infty \int_0^t P\{i + 1 - j \text{ departures in } t \mid T_k = s\}$$

$$k\mu e^{-k\mu s} \frac{(k\mu s)^{i-k}}{(i - k)!} ds dG(t)$$

$$= \int_0^\infty \int_0^t \binom{k}{j} (1 - e^{-\mu(t-s)})^{k-j} (e^{-\mu(t-s)})^j$$

$$k\mu e^{-k\mu s} \frac{(k\mu s)^{i-k}}{(i - k)!} ds dG(t) .$$

We now can verify by a direct substitution into the equations $\pi_j = \sum_i \pi_i P_{ij}$ that the limiting probabilities of this Markov chain are of the form

$$\pi_{k-1+j} = c\beta^j \; , \; j = 0,1, \ldots .$$

Substitution into any of the equations $\pi_j = \sum_i \pi_i P_{ij}$ when $j > k$ yields that β is given as the solution of

$$\beta = \int_0^\infty e^{-k\mu t(1-\beta)} dG(t) .$$

The values $\pi_0, \pi_1, \ldots, \pi_{k-2}$, can be obtained by recursively solving the first $k - 1$ of the steady-state equations, and c can then be computed by using $\sum_0^x \pi_i = 1$.

If we let W_Q^* denote the amount of time that a customer spends in queue, then we can show, upon conditioning, that

$$W_Q^* = \begin{cases} 0 & \text{with probability } \sum_0^{k-1} \pi_i = 1 - \frac{c\beta}{1-\beta} \\ \text{Exp}\,(k\mu(1-\beta)) & \text{with probability } \sum_k^\infty \pi_i = \frac{c\beta}{1-\beta} \end{cases}$$

where $\text{Exp}\,(k\mu(1 - \beta))$ is an exponential random variable with rate $k\mu(1 - \beta)$.

4. THE FINITE CAPACITY M/G/k

In this section, we consider an M/G/k queuing model having finite capacity N . That is, a model in which customers, arriving in accordance with a Poisson process having rate λ , enter the system if there are less than N others present when they arrive, and are then serviced by one of k servers, each of whom has service distribution G . Upon entering, a customer will either immediately enter service if at least one server is free or else join the queue if all servers are busy.

Our objective is to obtain an approximation for W_Q , the average time an entering customer spends waiting in queue. To get started we will make use of the idea that if a (possibly fictitious) cost structure is imposed, so that entering customers are forced to pay money (according to some rule) to the system, then the following identity holds--namely,

> time average rate at which the system earns = average arrival rate of entering customers × average amount paid by an entering customer.

By choosing appropriate cost rules, many useful formulae can be obtained as special cases. For instance, by supposing that each customer pays \$1 per unit time while in service, yields

$$\text{average number in service} = \lambda(1 - P_N)E[S] .$$

Also, if we suppose that each customer in the system pays $\$x$ per unit time whenever its remaining service time is x, then we get

$$V = \lambda(1 - P_N)E\left[SW_Q^* + \int_0^S (S - x)\,dx\right] = \lambda(1 - P_N)[E[S]W_Q + E[S^2]/2]$$

where V is the (time) average amount of work in the system and where W_Q^* is a random variable representing the (limiting) amount of time that the nth entering customer spends waiting in queue.

The above gives us one equation relating V and W_Q and one approach to obtaining W_Q would be to derive a second equation. An approximate second equation was given by Nozaki-Ross in [3] by means of the following approximation assumption.

Approximation Assumption

Given that a customer arrives to find i busy servers, $i > 0$, at the time he enters service the remaining service times of the other customers being served are approximately independent each having the equilibrium service distribution.

Using the above assumption as if it was exactly true, Nozaki-Ross were able to derive a second relationship between V and W_Q which resulted in an expression for W_Q as a function of P_N. By approximating P_N by its known value in the case where the service distribution is exponential, Nozaki-Ross came up with the following approximation for W_Q:

$$W_Q = \frac{\dfrac{E[S^2]}{2E[S]} \displaystyle\sum_{j=k}^{N-1} \frac{(\lambda E[S])^j}{k!k^{j-k}} - (N - k)\, \frac{E[S](\lambda E[S])^N}{k!k^{N-k}}}{\displaystyle\sum_{j=0}^{k-1} \frac{(\lambda E[S])^j}{j!} + \sum_{j=k}^{N-1} \frac{(\lambda E[S])^j}{k!k^{j-k}}\,(k - \lambda E[S])} .$$

The idea of an approximation assumption to approximate various quantities of interest of the model M/G/k was also used by Tijms, Van Hoorn, and Federgruen [6]. They used a slightly different approximation assumption to obtain approximations for the steady state probabilities. Other approximations for the M/G/k are also given in Boxma, Cohen, and Huffels [1], and Takahashi [5].

REFERENCES

1. O.J. BOXMA, J.W. COHEN, N. HUFFELS (1979) *Oper. Res.* 27, 1115-1127.
2. F.P. KELLY (1979) *Reversibility and Stochastic Networks*, Wiley.
3. S.A. NOZAKI, S.M. ROSS (1978) *J. Appl. Probab.* 15,826-834.
4. S.M. ROSS (1980) *Introduction to Probability Models*, 2nd Edition, Academic Press.
5. Y. TAKAHASHI (1977) *J. Oper. Res. Soc. Japan* 20,150-163.
6. H.C. TIJMS, M.H. VAN HOORN, A. FEDERGRUEN (1979) Approximations for the steady state probabilities in the M/G/c queue. Research Report 49, Free University, Amsterdam; *Adv. in Appl. Probab.*, to appear.

QUEUEING NETWORKS AND THEIR COMPUTER SYSTEM APPLICATIONS: AN INTRODUCTORY SURVEY

Kenneth C. Sevcik

Computer Systems Research Group
University of Toronto

ABSTRACT

Queueing networks and their properties have been subjected to intensive study in the past ten years. The primary application considered has been the analysis of computer system and computer-communication network performance. We present a brief introductory survey of some of the relevant literature.

1. INTRODUCTION

The work of numerous researchers over the past few years has added significantly to our cumulative knowledge of the properties of queueing networks. While queueing networks are useful models in various contexts, their application to studying the performance of computer systems and computer-communication networks has received the most attention. This paper is intended to provide a brief introductory overview of some of the literature in the area for someone with exposure to computer systems and queueing theory. In order to keep the survey brief, we only indicate the topics of each paper. The sources referenced should be examined to obtain a better understanding of the material. Books that relate to much of the following material include [Kell79, Klei75, Klei76, Koba78, GeMi80, SaCh81, Cour77].

In section 2, we discuss the properties of a special class of queueing networks, known as separable queueing networks. Separable networks are of particular importance because the equilibrium probability of a system state has a special form that facilitates efficient computation of certain performance measures. In section 3, we consider the application of queueing networks to the study of computer system performance. The challenge lies in relating the known characteristics of computer systems to the assumptions that lead to separability in queueing networks. Finally, in section 4, we examine the study of computer-communication networks using both queueing networks and related analytic techniques.

2. SEPARABLE QUEUEING NETWORKS

Early studies of queueing networks involved the interconnection of service centers where

M. A. H. Dempster et al. (eds.), Deterministic and Stochastic Scheduling, 211–222.

(1) customers were statistically indistinguishable,

(2) external arrival streams were Poisson,

(3) service demands at each service center were exponentially distributed, and

(4) fixed probabilities governed the selection of the next service center to visit upon completion of service at one center.

Jackson [Jack63] treated *open* networks, where external arrivals occur, and the number of customers within the network at a given time is not bounded. Upon completion of service at a service center (at least some of them), a customer may leave the network. Gordon and Newell [GoNe67] treated *closed* networks, where a fixed population of customers circulate continually among the service centers and there are no departures or external arrivals. Both Jackson and Gordon and Newell found that the equilibrium probability distribution of system state (where it exists) has a "product form" in which there is one factor corresponding to each service center. Jackson observed that the distribution is that which would result if the service centers were each independently subjected to a Poisson arrival stream of the appropriate intensity. (In the case of closed networks, the probabilities of various system states are normalized to guarantee that they sum to one.) Because each service center is represented in a single factor of the state probability distribution, networks with product form solutions have come to be known as *separable* networks.

Early computer system modelling studies were carried out by Scherr [Sche65] who used a machine-repairman model (which is a degenerate queueing network model) and Gaver [Gave67] among others. The computational techniques used for calculating the models' performance measures, however, precluded treating realistic closed models in which there are several service centers and a significant number of customers. A breakthrough that substantially increased the practical utility of queueing networks for modelling computer systems was the development of efficient computational algorithms by which performance measures could be calculated without explicit examination of all feasible system states [Buze73]. Such computational algorithms will be discussed further in a later section.

Recently, the generality of known separable networks has been greatly expanded. Customers may be associated with various *classes*, distinguishable by their routing patterns within the network and their mean service demands at various service centers [BCMP75, Kell76]. As they move among service centers, customers may change class. *Routing chains* are composed of classes among which a single customer may move. In addition, the distributions of service requirements need not be exponential at service centers where the scheduling discipline is *symmetric* [Kell79, Barb76] (i.e., satisfies *station balance* [ChHT77]). Arrival rates of customer classes, service rates at service centers, and routing patterns may depend on certain aspects of system state [KoRe75,Tows80]. (These dependencies are summarized by Sauer [Saue81]). Finally, the population in each class need not necessarily be either fixed or unbounded, but instead can be allowed to range between arbitrary upper and lower bounds [ReKo73,ReKo76]. In fact, constraints on feasible combinations of populations for various classes may be quite general [Lam77]. Separability arises from the ability to decompose the system's global balance equations into *independent balance equations* [Whit69] or *local balance equations* [Chan72]. Separability can also be demonstrated by showing that the Markov process representing the system state is dynamically reversible [Kell79].

Separable queueing networks have the property that any subnetwork can be analysed in isolation, and the subnetwork's behavior with respect to the rest of the system can be represented by a *flow equivalent server.* The queueing characteristics at service centers not in the subnetwork are identical for the original network and for the reduced network containing the flow equivalent server. Such *hierarchical decomposition* is an important technique in dealing with queueing network models [ChHW75,Cour77,Vant78].

Although the product form solution for separable networks gives the appearance that each center operates independently under Poisson arrivals, in fact, the flows of customers among service centers (called transition processes here) can be quite complex even in simple queueing networks. It has long been known that merging independent Poisson streams results in another Poisson stream (whose rate is the sum of the rates of the contributing streams). Similarly, splitting a Poisson stream by routing each event independently to output stream i with probability p_i results in an output stream that is Poisson (with rate p_i times the rate of the original stream). Finally, if a service center with a single server of rate μ with exponentially distributed service times is fed by a Poisson stream of rate λ, then the output stream is also Poisson (with rate $\min(\mu, \lambda)$). In some queueing networks, these three properties are sufficient to guarantee that all transition processes are Poisson. Such networks must satisfy a very restrictive property, however. They must contain no feedback routing cycles by which a single job may visit the same service center more than once. Whenever a job has the potential of reaching the same service center more than once, the flows become more complex. The simplest example is a single server queue with Poisson arrivals, exponential service times, and Bernoulli feedback (i.e., each job completing service re-enters the queue immediately with a fixed probability). This configuration was first studied by Burke [Burk56, Burk76] and later investigated further by others [DiMS80, LaPS79]. The somewhat surprising conclusion is that the input process (including fedback customers), the output process (including all service completions), and the feedback process are all non-Poisson, yet the departure process (excluding fedback customers) is once again Poisson. More generally, a queueing network represented by an arbitrary graph can be decomposed into subnetworks such that two service centers, i and j, are in the same subnetwork if and only if jobs may progress both from i to j and from j to i. If external arrival processes are Poisson at each node, then transition processes between subnetworks (as well as departure processes from the entire network) *are* Poisson, while transition processes within the subnetworks are not in general Poisson [WaVa78, LaPS79].

The equilibrium queue length distributions (where they exist) are of primary interest in the analysis of queueing networks. Distributions of queue length as observed at input and output instants, (when a job completes service at a center and/or joins the queue at a center) however, are also of interest. In those networks known to have product form solutions, the distributions of network state at input and output instants are conveniently related to equilibrium distributions. In open networks, the network state distribution at input and output instants is identical to the equilibrium state distribution. In closed networks, the network state distribution at instants when some job moves from one service center to another is identical to the equilibrium distribution of the same network but with one job (the one in transition) removed. In mixed networks, where some classes (or routing chains) are open while others are closed, customers in open chains see the equilibrium state distribution of the network at arrival instants, and customers in closed chains see the equilibrium state distribution with themselves removed. Various results on network state distributions at input and output instants have been obtained in several studies [Coop80, Kell79, LaRe80, SeMi81, Mela79a]. Parallel results in the framework of Operational Analysis were obtained by Buzen and Denning [BuDe80].

Recognizing that not all transition prócesses within queueing networks are Poisson, there have been a number of investigations of these processes [Burk72, Munt73, Dale76, DiCh74, Disn75, Mela79b]. Approximate characterizations of these transition processes have proven useful in certain applications [Kuhn76, SLTZ77, Albi80].

A performance measure of interest in applications of queueing network models is sojourn times, or the time spent by a customer in the network. Even in feed-forward networks (in which no job visits the same service center more than once), the delays incurred at successive service centers are not necessarily independent [Lemo77, Mitr79, SiFo79, Mela79c, WaVa79, Disn80]. This complicates the determination of the distribution of sojourn times [Burk69, Wong78a, Chow78]. While determination of the exact distribution is very complex, the results obtained by making simplistic independence assumptions are often quite accurate [LaSe78, Kies80, SaLa81].

3. COMPUTER SYSTEM APPLICATIONS

Applications of queueing network models in investigations of computer system performance reveal some important issues of practicality and relevance. First, computational algorithms for calculating performance measures based on separable queueing network models must produce answers of sufficient accuracy with an acceptable amount of computational effort. Second, for those systems that cannot be modelled adequately by separable queueing networks, approximate solution techniques leading to answers of useful accuracy are needed.

In principal, every closed queueing network possessing an equilibrium solution can be solved by the global balance technique, which involves solving a set of L simultaneous linear equations, where L is the number of distinct system states. Practically, however, models of typical systems involve many thousands of system states if not millions or more. Thus, even highly specialized linear equation solution packages designed especially for queueing network model analysis can only treat systems of limited size and complexity [Stew78].

Applications of queueing network models to computer systems only became truly practical with the discovery of efficient computational algorithms that exploit the product-form solution of separable networks. The first such algorithm (published by Buzen [Buze73]) treated single-class networks with exponential servers. Soon the algorithm was generalized to cover multi-class separable networks [MuWo73, ChHW75, ReKo75]. These algorithms have become known as *convolution algorithms* due to the vector and multi-dimensional matrix convolutions on which they are based. For separable networks representing a system of typical complexity, the convolution algorithm can calculate in a few hundred operations the same performance measures that would require many millions of operations to obtain by the global balance solution technique.

An alternative solution algorithm, called *mean value analysis* has recently been developed [ReLa80]. Mean value analysis requires essentially the same computational effort as does convolution, but it is easier to understand and is somewhat less susceptible to certain numerical instabilities. The exposition of mean value analysis instigated the creation of several other algorithms, each combining features of both convolution and mean value analysis, attempting to retain the advantages of each [ChSa80]. The algorithms mentioned above have been incorporated into software packages that are now widely used for computer system modelling studies. Most of the proposed algorithms fall short of treating the entire known class of separable queueing networks, however. Very recently, Sauer has proposed ways of extending the known computational algorithms to handle networks with various kinds of state-dependent arrival, service and routing properties, as well as mixtures of open and closed classes in a single network [Saue81].

In some situations involving systems with many devices or customers or classes of customers, even convolution and mean value analysis calculations can be prohibitively expensive. In such cases, one may resort to bounding techniques to ascertain ranges in which system performance is known to lie. *Asymptotic bound analysis* [MuWo75, DeBu78] considers performance at very light and very heavy load, while *balanced job bounds* [ZSEG81] relate system performance to that of systems in which all service centers are equally utilized.

When the asymptotic and balanced job bounds are not sufficiently tight, a new computational algorithm proposed by Chandy and Neuse can be used [ChNe81]. For models of large systems, their algorithm produces good approximate answers with much less computational effort than is required by the convolution or mean value analysis algorithms. Zahorjan has investigated various other ways of obtaining approximate solutions for separable networks at reduced computational cost [Zaho80].

For some systems, separable networks do not provide adequate models [Wolf77, Lazo77, SaCh79]. For example, priority scheduling disciplines or even first-come-first-served service where service times are highly variable violate the assumptions that lead to product-form solutions. Because direct global balance solution is impractical for most such systems, a number of approximate solution techniques have been developed. Most are based on a combination of decomposition, stage representations of service time distributions [Cox55, Whit74], and global balance solution of smaller systems. An example of such an approach is described by Marie [Mari79]. Another approach, based on diffusion approximations, is described by Gelenbe [Gele75]. Other similar techniques have been surveyed elsewhere [ChSa78, Trip79, SaCh80].

Many computer operating systems employ preemptive priority scheduling disciplines for processor dispatching. If the system uses the same preemptive priorities for allocating all system devices, good approximate performance estimates can be obtained by solving for each class in turn in order of priority, and reducing the processing capacity of each device as seen by a particular priority class to reflect the load placed on the device by higher priority classes. If, however, the cpu is scheduled based on preemptive priorities while the io devices are scheduled first-come-first-served (as is very often the case) then an iterative solution technique is required to investigate the interactions among the priority classes [Sevc77, ChYu79].

Memory space is an important computer system resource that must be treated differently than processors and storage devices. Queueing network models have been used to investigate the control of multiprogramming level in paging systems with the objective of avoiding thrashing [DeGr75, DKLP76, GrDe77, GeKu78]. The use of queueing network models as the basic analytic technique underlying a performance model for data base systems has also been proposed [Sevc81].

Most of the development of queueing network theory has been based on stochastic processes and their associated assumptions. In computer system contexts, however, those assumptions that can be tested at all are often blatantly violated. Operational analysis is an alternative framework, more closely related to computer systems, for establishing relationships among measurable quantities based on testable assumptions [DeBu78, BuDe80]. The formulae obtained from operational analysis are identical in form to those previously obtained by stochastic analysis (although the interpretations of the variables differ slightly). For this reason, there has been some controversy over the extent to which operational analysis represents more powerful results rather than just a reformulation of previously known results [SeKl79, Brya80].

4. COMPUTER-COMMUNICATION NETWORK APPLICATIONS

Queueing network models have also been used to investigate the performance of store-and-forward packet-switched computer-communication networks. Service centers are used to represent communication lines and concentrators as well as processors. Some of the problems that have been investigated involve topology design, routing, capacity selection, flow control, congestion control and distribution of end-to-end delay [Klei76, KoKo77, Reis78, Wong78b, LaWo81]. Communication networks may involve hundreds of service centers and many classes of customers (one for each logical communication path in the network). Exact solution of models of such networks is still feasible, however, by exploiting the fact that each class of customers typically visits only a small fraction of the service centers in the network [LaLi81].

Networks that allocate the transmission medium based on contention (such as the ALOHA network and numerous local area networks) give rise to interesting analytic problems (for which queueing networks provide little help) [Tane81, MoKl80].

5. CONCLUSIONS

Applications to computer systems and computer-communication networks have motivated extensive progress in the theory of queueing networks. The books and papers in the list of references reflect the breadth of studies relating to queueing network models and their applications.

REFERENCES

Albi80 S.L. Albin, Approximating superimposition arrival processes of queues, report, Bell Laboratories (April 1980).

Barb76 A.D. Barbour, Networks of queues and the method of stages, *Adv. Appl. Prob 8* (1976), 584-91.

BCMP75 F. Baskett, K.M. Chandy, R.R. Muntz and F. Palacios-Gomez, Open, closed and mixed networks of queues with different classes of customers, *J.ACM 22*, 2 (April 1975), 248-60.

Brya80 R.M. Bryant, On Homogeneity in M/G/1 queueing systems, *Performance Evaluation Review 9*, 2 (Summer 1980), 199-208.

Burk56 P.J. Burke, The Output of a queueing system, *Oper. Res. 4*, (1956), 699-714.

Burk69 P.J. Burke, The Dependence of sojourn times in tandem M/M/s queues, *Oper. Res. 17* (1969), 754-5.

Burk72 P.J. Burke, Output processes and tandem queues, Proc. Symp. on Comp. Comm. Networks and Teletraffic, Polytechnic Inst. of Brooklyn (April 1972), 419-28.

Burk76 P.J. Burke, Proof of a conjecture on the interarrival-time distribution in an M/M/1 queue with feedback, *IEEE-TC-24*, 5 (1976), 175-6.

Buze73 J.P. Buzen, Computational algorithms for closed queueing networks with exponential servers, *Comm. ACM 16*, 9 (September 1973), 527-31.

BuDe80 J.P. Buzen and P.J. Denning, Operational treatment of queue distributions and mean-value analysis, *Computer Performance 1*, 1 (1980).

Chan72 K.M. Chandy, The Analysis and solutions for general queueing networks, Proc. 6th Ann. Princeton Conf. on Inf. Sci and Systems (1972), 224-28.

ChHW75 K.M. Chandy, U. Herzog, and L.S. Woo, Parametric anaysis of queueing networks, *IBM J.R. & D. 19*, 1 (January 1975), 43-49.

ChHT77 K.M. Chandy, J.H. Howard and D.F. Towsley, Product form and local balance in queueing networks, *J.ACM 24*, 2 (April 1977), 250-63.

ChNe81 K.M. Chandy and D. Neuse, A Heuristic algorithm for queueing network models of computer systems, Proc. SIGMETRICS Conf. on Meas. and Mod. of Comp. Syst., Las Vegas (September 1981).

ChSa78 K.M. Chandy and C.H. Sauer, Approximate methods of analysis of queueing network models of computer systems, *Computing Surveys 10*, 3 (September 1978), 263-80.

ChSa80 K.M. Chandy and C.H. Sauer, Computational algorithms for product form queueing networks, *Comm. ACM 23*, 10 (October 1980).

Chow78 W.M. Chow, The Cycle time distribution of exponential central server models, *Proc. AFIPS NCC 43*, (1978).

ChYu79 W.M. Chow and P.S. Yu, An Approximation technique for central server queueing models with a priority dispatching rule, IBM Research Report, Yorktown Heights (1979).

Coop80 R.B. Cooper, *Introduction to Queueing Theory*, second edition, MacMillan (1980).

Cour77 P.J. Courtois, *Decomposability: queueing and computer system applications*, Academic Press (1977).

Cox55 D.R. Cox, A Use of complex probabilities in the theory of stochastic processes, *Proc. Cambridge Phil. Soc 51* (1955), 313-19.

Dale76 D.J. Daley, Queueing output processes, *Adv. Appl. Prob. 8* (1976), 395-415.

DeGr75 P.J. Denning and G.S. Graham, Multiprogrammed Memory management, *Proc. IEEE 63*, 6 (June 1975), 924-39.

DKLP76 P.J. Denning, K.C. Kahn, J. Leroudier, D. Potier, and R. Suri, Optimal multiprogramming, *Acta Informatica 7*, 2 (1976), 197-216.

DeBu78 P.J. Denning and J.P. Buzen, The Operational analysis of queueing network models, *Computing Surveys 10*, 3 (September 1978), 225-61.

DiCh74 R.L. Disney and W.P. Cherry, Some topics in queueing network theory, *Mathematical Methods in Queueing Theory*, A.B. Clarke (ed.), Springer-Verlag (1974).

Disn75 R.L. Disney, Random flow in queueing networks: a review and critique, *Trans. Amer. Inst. Industr. Engr. 7* (1975), 268-88.

Disn80 R.L. Disney, Queueing networks, Report VTR-7923, Dept. of Ind. Eng. and Oper. Res., Virginia Polytechnic Institute (August 1980).

DiMS80 R.L. Disney, D.C. McNickle, and B. Simon, The M/G/1 queue with instantaneous Bernoulli feedback, *Nav. Rsch. Log. Qtrly. 27*, 4 (December 1980), 635-44.

Gave67 D.P. Gaver, Probability models of multiprogramming computer systems, *J.ACM 14*, 3 (1967), 423-28.

Gele75 E. Gelenbe, On Approximate computer system models, *J.ACM 22*, 2 (April 1975), 261-69.

GeKu78 E. Gelenbe and A. Kurinckx, Random injection control of multiprogramming in virtual memory, *IEEE Trans. Softwr. Eng. 4*, 1 (1978), 2-17.

GeMi80 E. Gelenbe and I. Mitrani, *Analysis and Synthesis of Computer Systems*, Academic Press (1980).

GoNe67 W.J. Gordon and G.F. Newell, Closed queueing systems with exponential servers, *Oper. Res. 15* (1967), 254-65.

GrDe77 G.S. Graham and P.J. Denning, On the relative controllability of memory policies, *Computer Performance*, K.M. Chandy and M. Reiser (editors), Elsevier North-Holland (1977), 411-28.

Jack63 J.R. Jackson, Jobshop-like queueing systems, Mgmt. Sci. 10 (1963), 131-42.

Kell76 F.P. Kelly, Networks of queues, *Adv. Appl. Prob. 8* (1976), 416-32.

Kell79 F. P. Kelly, *Reversibility and Stochastic Networks*, John Wiley and Sons, Toronto (1979).

Kies80 P.C. Kiessler, A Simulation analysis of sojourn times in a Jackson network, Report VTR-8016, Dept. of Ind. Eng. and Oper. Res., Virginia Polytechnic Institute and State University (December 1980).

Klei75 L. Kleinrock, *Queueing Systems: Volume I - Theory*, Wiley-Interscience (1975).

Klei76 L. Kleinrock, *Queueing Systems: Volume II - Computer Applications*, Wiley-Interscience (1976).

KoRe75 H. Kobayashi and M. Reiser, On Generalization of job routing behavior in a queueing network model, IBM Research Report RC-5252, Yorktown Heights (1975).

KoKo77 H. Kobayashi and A. Konheim, Queueing models for computer communication systems, *IEEE Trans. Comm. Com-25*, 1 (January 1977), 2-29.

Koba78 H. Kobayashi, *Modeling and Analysis: An Introduction To System Performance Evaluation Methodology*, Addison-Wesley (1978).

Kuhn76 P. Kuhn, Analysis of complex queueing networks by decomposition, 8th International Teletraffic Congress, Melbourne, Australia (November 1976), 236-1-8.

LaPS79 J. Labetoulle, G. Pujolle, and C. Soula, Distribution of the flows in a general Jackson network, Report 341, IRIA Laboria (1979).

Lam77 S.S. Lam, Queueing networks with population size constraints, *IBM J.R. & D. 21*, 4 (July 1977), 370-78.

LaLi81 S. S. Lam and Y. K. Lien, A Tree convolution algorithm for the solution of queueing networks, Report TR-165, University of Texas at Austin (January 1981).

LaWo81 S.S. Lam and J.W. Wong, Queueing network models of packet-switching networks, Report CS-81-06, Dept. of Comp. Sci., Univ. of Waterloo, (February 1981).

LaRe80 S.S. Lavenberg and M. Reiser, Stationary state probabilities at arrival instants for closed queueing networks with multiple types of customers, *J. of Appl. Prob.* (December 1980).

Lazo77 E.D. Lazowska, The Use of percentiles in modelling cpu service time distributions, *Computer Performance*, K.M. Chandy and M. Reiser (editors), Elsevier North-Holland (1977), 53-66.

LaSe78 E.D. Lazowska and K.C. Sevcik, Exploiting decomposability to approximate quantiles of response times in queueing networks, *Proc. 2nd Int. Conf. on Operating Systems*, IRIA, Paris (October 1978).

Lemo77 A.J. Lemoine, Networks of queues -- a survey of equilibrium analysis, *Mgmt. Sci. 24*, 4 (1977), 464-81.

Mari79 R. Marie, An Approximate analytical method for general queueing networks, *IEEE Trans. on Softwr. Eng. SE-5*, 5 (September 1979).

Mela79a B. Melamed, On Markov jump processes imbedded at jump epochs and their queueing theoretic applications, Dept. of Ind. Eng. and Mgmt. Sci.,

Northwestern University (August 1979).

Mela79b B. Melamed, Characterizations of Poisson traffic streams in Jackson queueing networks, *Adv. Appl. Prob. 11* (1979), 422-38.

Mela79c B. Melamed, Sojourn times in queueing networks, Dept. of Ind. Eng. and Mgmt. Sci., Northwestern Univ. (August 1979).

Mitr79 I. Mitrani, A Critical note on a result by Lemoine, *Mgmt. Sci. 25*, 10 (October 1979), 1026-27.

MoKl80 M.Molle and L. Kleinrock, Virtual time CSMA: a new protocol with improved delay characteristics, UCLA (November 1980).

Munt73 R.R. Muntz, Poisson departure processes and queueing networks, Proc. 7th Ann. Princeton Conf. (March 1973), 435-40.

MuWo73 R.R. Muntz and J.W. Wong, Efficient computational procedures for closed queueing networks, Note 17, Computer Science Dept., UCLA (June 1973).

MuWo75 R.R. Muntz and J. Wong, Asymptotic properties of closed queueing network models, Proc. of 8th Annual Princeton Conf. on Inf. Sci. and Systems, Preinceton University (March 1974).

ReKo73 M. Reiser and H. Kobayashi, Numerical solution of semiclosed exponential server queueing networks, Proc. 7th Asilomar Conf. and Cir. and Syst., Asilomar (November 1973).

ReKo75 M. Reiser and H. Kobayashi, Queueing networks with multiple closed chains: theory and computational algorithms, *IBM J.R. & D. 19*, 3 (May 1975).

ReKo76 M. Reiser and H. Kobayashi, On the convolution algorithm for separable queueing networks, Proc. Int. Symp. on Comp. Perf. Mod. Meas. and Eval. Cambridge, Massachusetts (March 1976), 109-17.

ReLa80 M. Reiser and S.S. Lavenberg, Mean value analysis of closed multichain queueing networks, *J. ACM 27*, 2 (April 1980), 313-22.

Reis78 M. Reiser, A Queueing network analysis of computer communication networks with window flow control, IBM Research Report RC-7218, Yorktown Heights (July 1978).

SaLa81 S. Salza and S. S. Lavenberg, Approximating response time distributions in closed queueing network models of computer system performance, IBM Research Report RC-8735, Yorktown Heights (March 1981).

SaCh79 C.H. Sauer and K.M. Chandy, The Impact of distributions and disciplines on multiprocessor systems, *Comm. ACM 22*, 1 (January 1979), 25-34.

SaCh80 C.H. Sauer and K.M. Chandy, Approximate solution of queueing models, *IEEE Computer 13*, 4 (April 1980), 25-32.

SaCh81 C.H. Sauer and K.M. Chandy, *Computer System Performance Modeling*, Prentice Hall (1981).

Saue81 C.H. Sauer, Computational algorithms for state-dependent queueing networks, IBM Research Report RC-8698, IBM Yorktown Heights (February 1981).

Sche67 A.L. Scherr, *An Analysis of Time-Shared Computer Systems*, M.I.T. Press, Cambridge, Massachusetts (1967).

Sevc77 K.C. Sevcik, Priority scheduling disciplines in queueing network models of computer systems, *Proc. IFIP Congress 77* (1977), 565-70.

Sevc81 K.C. Sevcik, Data base system performance prediction using an analytical model, *Proc. 8th Conf. On Very Large Data Bases*, Cannes (September 1981).

SLTZ77 K.C. Sevcik, A.I. Levy, S.K. Tripathi and J. Zahorjan, Improved approximations of aggregated queueing network subsystsms, *Computer Performance*, K.M. Chandy and M. Reiser (editors), Elsevier North-Holland (1977), 1-22.

SeKl79 K.C. Sevcik and M.M. Klawe, Operational analysis versus stochastic modelling of computer systems, *Proc. Computer Science and Statistics: 12th Annual Symposium on the Interface*, Waterloo, Canada (1979).

SeMi81 K.C. Sevcik and I. Mitrani, The Distribution of queueing network states at input and output instants, *J.ACM 28*, 2 (April 1981).

SiFo79 B. Simon and R.D. Foley, Some results on sojourn times in acyclic Jackson networks, *Mgmt. Sci. 25*, (1979), 1027-34.

Stew78 W.J. Stewart, A Comparison of numerical techniques in Markov modelling, *Comm. ACM 21*, 2 (February 1978), 144-51.

Tane81 A. Tanenbaum, *Computer Networks*, Prentice Hall (1981).

Tows80 D.F. Towsley, Queueing network models with state-dependent routing, *J.ACM 27*, 2 (April 1980), 323-37.

Trip79 S.K. Tripathi, On Approximate solution techniques for queueing networks models of computer systems, Report CSRG-106, Computer Systems Research Group, Univ. of Toronto (September 1979).

Vant78 H. Vantilborgh, Exact aggregation in exponential queueing networks, *J.ACM 25*, 4 (October 1978).

WaVa78 J. Walrand and P. Varaiya, When is a flow in a queueing network Poissonian?, Memorandum ERL-M78/59, Electronics Research Laboratory, Univ. of California at Berkeley (August 1978).

WaVa79 J. Walrand and P. Varaiya, Sojourn times and the overtaking condition in Jacksonian networks, Memorandum ERL-M79/35, Electronics Research Laboratory, Univ, of California at Berkeley (May 1979).

Whit74 W. Whitt, The Continuity of queues, *Adv. Appl. Prob. 6* (1974), 175-83.

Whit69 P. Whittle, Nonlinear migration processes, *Internat. Stat. Inst. Bull. 42*, Book 1 (1969), 642-47.

Wolf77 R.W. Wolff, The Effect of service time regularity on system performance, *Computer Performance*, K.M. Chandy and M. Reiser (editors), Elsvier North-Holland (1977), 297-304.

Wong78a J.W. Wong, Distribution of end-to-end delay in message-switched networks, *Computer Networks 2*, 1 (February 1978), 44-49.

Wong78b J.W. Wong, Queueing network modeling of computer communication networks, *Computing Surveys 10*, 3 (September 1978), 343-52.

Zaho80 J. Zahorjan, The Approximate Solution of Large Queueing Network Models, Report CSRG-122, Computer Systems Research Group, Univ. of Toronto (1980).

ZSEG81 J. Zahorjan, K.C. Sevcik, D.L. Eager, and B.I. Galler, Balanced job bound analysis of queueing networks, to appear in *Comm. ACM.*

STATIONARY PROPERTIES OF TIMED VECTOR ADDITION SYSTEMS

Erol Gelenbe

Université de Paris Sud
INRIA, Rocquencourt

ABSTRACT

Vector addition systems, initially introduced by Karp and Miller, have been very successfully used for the description of complex concurrent behaviour. These models are essentially the same as Petri nets. Our purpose is to develop a quantitative "measurement" oriented theory of such systems in order to complement the existing qualitative theory. This paper is a first step in that direction.

1. INTRODUCTION

1.1. General considerations

Complex concurrent systems are conveniently described by vector addition systems (VAS) which originated with the work of Karp and Miller on parallel program schemata [1]. These models are essentially equivalent to the nets introduced by C.A. Petri which were originally suggested as models for management systems, but which are now very often effectively used to describe computer operating system behaviour. Petri nets and VAS are frequently used to prove qualitative properties of complex concurrent systems, for instance in the area of computer communication protocols, such as the non-existence of deadlock between parallel processes. They are also used as formal specification tools, as models in the design of large hardware/software systems, and for building standards of communication protocols.

It would be most convenient if essentially the same models could incorporate properties which would make them suitable for

M. A. H. Dempster et al. (eds.), Deterministic and Stochastic Scheduling, 223–232.

performance evaluation. Such an extension would have obvious practical advantages for the system designer. Indeed such extensions have been suggested by Sifakis [2], Ramamoorthy and Ho [3] and others, but have not led to the development of a substantial quantitative theory useable for performance evaluation and prediction.

In the area of performance evaluation, probabilistic queueing network models (see, for instance, [4]) have gained wide popularity. More recently, a more direct relationship between models of this type and measurement data was made via "operational models" introduced by Buzen and developed by Buzen and Denning [5], without probabilistic assumptions. However, a direct link between the formalisms used in queueing networks and those used in Petri nets has not been available, except for the most obvious analogy which would associate customers and servers on the one hand with tokens and transitions on the other hand.

This paper is an attempt to develop a "measurement" oriented theory within a formal context very close to that of VAS. The existence of such a theory would allow the analysis of both qualitative and quantitative aspects within one formal framework. Indeed, conventional VAS are imbedded in the model we use.

1.2. The formal model

Let $\underline{k}(t)$ be an ℓ-dimensional vector of natural numbers defined for each real valued $t \geq 0$. $\underline{k}(t)$ will be a left continuous function of t. $\underline{k}(0) = \underline{k}_0$ is given.

Let $T = \{0 = t_0 < t_1 < t_2 < \ldots < t_i < t_{i+1} < \ldots\}$ be a countable infinite subset of the positive real line.

We shall say that T is the set of transition times of $K = \{\underline{k}(t),\ t \geq 0\}$; namely, we suppose that

$$\underline{k}(t) = \underline{k}(t_i^+) \text{ for } t \in]t_i, t_{i+1}], \text{ for each } i = 0,1,2,\ldots .$$

This means that changes, or "jumps", in K can only occur at the instants contained in T. This, of course, does not imply that there will be a jump at each instant which is an element of T.

A subset S of $\{1,2,\ldots,\ell\}$ will be called a *cut*, and for a given S we shall define for each $t \geq 0$,

$$N(t) = \Sigma_{i \in S}\ k_i(t),$$

where $k_i(t)$ is the i-th component of the vector $\underline{k}(t)$. The cut S will also give rise to a restriction T_S of T:

$$T_S = \{t_i : t_i \in T \text{ and } N(t_i^+) \neq N(t_i)\}.$$

We can view T_S as the set of jump instants of $\{N(t),\ t \geq 0\}$, or as the jump instants in T which can be detected by observing N(t).

1.3. Relationship with vector addition systems

The relationship between the model described in the preceding section and VAS is obvious.

A VAS is defined by an ℓ-dimensional vector of natural numbers $\underline{k}_0$, and by a finite set $R = \{\underline{r}_1,\ldots,\underline{r}_k\}$ of integer valued ℓ-dimensional vectors. From $\underline{k}_0$ we can construct any ℓ-dimensional vector as follows:

$$\underline{k}^0 = \underline{k}_0,$$

$$\underline{k}^m = \underline{k}^{m-1} + \underline{r}^m, \quad \underline{r}^m \in R,$$

as long as each $\underline{k}_i$, $1 \le i \le m$, is composed of natural numbers. That is, we use the elements of R to transform $\underline{k}_0$ by successive additions with the only restriction that each of the resulting vectors remain in $\mathbb{N}^\ell$ (the ℓ-dimensional space of natural numbers).

With respect to the model presented in Section 1.2, we can simply consider that

$$\underline{k}(t_m^+) = \underline{k}^m.$$

Thus our model differs from a VAS simply by the introduction of *time*.

1.4. Relationship with Petri nets

Consider the Petri net shown in Figure 1. Arrows represent the flow of tokens among the four places or transitions. The incoming arrow to transition no. 1 indicates that it may receive tokens from the "outside world", while tokens may go to the "outside world" from transitions nos. 2 and 4. The system may be represented by the vector

$$\underline{k} = (k_1, k_2, k_3, k_4),$$

where k_i is the number of tokens present at position i. In Figure 1 we show the initial vector which is

$$\underline{k}_0 = (2,3,1,2) = \underline{k}(0),$$

and the numbers in parentheses next to each transition represent the number of tokens necessary to fire that transition. Thus $\underline{k}_0$ is a stable state of the system.

Suppose now that at time $t_1 > 0$ one token arrives to transition 1 which will then fire and release its three tokens to each of 2,3,4 which will in turn fire, etc., leading to a new stable state

$$\underline{k}(t_1^+) = (0,1,1,0).$$

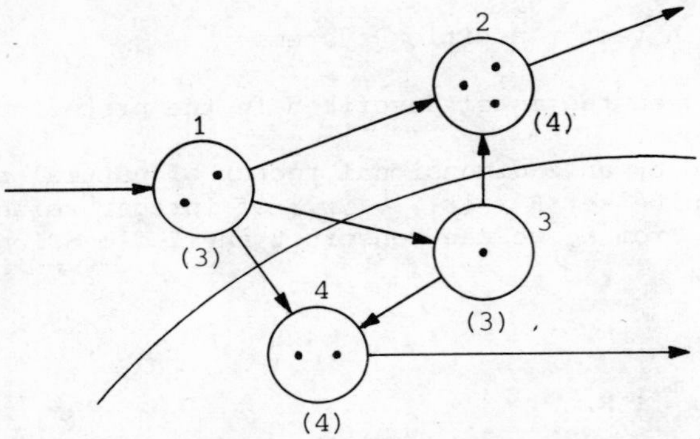

Figure 1. Example of a Petri net.

Let the cut S be the set {3,4} as shown in Figure 1. Then

$$N(0) = 3$$

and

$$N(t_1^+) = 1.$$

An external arrival of a token at time t_2 will provoke a state transition of the system to

$$\underline{k}(t_2^+) = (1,1,1,0)$$

and so on. Thus the formal model described in Section 1.2 differs from a conventional Petri net only in the introduction of the time parameter t, as was the case with respect to VAS.

1.5. Relationship with probabilistic queueing models

Consider the probabilistic network of queues shown in Figure 2. Customers arriving to the system are routed initially to service station i, $1 \le i \le \ell$, with probability q_i, and after some queueing and service time at that station will either leave the system with probability $p_{i,\ell+1}$ or enter queue j with probability $p_{i,j}$, and so on.

We shall denote by ω a complete history of the system and Ω will be the set of all such histories. Any ω can be constructed as follows, supposing that these histories begin at time t = 0.

Let successive customers arrive at instants

$$A = \{0 \le a_1 \le a_2 \le a_3 \le \ldots \le a_i \le a_{i+1} \le \ldots\}$$

and for the i-th customer in the sequence let

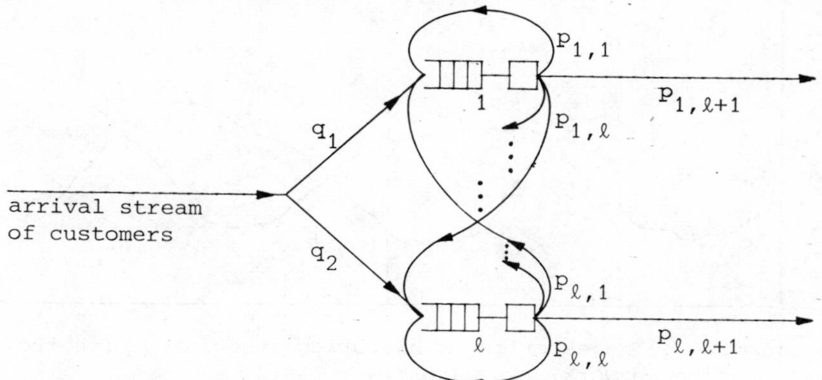

Figure 2. Probabilistic network of queues.

$$Y_i = \{Y_{i1}, Y_{i2}, \ldots, Y_{ij}, Y_{i,j+1}, \ldots\}$$

be the sequence of service stations visited, at which it will receive services of duration

$$Z_i = \{Z_{i1}, Z_{i2}, \ldots, Z_{ij}, Z_{i,j+1}, \ldots\}.$$

Then each ω will be completely defined if we specify the three sets $A, \{Y_i\}, \{Z_i\}$.

For a given ω, however, we may only be interested in the movements of the customers in the system, in which case we can construct the function

$$\underline{k}(\omega,t) = (k_1(\omega,t), \ldots, k_\ell(\omega,t)), \quad t \geq 0,$$

where $k_j(\omega,t)$ is the number of customers in queue j at time t for history ω. Similarly, from history ω we can construct the jump instants

$$0 \leq t_1(\omega) \leq t_2(\omega) \leq \ldots \leq t_i(\omega) \leq \ldots .$$

Thus the $\underline{k}(t)$ constructed in Section 1.2 corresponds to the $\underline{k}(\omega,t)$ of this section.

In the probabilistic framework, one most often works directly with probabilistic distributions, rather than with sample paths, even though the probability distributions are initially assigned to each history ω in Ω.

Thus for some time $\tilde{t}$ one tends to compute quantities related to $p(\underline{k}(\tilde{t}) \in A)$, which is the probability that the (random) process $\underline{k}(t) = \{\{\underline{k}(\omega,t): \omega \in \Omega\}, t \geq 0\}$ is in the subset A of its set of possible states at time $\tilde{t}$; but this is in fact

(1) $p(\underline{k}(\tilde{t}) \in A) = \Sigma_{\omega \in \Omega(\tilde{t},A)}\, p(\omega)$

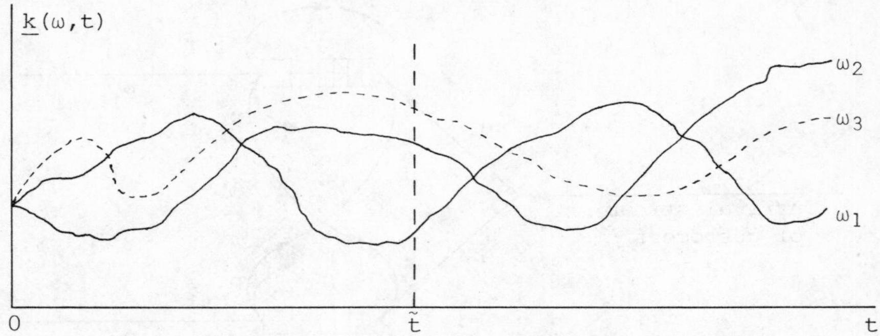

Figure 3. Trajectories or histories associated with the queueing network.

where

(2) $\Omega(\tilde{t},A) = \{\omega | \underline{k}(\omega,\tilde{t}) \in A\}$.

Thus, results concerning $\underline{k}(\omega,t)$ for one ω or for a family of ω's may be carried over to the probability distributions of a probabilistic model via (1) and (2).

2. PROPERTIES OF FLOWS

Let us now return to the properties of the formal model defined in Section 1.2 as follows:

- $\underline{k}(t) \in \mathbb{N}^\ell$ (the ℓ-dimensional space of natural numbers) defined for each t; $\underline{k}(t) = (k_1(t),\ldots,k_\ell(t))$;
- $T = \{0 < t_1 < t_2 < \ldots < t_i < t_{i+1} < \ldots\}$, the set of jump instants of $\underline{k}(t)$;
- S a subset of $\{1,\ldots,\ell\}$ which we call a cut, and which gives rise to

  $N(t) = \Sigma_{i \in S}\ k_i(t),\quad t \geq 0$

  and to $T_S = \{t_i\colon t_i \in T \text{ and } N(t_i^+) \neq N(t_i)\}$.

All these quantities are to be viewed as being *deterministic*. One may also view $\{\underline{k}(t),\ t \geq 0\}$ as being in fact identical to some sample path $\{\underline{k}(\omega,t),\ t \geq 0\}$ associated with a given history ω of the random process $\{\underline{k}(t),\ t \geq 0\} = \{\{\underline{k}(\omega,t)\colon \omega \in \Omega\},\ t \geq 0\}$.

2.1. Flows associated with the cut S

$T_S \subseteq T$ is obtained by considering only those time instants at which N(t) changes value. We shall define two subsets Λ and Γ of T_S such that

$$\Lambda \cap \Gamma = \emptyset, \quad \Lambda \cup \Gamma = T_S,$$

such that

$$\Lambda = \{t_i : t_i \in T_S \text{ and } N(t_i^+) > N(t_i)\},$$

$$\Gamma = \{t_i : t_i \in T_S \text{ and } N(t_i^+) < N(t_i)\},$$

and write

$$\Lambda = \{0 < a_1 < a_2 < \ldots < a_i < a_{i+1} < \ldots\},$$

$$\Gamma = \{0 < d_1 < d_2 < \ldots < d_i < d_{i+1} < \ldots\},$$

which we shall view as the *arrival* and *departure* instants, respectively, to and from the cut S.

2.2. Flow assumptions

Some restrictive assumptions will now be made concerning the flows which we consider.

(a) *Initial condition on* $N(t)$:

$$N(0) = 0.$$

(b) *One step transitions of* $N(t)$:

$$N(t_i^+) = N(t_i) \pm 1 \quad \text{for any } t_i \in T_S.$$

(c) *Causal flows*:

$$\Sigma_{i=1}^{\infty} 1(a_i < t) \geq \Sigma_{i=1}^{\infty} 1(d_i < t) \quad \text{for all } t \geq 0.$$

(d) *Stationary flows*: there exist real numbers $0 < \lambda < \infty$, $0 < \gamma < \infty$ such that

$$\lim_{i\uparrow\infty} \frac{1}{i}|a_i - \frac{i}{\lambda}| = 0, \quad \lim_{i\uparrow\infty} \frac{1}{i}|d_i - \frac{i}{\gamma}| = 0.$$

Under these assumptions, we can now present some useful general results. They are given here without proof.

3. PROPERTIES OF ARRIVAL AND DEPARTURE INSTANT MEASURES

From a theoretical standpoint, Λ and Γ define sequences of significant instants of the formal system. Thus properties of observations of the system at those instants are of interest. If one takes a practical viewpoint arising from the necessity to carry out measurements, the Λ or Γ instants can be used to trigger data collection.

Our first result states that the sequences Λ and Γ are equivalent with respect to the measures they offer from observations of the state of the cut S, namely of $\{N(t), t \geq 0\}$.

Let us define for each $t \geq 0$ and natural number $n \geq 0$,

$$U(n,t) = \Sigma_{i=1}^{\infty} 1(N(a_i) = n, a_i < t)/\Sigma_{i=1}^{\infty} 1(a_i < t),$$

$$V(n,t) = \Sigma_{i=1}^{\infty} 1(N(d_i^+) = n, d_i < t)/\Sigma_{i=1}^{\infty} 1(d_i < t).$$

We can say that $U(n,t)$ is the relative frequency of observing that $N(t) = n$ at arrival instants, and that $V(n,t)$ is the corresponding quantity for departure instants.

THEOREM 3.1. *Under assumptions* 2.2(a)-(d),

$$\lim_{t\uparrow\infty} |U(n,t)-V(n,t)| = 0$$

if $\lambda = \gamma$.

Analogous theorems are often established for various queueing theoretic models under probabilistic assumptions.

4. TIME AVERAGES AND EXCURSIONS

A well-known result which reappears in various contexts is the theorem first proved by J.D.C. Little which relates average response times in queueing systems to the average number of customers in the system.

In our formal model, though $N(t)$ may be viewed as being the "number of customers in the sub-system S", there is no specific relationship established between arrival and departure instants of a given "customer". If such a relationship can be established, then the result below can be viewed as a form of Little's theorem, proved in the context of our formal model. Otherwise it should merely be viewed as a general abstract property relating the flows Λ and Γ to the time average behaviour of $N(t)$, $t \geq 0$.

Let us define

$$T(t) = \Sigma_{i=1}^{\infty} (d_i - a_i) 1(d_i < t)/\Sigma_{i=1}^{\infty} 1(d_i < t).$$

REMARK 4.1. *By assumption* 2.2(c), $T(t) \geq 0$ *for all* $t \geq 0$.

Proof. Proceed by contradiction. If this is not true there must be at least one d_j, $d_j < t$, such that $d_j < a_j$. But then for some small enough $\varepsilon > 0$,

$$\Sigma_{i=1}^{\infty} 1(d_i < d_j + \varepsilon) = j$$

while

$$\Sigma_{i=1}^{\infty} 1(a_i < d_j+\varepsilon) < j$$

where it suffices to choose $d_j+\varepsilon < a_j$. But this contradicts the assumption 2.2(c) for $t = d_j+\varepsilon$. Therefore the remark is true. □

THEOREM 4.2. *Under assumptions* 2.2(a)-(d) *and when* $\lambda = \gamma$ *it follows that*

$$\lim_{t\uparrow\infty} \frac{1}{t}\left|\frac{1}{t} \int_0^t N(\tau)d\tau - \lambda T(t)\right| = 0.$$

Notice that this is not quite the usual statement of Little's theorem.

5. CONCLUSIONS

The formal model presented in this paper constitutes a first step towards a measurement oriented theory of a class of models which are now widely accepted to represent complex concurrent systems functioning under predetermined or dynamic schedules. It may also be viewed as a tool for analysing the sample path behaviour of stochastic queueing network models.

The theorems we present are indicative of the type of general results which may be proved in this formal context. These theorems can be carried through to stochastic processes for which each sample path satisfies our restrictive conditions: the proof of this must be carried out for each stochastic model considered and will often be a non-trivial task. Conversely, there is no reason to assume that *proofs* of similar results in certain stochastic queueing models carry over to our formal framework, unless they can be shown to hold for each sample path which should then be verified with respect to our assumptions.

It will be interesting to see to what extent the approach we propose can lead to effective computations of useful performance measures.

ACKNOWLEDGMENT

This work was supported by a contract from the Agence de Développement des Applications de l'Informatique.

REFERENCES

1. R.M. KARP, R.E. MILLER (1966) Properties of a model for parallel computations. *SIAM J. Appl. Math.* 14,1390-1411.

2. J. SIFAKIS (1976) Thèse de Docteur-Ingénieur, Grenoble.
3. C.V. RAMAMOORTHY, G.S. HO (1980) Performance evaluation of asynchronous concurrent systems using Petri nets. *IEEE Trans. Software Engrg.* SE-6,440-449.
4. E. GELENBE, I. MITRANI (1980) *Analysis and Synthesis of Computer System Models*, Academic Press, New York.
5. J.P. BUZEN, P.J. DENNING (1980) Measuring and calculating queue length distributions. *Computer*, IEEE, 13.4,33-44.

THE MULTIPRODUCT LOT SCHEDULING PROBLEM

Linus Schrage

University of Chicago
Chicago

Université Catholique de Louvain
Louvain-la-Neuve

European Institute for Advanced Studies in Management
Brussels

ABSTRACT

An NP-hard problem of considerable practical interest is the multi-product lot scheduling problem. In its simplest form there are P products to be scheduled on a single machine over a finite interval (0, T). Associated with each product i is a demand schedule D_t, a per unit time holding cost h_t, and a changeover cost vector c_{ji} which is the cost of starting production on i if the machine previously produced product j. In practical problems one might wish to treat the D_t as random variables, although this feature is typically disregarded by solution procedures. Example situations might be a television manufacturer who produces several different styles and sizes of televisions on a single line or a chemical processor who produces several different chemicals in batches on a single expensive machine. We briefly summarize previous approaches to this problem starting with the work of Manne, Dzielinski, Gomory, Lasdon and Terjung and then analyze LP-like approximations to this model and provide bounds on the closeness of the LP solution to the exact IP solution as the problem size gets large.

M. A. H. Dempster et al. (eds.), Deterministic and Stochastic Scheduling, 233–244.

INTRODUCTION

The lot scheduling problem is a generalization of the one machine sequencing problem. One wishes to determine which products should be produced when on a single machine so as to meet a specified demand schedule at minimum cost. The relevant costs are typically setup or changeover costs and inventory related costs.

Practical variations of the problem arise in the scheduling of appliances (e.g., TV's) on assembly lines or batches of different chemicals through an expensive chemical processor, or tires in a tire factory.

Analysis of lot scheduling problems have historically proceeded along one of two lines: a) the constant demand, constant costs, continuous time case and b) the dynamic demand and cost, discrete time case. Our analysis will be devoted mainly toward the dynamic demand, discrete time case. Elmagrahby (1978) presents a recent survey of the continuous time case.

We will start by considering single product dynamic lot scheduling problems. Happily, it appears that single product lot scheduling algorithms can be used as subroutines in multiproduct lot scheduling algorithms. Unhappily, many of the interesting single product lot scheduling problems are already NP-hard without introducing the complication of additional products. It is the multiproduct problem which is of real practical interest.

DEFINITION OF THE SINGLE PRODUCT LOTSIZING PROBLEM

A single product dynamic lotsizing problem is described by the following set of parameters:

T = number of periods; the periods are indexed with $t = 1, 2, \ldots, T$,

E_t = maximum allowed inventory at the end of period t,

A_t = production capacity in period t,

S_t = setup cost incurred in period t if there was production of the product in period t but no production of the same product in period $t-1$,

R_t = an overhead or rental charge which is incurred, possibly in addition to S_t, if there is positive production of the product in period t,

C_t = production cost per unit produced in period t,

H_t = end of period holding cost per unit applied to positive end of period inventory,

B_t = end of period backlogging cost per unit applied to negative end of period inventory,
W_t = production capacity wasted (or lost to) on setup if the product is produced in t but not in t-1,
D_t = demand in period t.

It is worthwhile dwelling on the distinction between S_t and R_t. Many models, such as the Wagner-Whitin (1959) dynamic lotsize model, implicitly assume that S_t is zero. The so-called setup cost in these models really corresponds to R_t. In practical problems it is frequently the case that one need not incur an additional setup in period t if there was production in period t-1.

One additional problem characteristic of interest is the type of policy allowed. We distinguish two cases: i) all or nothing production policies, denoted 0/A, and ii) general production, denoted G. Under a 0/A policy the only production level allowed in period t is either 0 or A_t, whereas under a G policy any production level between 0 and A_t is allowed. There are two motivations for considering 0/A policies: i) computational simplification and ii) 0/A policies are frequently used in practice, e.g., if setups are done only at the beginning of a shift or outside of regular production hours.

COMPUTATIONAL DIFFICULTY OF ONE MACHINE LOTSIZING

The computational complexity of the lotsizing problem in the NP-hardness sense was first studied by Florian, Lenstra, and Rinnooy Kan (1980). A comprehensive, recent analysis of the computational complexity of lotsizing appears in Bitran and Yanasse (1981). The tabulations in Table 1 are largely extensions of the latter paper to include backordering, $S_t > 0$, and 0/A policies. The difficulty of a problem depends closely upon how the various parameters change over time. The possible changes over time we will consider are defined below.

Z: for zero: the indicated parameter is zero in every period,
C: the indicated parameter is constant from period to period,
NI: the indicated parameter is nonincreasing with time,
ND: the indicated parameter is nondecreasing with time,
∞: the indicated parameter is infinite in every period,
G: general: no assumptions are made regarding how the parameter changes over time.

The generality of these assumptions are described in Figure 1. G is the most general while Z and ∞ are the most restrictive. Any algorithm for a specific problem type can also be applied to a problem type having more restrictive assumptions.

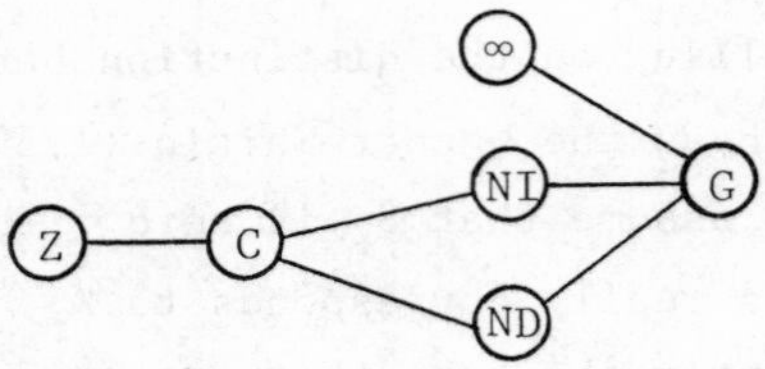

Figure 1 Generality of Restrictions.

Problem Type	E_t	A_t	S_t	R_t	C_t	H_t	B_t	W_t	D_t	Policy	Difficulty
p1	∞	NI	Z	ND	ND	Z	∞	Z	G	G	T
p2	∞	∞	G	G	G	G	∞	G	G	G	T^2
p3	∞	G	Z	C	C	Z	∞	Z	G	G	T^2
p4	G	C	G	G	G	G	G	Z	G	O/A	T^2
p5	G	C	G	G	G	G	G	C	G	O/A	T^3
p6	C	∞	G	Z	G	G	∞	Z	G	G	T^3
p7	∞	C	Z	NI	NI	G	∞	Z	G	G	T^3
p8	∞	C	G	G	G	G	∞	Z	G	G	T^4
p9	∞	ND	Z	NI	NI	G	∞	Z	G	G	T^4

Table 1 Computational Difficulty of One Product Lotsizing.

Notice in particular that if we restrict ourselves to O/A policies, problems p4 and p5, then quite general problems can be solved. We will be particularly interested in O/A policies.

Many lot scheduling problems can be shown to be NP-hard. A summary of current results in this regard is shown in Table 2. Notice that 2 and 3 product lotsizing problems become hard very quickly. Problem N11, for example, could be solved in T^2 time if the number of products was one instead of three.

Problem ID	No Products	E_t	A_t	S_t	R_t	C_t	H_t	B_t	W_t	D_t	Policy
N1	1	∞	G	Z	C	Z	Z	∞	Z	C	G
N2	1	∞	NI	Z	C	NI	Z	∞	Z	G	G
N3	1	∞	ND	Z	C	ND	Z	∞	Z	G	G
N4	1	∞	ND	Z	ND	Z	Z	∞	Z	G	G
N5	1	∞	NI	Z	NI	Z	Z	∞	Z	G	G
N6	1	∞	NI	Z	C	Z	G	∞	Z	G	G
N7	1	∞	NI	Z	C	ND	C	∞	Z	G	G
N8	2	∞	ND	Z	C	Z	Z	∞	Z	G	G
N9	2	∞	NI	Z	C	Z	Z	∞	Z	G	G
N10	2	∞	C	Z	NI	NI	G	∞	Z	G	G
N11	3	∞		Z	C	Z	C	∞	Z	G	G

Table 2 NP-hard Lotsizing Problems.

ALGORITHM FOR CAPACITATED, ONE PRODUCT, ONE MACHINE TYPE LOT-SIZING PROBLEM

If A_t and W_t are constant over time and we restrict ourselves to O/A policies, then a straightforward dynamic programming algorithm can be used to solve the one product lotsizing problem. Additionally, several machines can be accommodated as long as they are identical. Without loss we can assume that each machine has unit capacity per time period and that the demands D_t are measured in units of machine capacity.

Define:

M = number of machines,

L (u,p,n,t) = minimum cost until the end of period t if from 1 to t we used u setups, p full production periods (no setup), and n machines were in use at the end of period t,

$$V_t = p + u*(1-W) - \sum_{i=1}^{t} D_i = \text{period } t \text{ ending inventory.}$$

The dynamic programming recursion is

$$L(u,p,n,t+1) = H_{t+1} * \max(0, V_{t+1}) + B_{t+1} * \max(0, -V_{t+1}) + \min\left[\min_{i=0 \text{ to } n-1} (S_{t+1} * (n-i) + L(u-n+i, p-i, i, t)), \min_{i=n \text{ to } M} (L(u, p-n, i, t)\right] + n * R_{t+1}.$$

Here i indexes over the number of machines used in period t.

L (u,p,n,t) must be calculated for all possible values of u,p,n,t. The maximum possible values for u and p are MT/2 and MT respectively. Thus, of the order (MT)(MT)MT quantities must be calculated. Each of these involves finding the minimum of M + 1 quantities. Thus, the total work is of the order $T^3 M^4$. If a setup does not waste any capacity, then u,p can be replaced by u+p and the total work is proportional to $T^2 M^3$. The above recursion is essentially the one used by Lasdon and Terjung (1971) in their tire company application.

MULTIPRODUCT, ONE MACHINE LOTSIZING

One of the fundamental models for multiproduct lotsizing is due to Manne (1958), Dzielinski and Gomory (1965), and Lasdon and Terjung (1971). This model assumes that setup times and costs are sequence independent. That is, the setup effort for product j is independent of which product was run previously. As suggested by Manne, the method is essentially as follows: 1) Generate every possible feasible single product schedule for each product and then 2) Minimize the total cost of schedules selected subject to a) selecting exactly one schedule for each product and b) resource requirements of the schedules selected do not together exceed the resource availability each period. The contributions of Dzielinski, Gomory, Lasdon and Terjung were to show that algorithms for one product lotsizing can be exploited to avoid the necessity of generating all feasible schedules. We will call the general approach based on all of the above the MDGLT approach.

We will show how the MDGLT approach can be extended to the multiproduct case in which setup costs are sequence dependent. In particular, we consider the case where products can be partitioned into classes. Within a class, setup costs are sequence independent. As with the MDGLT method the problem is formulated as an integer program. We show that if the number of classes plus the number of time periods is small relative to the number of products, then this formulation is a good one in a certain well defined sense.

Displaying the linear 0/1 integer program for the sequence dependent multiproduct lotsizing problem requires us to introduce the following notation:

Sets

$Q(p)$ = the set of all possible feasible schedules for product p. The presumption is that we will not need to explicitly generate all members of $Q(p)$.

Parameters

S_{uv} = cost of a changeover from class u to class v.

U_k = cost of schedule k. This cost includes all costs except the sequence dependent setup cost component.

F_{kut} = 1 if schedule k is a schedule which begins production of a class u product in t, else 0.

G_{kut} = 1 if schedule k is a schedule which ends production of a class u product in t-1, else 0.

Variables

Z_k = 1 if schedule k is chosen, else 0.

X_{uvt} = 1 if a changeover from class u to v occurs at the beginning of period t, else 0.

Y_{ut} = 1 if the machine is idle in class u during period t, else 0.

The integer programming formulation is then:

$$\text{Min} \sum_p \sum_{k \in Q(p)} U_k Z_k + \sum_u \sum_v \sum_t S_{uv} X_{uvt}$$

s.t.

1) $\sum_{k \in Q(p)} Z_k = 1$ for every product p,

2) $\sum_v X_{vut} + \sum_p \sum_{k \in Q(p)} G_{kut} Z_k + Y_{ut-1}$

$$= \sum_v X_{uvt} + \sum_p \sum_{k \in Q(p)} F_{kut} Z_k + Y_{ut} \quad \text{for every class u and period t,}$$

$Z_k = 0$ or 1 for all k,

X_{uvt} and $Y_{ut} = 0$ or 1 for all u,v, and t.

QUALITY OF THE MULTIPRODUCT LP/IP

The IP problem tends to be easy to solve if when it is solved as an LP with the integrality constraints relaxed, many of the variables naturally take on integer values.

Let P be the number of products. Thus, P equals the number of constraints of the form $\Sigma Z_k = 1$. Let N = the number of other constraints. Typically, N = the product of the number of time periods and the number of classes. We can make the following statement regarding the integrality or goodness of the IP solution.

Theorem: The number of products with fractional schedules in the LP optimum is less than or equal to N.

Proof: Let f equal the number of products with fractional schedules in the LP optimum. A product with a fractional schedule must have at least 2 schedule variables, Z_k, nonzero. All other products will have exactly one schedule variable nonzero. Thus, the number of nonzero schedule variables is greater than or equal to 2f + (p-f). In a basic feasible LP solution the number of nonzero variables is less than or equal to P + N. Thus, $2f + p - f \leq P + N$, so $f \leq N$, as the theorem states.

Thus, if the number of products is large relative to the number of time periods plus classes, then we can expect most of the schedule variables to take on integer values in the LP optimum.

One further observation one can make about the goodness of the LP solution is that its cost at least equals the sum of the costs of the individual optima when each product is considered by itself. It is very easy to give formulations which do not have the above feature. Consider the following formulation which is typical of formulations occasionally proposed for lot-sizing problems. Define:

Parameters

S_{ij} = cost of switching from product i to j in any period, $S_{ii} = 0$,

h = holding cost applied/unit of ending inventory of any product in any period,
c = cost/unit of producing any product in any period,
D = demand in each period for each product, i.e., all products are identical except for changeover costs,
T = number of periods.

Decision variables

X_{jt} = 1 if product j is produced in period t, else 0,

Z_{ijt} = 1 if a switch from product i to j is made at the start of period t, else 0,

I_{jt} = inventory of product j at the end of period t.

No backlogging is permitted.
The formulation is:

$$\text{Min} \sum_j \sum_t C_j X_{jt} + \sum_j \sum_t h I_{jt} + \sum_i \sum_j \sum_t S_{ij} Z_{ijt}$$

s.t.

$$I_{jt-1} + X_{jt} - I_{jt} = D \qquad \text{for all products j and periods t,}$$

$$X_{jt} \le \sum_i Z_{ijt} \qquad \text{for all products j and periods t,}$$

$$\sum_i Z_{ijt} = \sum_r Z_{jrt+1} \qquad \text{for all products j and periods t,}$$

$$\sum_i \sum_j Z_{ij1} = 1 \qquad \text{(initialization),}$$

$$X_{jt} \text{ and } Z_{ijt} = 0 \text{ or } 1.$$

The appeal of this formulation is that it does not require one to generate single product schedules outside the formulation. This formulation solves the entire problem. The user does not have to solve part of the problem.

If, however, the integrality constraints are relaxed to $0 \le X_{jt} \le 1$ and $0 \le Z_{ijt} \le 1$ and the resulting LP is solved we can expect that the LP solution is a poor approximation to the IP solution. As T gets large we will find that the LP solution is $Z_{iit} = 1/P$, $Z_{ijt} = 0$ for $i \neq j$ and $X_{jt} = D$. If only 0/A policies are allowed, then this is a poor approximation to the integer optimum because as T gets large no holding or setup costs are incurred.

INTEGRATING THE SINGLE PRODUCT LOTSIZING MODEL INTO THE MULTI-PRODUCT IP MODEL

The difficulty with the multiproduct IP as formulated is that it requires the generation of all feasible schedules for each product. Dzielinski and Gomory (1965) discussed how algorithms for single product lotsizing could be exploited in the multiproduct situation so that only "interesting" single product schedules need be generated. We now discuss how this idea can be extended to the sequence dependent case.

When the problem is solved as an LP, let λ_{ut} be the dual price on the constraint:

$$\sum_{v} X_{uvt} + \sum_{p} \sum_{k \in Q(p)} F_{kut} Z_k + Y_{ut}$$

$$- \sum_{v} X_{vut} - \sum_{p} \sum_{k \in Q(p)} G_{kut} Z_k - Y_{ut-1} = 0.$$

The multiproduct problem is solved then essentially as follows.

0) Start with an initial arbitrary set of schedules such that there is at least one schedule for each product.

1) Solve the LP and get the dual prices λ_{ut}.

2) For each product p:
   2.1.) Add a rental surcharge $\hat{R}_{pt} = R_{pt} + \lambda_{ut}$, where u is the class of product p.

   2.2.) Use a one-product lotsizing algorithm on product p but using $\hat{R}_{pt}$ instead of R_{pt}.

   2.3.) Call the resulting schedule k, calculate U_k (using the actual rental charge R_{pt}) and add it to the LP.

   2.4.) Go to (1).

The iteration stops when none of the dual prices change in step 1.

REFERENCES

P. AFENTAKIS, B. GAVISH, U. KARMARKAR (1980) Exact solutions to the lot-sizing problem in multistage assembly systems. Technical Report, University of Rochester.

K.R. BAKER, P.S. DIXON, M.J. MAGAZINE, E.A. SILVER (1978) An algorithm for the dynamic lot-size problem with time-varying production capacity constraints. *Management Sci.* 24,1710-1720.

G.R. BITRAN, H.H. YANASSE (1981) Computational complexity of the capacitated lot size problem. Technical Report, Sloan School, M.I.T.

J. BLACKBURN, R. MILLEN (1979) Selecting a lot-sizing technique for a single-level assembly process: part I - analytical results. *Production and Inventory Management*, 3rd Quarter.

J. BLACKBURN, R. MILLEN (1979) Selecting a lot-sizing technique for a single-level assembly process: part II - empirical results. *Production and Inventory Management*, 4th Quarter.

W. CROWSTON, M. WAGNER, J. WILLIAMS (1973) Economic lotsize determination in multi-stage assembly systems. *Management Sci.*

C. DELPORTE, J. THOMAS (1977) Lotsizing and sequencing for N products on one facility. *Management Sci.* 23,1070-1079.

M.A.H. DEMPSTER, M.L. FISHER, L. JANSEN, B.J. LAGEWEG, J.K. LENSTRA, A.H.G. RINNOOY KAN (1981) Analytical evaluation of hierarchical planning systems. *Oper. Res.*, to appear.

B. DZIELINSKI, R. GOMORY (1965) Optimal programming of lot sizes, inventories and labor allocations. *Management Sci.* 11,874-890.

S. ELMAGHRABY (1978) The economic lot scheduling problem (ELSP): review and extensions. *Management Sci.* 24,587-598.

M. FLORIAN, M. KLEIN (1971) Deterministic production planning with concave costs and capacity constraints. *Management Sci.* 18, 12-20.

M. FLORIAN, J.K. LENSTRA, A.H.G. RINNOOY KAN (1980) Deterministic production planning: algorithms and complexity. *Management Sci.* 26,669-679.

A.M. GEOFFRION, G.W. GRAVES (1976) Scheduling parallel production lines with changeover costs: practical application of a quadratic assignment/LP approach. *Oper. Res.* 24,595-610.

S. GORENSTEIN (1970) Planning tire production. *Management Sci.* 17, 72-82.

R.H. KING, R.R. LOVE, JR. (1980) Coordinating decisions for increased profits. *Interfaces* 10,4-19.

M.R. LAMBRECHT, J. VANDER EECKEN (1978) A facilities in series capacity constrained dynamic lot-size model. *European J. Oper. Res.* 2,42-49.

M.R. LAMBRECHT, J. VANDER EECKEN (1978) Capacity constrained single-facility dynamic lot-size model. *European J. Oper. Res.* 2,1-8.

M.R. LAMBRECHT, H. VANDERVEKEN (1979) Heuristic procedures for the single operation, multi-item loading problem. *AIIE Trans.* 11,319-326.

M.R. LAMBRECHT, H. VANDERVEKEN (1979) Production scheduling and sequencing for multi-stage production systems. *OR Spektrum* 1,103-114.
L.S. LASDON, R.C. TERJUNG (1971) An efficient algorithm for multi-item scheduling. *Oper. Res.* 19.4.
S.F. LOVE (1972) A facilities in series inventory model with nested schedules. *Management Sci.* 18,327-338.
A.S. MANNE (1958) Programming of economic lot sizes. *Management Sci.* 4,115-135.
T. MORTON, J. MENSCHING (1978) A finite production rate dynamic lot size model. Working Paper 59-77-78, Graduate School of Industrial Administration, Carnegie-Mellon University.
L.G. SCHWARZ, L. SCHRAGE (1975) Optimal and system myopic policies for multi-echelon production/inventory assembly systems. *Management Sci.* 21.
S. SETHI, S. CHAND (1980) Multiple finite production rate dynamic lot size inventory models. Technical Report, Purdue University.
E.A. SILVER, H. MEAL (1973) A heuristic for selecting lot-size quantities for the case of a deterministic time varying demand rate and discrete opportunities for replenishment. *Production and Inventory Control Management*, 2nd Quarter, 64-74.
E.A. SILVER (1976) A simple method of determining order quantities in joint replenishments under deterministic demand. *Management Sci.* 22,1351-1361.
J.A.E.E. VAN NUNEN, J. WESSELS (1978) Multi-item lot size determination and scheduling under capacity constraints. *European J. Oper. Res.* 2,36-41.
H.M. WAGNER, T.M. WHITIN (1959) Dynamic version of the economic lot size model. *Management Sci.* 15,86-96.
W. ZANGWILL (1968) Minimum concave cost flows in certain networks. *Management Sci.* 14,429-450.

AN INTRODUCTION TO PROOF TECHNIQUES FOR BIN-PACKING APPROXIMATION ALGORITHMS

E. G. Coffman, Jr.

Bell Laboratories, Murray Hill, NJ

I. COMBINATORIAL MODELS

Introductory Remarks - Upon introducing a seemingly small change or generalization into either the model or an approximation algorithm for a previously studied bin packing problem, it has been common to find that very little of the analysis of the original problem can be exploited in analyzing the new problem. Such experiences may suggest that the mathematics of bin packing does not contain a central, well-structured theory that provides powerful, broadly applicable techniques for the analysis of approximation algorithms. It would be difficult to repudiate completely such an observation, but we hope to show that the extent to which it is true is largely inevitable. Specifically, our aim in this brief tutorial extension of the survey in [GJ] will be to explain somewhat informally certain techniques that do enjoy a moderately broad applicability, while making it clear where the novelty and perhaps ingenuity of the approach to an individual problem may be required.

For purposes of illustration we confine ourselves to approximation algorithms for three of the fundamental optimization problems: P1, minimizing the common capacity of a set of m bins required to pack a given list of pieces; P2, minimizing the number of unit capacity bins needed to pack a given list of pieces; and P3, minimizing the height of a unit-width rectangle needed to pack a collection of squares whose sides must be parallel to the bottom or sides of the enclosing

M. A. H. Dempster et al. (eds.), Deterministic and Stochastic Scheduling, 245–270.

rectangle. Although the set of techniques to be discussed, as well as the problem set chosen, are quite small, the literature will verify that they have been quite commonly applied with success, and therefore provide an important background for the newcomer to this class of problems.

We shall concentrate only on defining the structure of certain proofs; the details that we suppress are left to the interested reader, who may consult the appropriate references in [GJ]. In this connection, we should also point out that, with the exception of the NFD rule for P2, we do not examine the achievability of bounds or performance ratios. Our concern focuses simply on proofs that certain numbers or expressions provide upper bounds on performance ratios.

We shall use the notation defined in [GJ] with one minor simplification. We let $s_i = s(p_i)$ denote the size of p_i. Also, with respect to any given bin, we introduce the term level to refer to the sum of piece sizes in that bin.

The algorithms to be considered all pack pieces in sequence as they are drawn from some given list, L. For P1 we look at the Lowest-Fit (LF) rule whereby each piece is put into a bin currently of lowest level. If L is first put in order by decreasing piece size, then we have the LFD rule. For P2 we consider the Next-Fit (NF) and First-Fit (FF) rules. When packed by the NF rule, each piece is put into the same bin receiving the previous piece, if it will fit; if it does not fit (or if it is the very first piece), it is put into a new, empty bin. With FF each piece is packed into the first bin of the sequence $B_1, B_2, \ldots$ into which it will fit. As before, the NFD and FFD rules first put L in decreasing order of piece size. For P3 we analyze a bottom-up rule by which each successive square is placed left-justified at the lowest possible level in the current packing. Specifically, we examine the BUD (Bottom-Up-Decreasing) rule which first orders L by decreasing square size. Further details of these algorithms can be found in [GJ] and the references made therein.

In the sequel we discuss a general problem reduction approach, area arguments, and weighting functions, in that order. Before getting into the first of these a remark on induction proofs is appropriate. Proofs by induction on the number of pieces or bins are natural to consider in proving worst-case bounds. But they have been effective only for the simplest and least effective algorithms. Although induction can often be used to establish certain useful properties of approximate packings, such packings seldom have the inductive structure necessary for parlaying these properties into a proof of the desired bound. This absence of inductive structure is illustrated in [Gr] for one of the better algorithms for P2 by exhibiting lists of pieces which have proper sub-lists that in fact require more bins to pack by the FFD rule.

A Problem Reduction Approach [Gr] - We describe here an approach that is organized around proofs by contradiction. One begins with the assumption that there exists a list L that violates the desired bound, and in order to restrict the class of possible counter-examples, it is common also to assume that L is smallest in some sense, e.g. there is no list of fewer pieces that violates the bound. The next step is to exploit the structure of the approximation algorithm in order to derive further properties that any candidate counter-example must possess. The intent (or hope) is that the resulting class of such candidates is so small that it can be easily dealt with, e.g. in an essentially enumerative manner, showing individually for each such candidate that it cannot violate the bound. Of course, in this generality the approach is difficult to appreciate, so let us consider the following example.

We consider P1 and the proof of the absolute bound

$$\frac{C_{LFD}(L)}{C_{OPT}(L)} \le \frac{4}{3} - \frac{1}{3m} \qquad (1)$$

where $C_{LFD}(L)$ represents the minimum, common bin capacity needed to pack L into m bins by the LFD rule, and $C_{OPT}(L)$ is similarly defined for an optimization rule. Let the pieces of L be indexed so that $s_1 \ge s_2 \ge \ldots \ge s_n$.

Now suppose L is a smallest list (in the sense of $|L|$) violating the bound (1). Let s be the size of a piece, p, in the LFD packing whose top is at the highest level, i.e. the minimum bin capacity $C_{LFD}(L)$ required by the LFD rule with m bins. By definition of the LFD rule every bin must be occupied up to a level not less than $C_{LFD}(L) - s$, the level on which the bottom of p rests. But it is readily verified that, for this level, we have the upper bound $C_{LFD}(L) - s \leq \frac{1}{m} \sum_{s_i \neq s} s_i$, which is achieved only when all bins not containing p attain the same level. Thus,

$$C_{LFD}(L) = s + [C_{LFD}(L)-s] \leq s + \frac{1}{m} \sum_{s_i \neq s} s_i$$

and

$$C_{LFD}(L) \leq s\left(1 - \frac{1}{m}\right) + \frac{1}{m} \sum_{i=1}^{n} s_i . \tag{2}$$

We now observe that we must have $s = s_n$. For suppose not, i.e. suppose the top of some piece other than the smallest piece p_n achieves a highest level in the LFD packing. If we discard p_n from L, we are left with a list L′ that is easily seen to require the same capacity according to the LFD rule (the packing for L′ is obtained simply by removing p_n from the packing for L). Also, it is clear that L′ requires a capacity that is at most $C_{OPT}(L)$ when using an optimization rule. Obviously, (1) would still be violated with L′; hence $L' \subset L$ would contradict our assumption that L is a smallest counterexample.

Next, since no rule can do better than to fill all bins to the same level, we must have

$$C_{OPT}(L) \geq \frac{1}{m} \sum_{i=1}^{n} s_i . \tag{3}$$

From (2), violation of (1) implies

$$\frac{4}{3} - \frac{1}{3m} < \frac{C_{LFD}(L)}{C_{OPT}(L)} \leq \frac{s_n\left(1 - \frac{1}{m}\right)}{C_{OPT}(L)} + \frac{\frac{1}{m}\sum_{i=1}^{n} s_i}{C_{OPT}(L)}$$

so that by (3)

$$1 + \frac{1}{3}\left(1 - \frac{1}{m}\right) < \frac{s_n\left(1 - \frac{1}{m}\right)}{C_{OPT}(L)} + 1$$

and hence

$$C_{OPT}(L) < 3s_n.$$

Through this inequality we have achieved a major reduction in our problem, for it says that an optimum packing for a smallest counter-example packs at most two pieces per bin. At this point we can effectively examine all such packings to verify, through simple transformations, that the corresponding LFD packings are in fact optimal. This contradiction then proves the theorem.

Area Arguments - There is a strong intuitive appeal to proofs using area arguments, stemming from their obvious geometrical representations. Note that the concept of area is not inherent to one-dimensional bin packing; it comes about from the way in which packings are usually drawn (see the figures in [GJ] for example). The bins and pieces are endowed with another (horizontal) dimension that does not change the dimensionality of the problem, since it takes the same value for all bins and pieces.

The basic argument consists of finding a lower bound to that fraction of the total area which must be occupied by the pieces in packings produced by a given approximation algorithm, and an upper bound to the area occupied by these pieces in an optimum packing. From simple expressions of these areas in terms of the performance measure, one obtains from their ratio a bound on relative performance.

As an illustration consider P2 and a proof of the asymptotic bound $R_{NF}^{\infty} = 2$. Note that, informally we can associate R_{NF}^{∞} with the smallest possible multiplicative constant, α, in a bound of the form $NF(L) \leq \alpha OPT(L) + \beta$, where NF(L) and OPT(L) denote the numbers of bins used in packing L by the NF and an optimization rule, respectively, and α and β are constants. (See [GJ].)

The basic observation is that the cumulative size of the pieces in any two consecutive occupied NF bins,

B_i and B_{i+1}, must exceed 1. This follows simply from the fact that the first piece packed in B_{i+1} did not fit into the unused capacity of B_i. Thus, we see that $\sum_{i=1}^{n} s_i$ must constitute more than 1/2 the total "area" of the first NF(L) - 1 bins. An optimization rule can do no better than to pack each bin completely full, so we have

$$\mathrm{OPT}(L) \geq \sum_{i=1}^{n} s_i. \tag{4}$$

Therefore, $[\mathrm{NF}(L)-1]/2 \leq \sum_{i=1}^{n} s_i \leq \mathrm{OPT}(L)$ and $\mathrm{NF}(L) \leq 2\mathrm{OPT}(L) + 1$, from which $R_{NF}^{\infty} = 2$ follows. A slight tightening of (4) to $\mathrm{OPT}(L) \geq \left\lceil \sum_{i=1}^{n} s_i \right\rceil$ suffices to prove with very little effort that 2 is in fact an absolute bound for the NF rule.

It is readily seen that the above argument can also be used to prove a bound of 2 for any of the algorithms NFD, FF, FFD. Indeed, it is generally true that some sort of area argument is virtually always successful in proving upper bounds for the types of algorithms applied to P2; however, such an argument can seldom be expected to provide a proof of tight (achievable) bounds. Prospects for proving tight bounds are much improved by considering a more general approach known as the weighting function approach. We shall return to a discussion of this method after an illustration of area arguments applied to 2-dimensional bin-packing [BCR].

Area arguments are very common among proofs of bounds for two-dimensional packing problems. Often they seem to be the only effective technique available. To illustrate this application of area arguments we consider the proof of the absolute bound of 2 for the BUD rule for packing squares. An example is shown in Fig. 1 where we have normalized the bin (strip) width to 1.

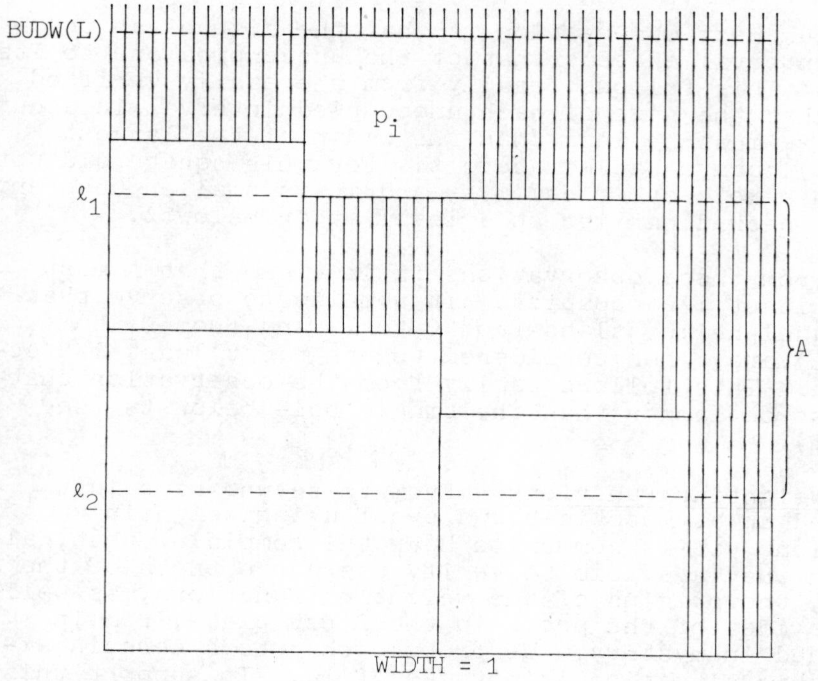

Figure 1 - Illustration of BUDW Square Packing

We begin by defining the horizontal cuts shown in the figure at levels ℓ_1 and ℓ_2. The first is drawn through the bottom of a top-most square (p_i) in the packing, and the second is drawn at a level $\ell_2 = s_i$, the height (and width) of p_i, and hence passes through the squares left-justified on the bottom of the bin. We want to show first that the area A between ℓ_1 and ℓ_2 must be at least 1/2 occupied. For this we consider an arbitrary horizontal cut at level ℓ through A. Clearly, such a cut can be partitioned into intervals $I_1, I_2, \ldots$ alternating between occupied and unoccupied regions of A.

Our first observation is that I_1 must correspond to an occupied region. This follows from the left-justification of squares and the consequent fact that I_1 must be defined by the width of some square whose bottom is at or below ℓ. Such a square cannot be wholly beneath ℓ, for some smaller square (e.g. p_i) packed later would subsequently have been packed into the hole defined by I_1.

Our next observation is that the size of an occupied interval exceeds that of the unoccupied one to its right. This follows readily from the easily verified fact that the size of each unoccupied interval in a cut below ℓ_1 must be less than the width of the largest (first) square packed above ℓ. For this square did not fit in a hole at or below ℓ, and it is smaller than any square packed earlier on a level at or below ℓ.

From these observations it is clear that A must be at least 1/2 occupied. It remains to observe that the equal-height slabs from 0 to ℓ_2 and from ℓ_1 to BUDW(L) are, when considered together, at least 1/2 occupied. This follows easily from the observation that p_i must be larger than the single hole below ℓ_2 (see Fig. 1).

Weighting Functions - From those who have just studied an asymptotic-bound proof using weighting functions, it is common to hear the complaint that, although they are able to verify the correctness of the proof, the origins of the weighting functions, as well as the idea of the proof in the first place, remain shrouded in mystery. We propose that to a certain extent this situation is not justified. To support this view we shall first introduce the method as a rather natural generalization of area arguments, and then discuss briefly the role and design of weighting functions using the NFD rule as a specific illustration.

We must concede at the outset, however, that we will be unable to completely eliminate the element of mystery from the design of weighting functions. But we will try to show how and why this is only to be expected. Although a variety of guidelines can be formulated, at present a successful search for weighting functions normally must rely at some point on ingenuity (or luck).

Also, our brief tutorial treatment runs the risk of considerable oversimplification. The number of problems we consider is very small, and we avoid the lengthier proofs altogether. Indeed, lengthy proofs seem to be more the rule than the exception. Furthermore, the frequently arduous case analyses accompanying the weighting function approach often contain subtle combinatorial arguments not touched upon in the present study. Thus, we emphasize that our goal concentrates on an informal explanation of the basis and use of weighting functions, without any serious attempt at formulating a completely general method.

We shall concentrate on problem P2 and, for purposes of discussion, let us suppose initially that we are considering a proof of an asymptotic bound α for approximation algorithm A. Modifying slightly our earlier definition, an area-argument approach requires first a proof that

$$\sum_{i=1}^{n} s_i \geq \frac{1}{\alpha} [A(L) - \beta] \qquad (5)$$

where $\beta > 0$ is a constant independent of L. Informally, within some bounded amount, the occupied area, $\sum_{i=1}^{n} s_i$, of any A-packing is at least $1/\alpha$ the total of A(L). On proving (5), we proceed as before to make use of the bound

$$\sum_{i=1}^{n} s_i \leq OPT(L) \qquad (6)$$

to obtain

$$A(L) \leq \alpha OPT(L) + \beta, \qquad (7)$$

thus completing the proof that α is an asymptotic bound.

To prove (5) one can begin as usual with an experimental study of A-packings, especially those for which the relative performance of A appears to be poor. Typically, in a non-trivial problem, one will encounter such packings having bins less than $1/\alpha$ full; clearly, if (5) is to hold the shortfall (the amount less than $1/\alpha$) in such bins has to be compensated in some way, or the cumulative shortfall has to be shown to be bounded by some constant. In our earlier NF example, recall that such bins were disposed of by showing that their levels were compensated by the levels of preceding bins. If a similar property applies to A-packings or if, for example, there can only be a bounded number of bins with levels less than $1/\alpha$, then the basic area argument will succeed. Unfortunately, observations of this simplicity are either unavailable or inadequate for the analysis of most of the more effective approximation rules.

For example, with the FF rule a little reflection will show that the major difficulties of an area argument will arise in trying to account for the cumulative shortfall in sequences of bins having single pieces whose sizes are in $\left[\frac{1}{2}, \frac{10}{17}\right)$; this is a shortfall that is not easily compensated for and one that cannot be bounded independently of the list length. Similarly, with the NFD rule, it is the lists that begin with sequences of pieces exceeding 1/2 in size that will correspond to the "hard part" of an area argument. For both these rules and most others, the failure of the simple area argument establishes a need for a more effective, general approach.

Such an approach is provided by the key observation that for (7) to hold we need not be restricted to the simple sum function in (5) and (6), so long as the inequalities can still be made to hold. In particular, it is natural to consider weighted sums $\sum w(s_i)s_i$, where w(s) is a weighting function of piece size; for with such functions we can adjust the weights (influence) of piece sizes in a way that reflects the difficulties encountered in an area argument. Clearly, if we choose $w(s) > 1$ over certain intervals, in general we must select $w(s) < 1$ over other intervals; hence we may be obtaining new difficulties in trade for the old ones of the area argument.

This can be illustrated simply by the weighting functions for the FF and NFD rules. In these cases, the approach is organized around proofs of the two inequalities [JDUGG],[BC]

$$\sum_{i=1}^{n} w(s_i)s_i \geq \frac{1}{\alpha}\,[A(L) - \beta] \tag{8}$$

and

$$\sum_{i=1}^{n} w(s_i)s_i \leq OPT(L) \tag{9}$$

where $\alpha_{FF} = \frac{17}{10}$, $\beta_{FF} = 1$, $\alpha_{NFD} = 1.691...$, and $\beta_{NFD} = 3$, using the obvious notation. The weighting functions selected in each case have the property that $w(s) \geq 1$ for $s > \frac{1}{2}$ and $w(s) \leq 1$ for $s \leq \frac{1}{2}$. In both cases, it is shown that with the given weighting function it is not possible for any algorithm to pack a bin with pieces whose cumulative weighted sum exceeds 1; it follows that for an optimization rule, (9) must hold. Next, an attempt is made to prove the first inequality by showing that $\sum_{i=1}^{n} w(s_i)s_i \geq \frac{1}{\alpha}$ for each bin packed by the approximation rule. In each case, there is no difficulty with the bins containing a piece in $\left(\frac{1}{2}, \frac{1}{\alpha}\right)$, which were troublesome in the area argument. But this has been traded for difficulties in dealing with bins containing certain other piece configurations whose weighted sum is less than one. However, the new problems are more tractable; specifically, it is possible with both weighting functions to show that the cumulative shortfall of the weighted sums in such bins is bounded by a constant: 1 for FF and 3 for NFD.

Although there is an apparent flexibility introduced by weighting functions, we cannot provide any systematic procedure for finding such functions that work. Indeed, there is no known a priori basis for determining whether <u>any</u> such function exists for a given bound. In searching for a weighting function one must be guided by the requirements of (8) and (9) and the specific structures of packings produced by the approximation algorithm. One attempts, in a trial-and-error approach, to use those piece configurations (partitions of the unit interval) that explain why a given function does not satisfy (8) or (9), as a basis for selecting a new, trial function.

We now illustrate our discussion of weighting functions with a more detailed outline of the proof of a tight asymptotic bound for the NFD rule. At the outset we have the critical problem of finding what appears to be a worst-case example. Again, there is no systematic, general procedure for finding such examples and some of them can be very difficult to find (see [JDUGG] for the worst-case examples of the FF rule). In general, one simply uses the structure of the algorithm to find a list that the approximation rule packs as poorly as possible (i.e. wastes as much bin capacity as

(a)
Optimum packing,
each bin having one piece of size
$\frac{1}{2} + \varepsilon, \frac{1}{3} + \varepsilon, \frac{1}{7} + \varepsilon, \ldots$

(b)
NFD packing

$\frac{1}{2} + \varepsilon \qquad\qquad 2 \times \left(\frac{1}{3} + \varepsilon\right) \qquad 6 \times \left(\frac{1}{7} + \varepsilon\right)$

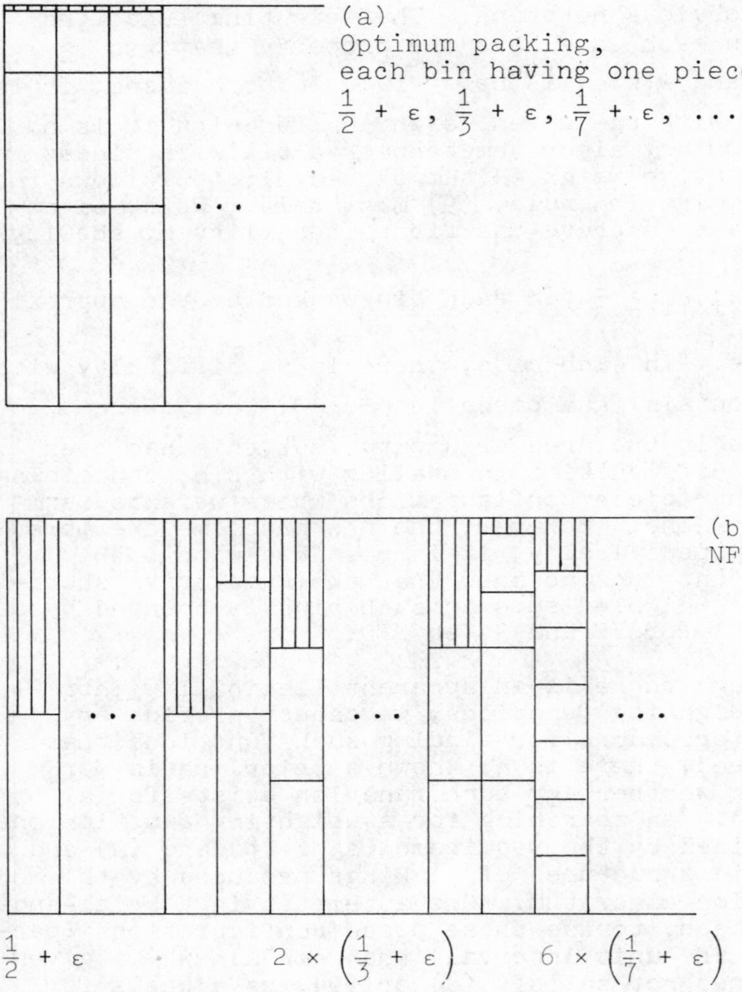

Figure 2 - Worst-Case for NFD

possible), whereas an optimum packing wastes little or
no space at all.

After such a study of the NFD rule one may arrive
at the example of Fig. 2. If one begins with an example performing not so badly as this one, there is
some hope (but certainly no assurance) that the precise
reasons why the subsequent attempt at an upper bound
proof fails will suggest the modifications needed to
obtain worse examples. Thus, in general the processes

of finding worst-cases and appropriate weighting functions may interact.

It is readily verified from the figure that in the limit $\varepsilon \to 0$, the performance ratio is $\gamma = 1 + \frac{1}{2} + \frac{1}{6} + \frac{1}{42} + \ldots \doteq 1.691\ldots$. Note that the sequence $\frac{1}{2}, \frac{1}{3}, \frac{1}{7}, \frac{1}{43}, \ldots$ is a "greedy" partition of $[0,1)$ restricted to unit fractions. That is, if $a_1, \ldots, a_i$ are the first i fractions, a_{i+1} is the largest unit fraction in the interval $\left(0, 1 - \sum_{j=1}^{i} a_j\right)$.

As with other weighting function proofs we now use a somewhat more convenient, normalized function, $W(s) = \gamma s w(s)$, with which we attempt to prove for some constant β the following normalization of (8) and (9):

$$\sum_{i=1}^{n} W(s_i) \geq NFD(L) - \beta \tag{10}$$

and

$$\sum_{i=1}^{n} W(s_i) \leq \gamma OPT(L). \tag{11}$$

Note that an area argument under this normalization corresponds to $W(s) = \gamma s$.

Next, consider the following preliminary guidelines as an aid in the design of an appropriate weighting function. Admittedly, they will not be definitive; in general, they may have to be augmented by considerable ingenuity.

1. First, recalling the difficulty encountered by an area argument, we try to simplify the proof of (10) by simplifying those cases where a bin has a single piece in the interval $\left(\frac{1}{2}, \frac{1}{\gamma}\right)$; for this, we seek a function that assigns to such pieces a weight such that

$$\frac{W(s)}{\gamma s} \geq 1 \quad \text{for} \quad s \in \left(\frac{1}{2}, \frac{1}{\gamma}\right).$$

2. Because of (11) the remainder of the function must compensate in other intervals by assigning weights less than γs. In view of the requirement for (10), we want to keep $\gamma s - W(s)$ sufficiently small, subject to the weights chosen for $s \in \left(\frac{1}{2}, \frac{1}{\gamma}\right)$. Of course, a very specific requirement is that the bin partitions in Fig. 2a each have a total weight not exceeding γ, and it is desirable that all bins in Fig. 2b have a total weight of at least 1.

3. Finally, it is obviously desirable to have a function with a simple, tractable form vis-a-vis the analysis supporting proofs of (10) and (11). For example, simpler arguments can usually be expected from functions with a finite or at least a denumerable range; failing this we can try for linear functions of piece size, and so on. In this respect there is an intuitive basis for exploiting the special role of unit fractions occurring in many bin-packing algorithms (see Fig. 2). For example, suppose bin B has k pieces and it is not the last non-empty bin. If the packing is done in an order of non-increasing piece size, then the (tight) lower bound on maximum piece size is given by the unit fraction $\frac{1}{k+1}$. (Selecting all k piece sizes to be just larger than $\frac{1}{k+1}$ thus maximizes the wasted space in the bin.) As a result, simplified proofs of (10) and (11) lead one to contrive functions $W(s)$ such that (i) the cumulative weight of k pieces of size $\frac{1}{k+1}$ is 1, for each $k \geq 1$, and (ii) either $W(s)$ or $W(s)/s$ does not distinguish between piece sizes that are contained in the same interval bounded by adjacent unit fractions, i.e. $\left(\frac{1}{k+1}, \frac{1}{k}\right]$ for some integer $k \geq 1$.

With these admittedly hazy ideas, and perhaps some preliminary trial-and-error, the following weighting function may emerge as a reasonable proposal, consistent with the guidelines

$$W(s) = \frac{1}{k}, \quad s \in \left(\frac{1}{k+1}, \frac{1}{k}\right], \quad k = 1,2,\ldots.$$

Happily, with this weighting function it is not difficult to verify (11); specifically, (11) follows from a proof that pieces just larger than $\frac{1}{2}, \frac{1}{3}, \frac{1}{7}, \frac{1}{43}, \ldots$ with weights $1, \frac{1}{2}, \frac{1}{6}, \frac{1}{42}, \ldots$ create that partition of the unit interval which maximizes the sum of weight. Moreover, in an attempted proof of (10) we see that successive bins containing k pieces, each having a size in $\left(\frac{1}{k+1}, \frac{1}{k}\right]$ have the desired sum of weights $k \frac{1}{k} = 1$. Unfortunately, the "mixed" bins of an arbitrary NFD packing (those having pieces from more than one interval $\left(\frac{1}{k+1}, \frac{1}{k}\right]$), can have sums of weights that are less than 1; and the problem of accounting for the cumulative shortfall in such bins seems to be no easier than the problem we faced in applying an area argument.

A conclusion from this trial is that we must increase the weighting function in some way. But we cannot increase the cumulative weight of the pieces in Fig. 2a, if (11) is to hold. So let us try to increase the weights of only those piece sizes not in the intervals $\left(\frac{1}{2}, 1\right], \left(\frac{1}{3}, \frac{1}{2}\right], \left(\frac{1}{7}, \frac{1}{6}\right], \ldots$, in a way that does not destroy the weight-maximizing property of the sequence in Fig. 2a.

One simple attempt at this produces the weighting function: For $s \in \left(\frac{1}{k+1}, \frac{1}{k}\right]$,

$$W(s) = \begin{cases} \frac{1}{k}, & k \in \{1,2,6,42,\ldots\}, \\ \frac{k+1}{k} s, & \text{otherwise}. \end{cases} \tag{12}$$

This happens to succeed, although the choice of $\frac{k+1}{k} s$ certainly cannot be claimed unique for this purpose. Inequality (11) continues to hold, and does so because the weight maximizing property of the partition in Fig. 2a obtains even with the increased weights given to piece sizes not in the intervals $\left(\frac{1}{2}, 1\right], \left(\frac{1}{3}, \frac{1}{2}\right], \left(\frac{1}{7}, \frac{1}{6}\right], \ldots$. In proving (10) it is still found that mixed bins can produce a shortfall, but it can now be shown without too much difficulty that the cumulative shortfall in such bins is bounded.

For practical reasons we have described a simplified weighting-function "search." In reality, the number of iterations can be expected to be painfully large, especially with approximation rules having a structure not so simple as that of NFD. Iteratively, the removal of difficulties (with certain piece configurations) based on one function tends to create other difficulties with a new function. Indeed, as indicated in [GJ], successful weighting functions have not been found for bounds thought to be tight for the performance of a number of approximation algorithms. In these cases, performance ratios have been adopted that are subject to a successful weighting function approach, but are larger than that achieved by any known example.

It should be pointed out in this regard that the weighting function approach has also been applied to problems other than P2. Among others, problem P1 and the problem of maximizing the number of pieces packed (see section 4 of [GJ]) have been successfully attacked using weighting functions.

II. PROBABILITY MODELS

Introductory Remarks - Although the worst-case bounds of a combinatorial analysis provide guarantees on performance, it is also very desirable to have measures of expected performance. In the next two sections we shall be analyzing a mathematical model in which piece sizes are assumed to be independent, identically distributed (i.i.d.) random variables whose probability distribution is given by $G(x)$. That is, using upper case for random variables, $G(x) = Pr\{S_i \leq x\}$, $1 \leq i \leq n$. The object of the analysis will be expected values as a function of $G(x)$ and the problem parameters.

The basis of our probability model is a Markov chain whose asymptotic analysis provides the desired results. The representation of packing processes by Markov chains is essentially limited to the on-line algorithms. The off-line algorithms, when circumstances permit their use, are more effective, but they involve orderings by piece size which destroy the crucial independence assumption.

In the next two sections we outline in some detail the analysis of the LF and NF rules for problems P1 and P2 defined earlier. As before we concentrate on the structure of proofs and the essential observations; the details suppressed are largely routine and can be found in the literature [CSHY],[CHS].

We shall also be considering lower bounds on the expected performance of optimization rules, so that a bound on the ratio of the expected values for approximation and optimization rules can be found. This comparison of expected values is certainly of interest, but so also is the expected value of the comparison (performance ratio). Unfortunately, no results on this substantially harder problem can be reported for the analysis of the LF and NF rules. Interesting results of this type, however, have been found for certain algorithms for the two-dimensional bin packing problem.

The LF Rule - Consider first the $m = 2$ bin case, and in the LF packing of L, let $h_i(L)$ denote the height or level of the top of p_i. $C_{LF}(L) = \max_{1 \le i \le n} \{h_i(L)\}$ denotes the minimum common capacity of the two bins required by the LF packing, and $C_{OPT}(L)$ denotes the corresponding minimum achievable capacity. We assume that $G(x)$ has a density g and finite first and second moments $E[S]$ and $E[S^2]$.

The study of minimum capacities reduces essentially to the analysis of the variations between adjacent piece heights in the two bins. Formally, in a packing produced by the LF rule for L in two bins, we define the variations, V_i, corresponding to p_i, $1 \le i \le n$, as

$$V_i = \begin{cases} S_i, & i = 1, \\ |V_{i-1} - S_i|, & 1 < i \le n. \end{cases} \tag{1}$$

Under the LF-rule, because of the recurrence in (1), it is obvious that the sequence $\{V_i\}$ forms a Markov chain with the distribution of the initial variation, V_1, given by $G(x)$. Thus, in terms of the transition kernel, $K(x,y)$, the distributions of successive variations are related by

$$F_{V_{i+1}}(y) = \int_0^\infty K(x,y)dF_{V_i}(x), \tag{2}$$

$$F_{V_1}(y) = G(y).$$

Based on the LF-rule we have $K(x,y) = \Pr[V_{i+1} \le y | V_i = x] = \Pr[|x-S_{i+1}| \le y]$ and hence

$$K(x,y) = G(x+y) - G(x-y). \tag{3}$$

Let V be a random variable having the stationary distribution $F_V(x) = \lim_{n\to\infty} F_{V_N}(x)$, assuming this limit exists.

<u>Theorem 1</u> [Fe]. Suppose $g(x)$ is strictly positive and continuous. Then F_V exists and has the density

$$f_V(y) = \frac{1}{E[S]}[1 - G(y)]. \tag{4}$$

<u>Proof</u>. The density f_V is found by taking the limit of (2) expressed in terms of densities. On substituting the density corresponding to (3) we obtain

$$f_V(y) = \int_0^\infty f_V(x+y)g(x)dx + \int_0^\infty f_V(x)g(x+y)dx. \tag{5}$$

Differentiating (5) we obtain a differential equation that is routinely solved. Further technical details can be found in [Fe]. □

For the minimum capacity we have routinely $C_{LF}(L) = \frac{1}{2}\sum_{i=1}^n S_i + V_n/2$ and hence, taking the expectation over all n-element lists, we have

$$E[C_{LF}(L)] = \frac{n}{2}E[S] + E[V_n]/2. \tag{6}$$

Of course, for arbitrary $G(x)$, $E[V_n]$ may be difficult to calculate. We shall consider an approximation of $E[V_n]$ after a brief discussion of the general case $m \ge 2$.

For arbitrary $m \geq 2$, consider the level, ℓ_i; on which p_i rests in the LF packing of L, and let $Z_j(i)$, $1 \leq j \leq m-1$, be the residual piece heights in the m-1 bins to which p_i is not assigned. We have

$$C_{LF}(L) = \ell_n + \max\left\{S_n, \max_j\{Z_j(n)\}\right\} \tag{7}$$

where

$$\ell_n = \frac{1}{m}\left[\sum_{i=1}^{n-1} S_i - \sum_{j=1}^{m-1} Z_j(n)\right]. \tag{8}$$

Although for fixed n, the $Z_j(n)$ are not i.i.d. random variables, it is readily seen that they become so in the limit $n\to\infty$, and their distribution converges to $F_V(x)$ as given in the $m = 2$ case. (Note that the piece height sequences in the m bins are asymptotically identical and independent renewal processes.) Thus, from (7) and (8) we can work out an approximation for $E[C_{LF}(L)]$ by treating the $Z_j(n)$ as i.i.d. random variables with the density in (4).

The quality of this approximation depends upon the rate of convergence to $F_V(x)$, which in turn will depend in general on $G(x)$. In the next section we shall illustrate a method of assessing this approximation which exploits the property of <u>geometric</u> convergence. Specifically, $\{V_i\}$ is said to converge geometrically if there exist positive constants a and b, $b < 1$, such that for all n sufficiently large

$$|K^{(n)}(x,y) - F_V(y)| \leq ab^n \tag{9}$$

where $K^{(n)}(x,y)$ is the n-step transition kernel.

Finally, expressions for $E[C_{LF}(L)]$ can be compared to the lower bound

$$E[C_{OPT}(L)] \geq \sum_{i=1}^{n} E[S_i] = \frac{nE[S]}{m} \tag{10}$$

to obtain bounds on the expected LF minimum capacities relative to the expected optima.

The NF Rule – The bin capacities are now fixed at 1, and the performance measure is the number of bins used. We proceed essentially as before. However, because of the greater difficulty in analyzing finite packings, we begin by considering infinite packing sequences. Later, we discuss the representation of finite problems as prefixes of infinite packings.

The sizes of the pieces to be packed are specified in an infinite sequence $\langle S_i, i \geq 1 \rangle$ of i.i.d. random variables, again with distribution $G(x)$. Our concern now focuses directly on bin-level sequences* $\langle L_i, i \geq 1 \rangle$, rather than on the level variations, produced by the NF rule. It follows easily from the structure of this rule, that the level sequence constitutes a Markov chain described by

$$F_{L_{i+1}}(y) = \int_0^1 Q(x,y)dF_{L_i}(x), \quad i \geq 1,$$

where $F_{L_1}(y) = Q(1,y)$, and $Q(x,y)$ is the one-step transition probability.

For the calculation of $Q(x,y)$ the following notation is convenient. Let W_i be the size of the first piece packed in bin B_i. Let X_n denote the sum of $n \geq 0$ independent, identically distributed piece-sizes, where X_o takes on the value zero only. By definition of the Next-Fit rule we must have $W_{i+1} > 1 - L_i$, $i \geq 1$. Hence $Q(x,y) = 0$ for all $y \leq 1-x$. Now $L_{i+1} \leq y$ if and only if for some $n \geq 0$, the sum of W_{i+1} and the next n piece sizes is no greater than y, but the sum of W_{i+1} and the next n+1 piece-sizes exceeds 1. Hence, using the Markov property of $\{L_i\}$ and a complete set of events,

$$Q(x,y) = \sum_{n=0}^{\infty} \Pr\{W_{i+1} + X_n \leq y,\ W_{i+1} + X_n + S > 1 | L_i = x\}. \tag{11}$$

---

*We retain the notation of [CSHY], in spite of its partial conflict with our list notation, because all other choices seem to have worse disadvantages.

The conditional distribution of W_{i+1} given that $L_i = x$ is simply $[G(w) - G(1-x)]/[1 - G(1-x)]$, for $1-x < w$, and 0 otherwise. Thus (11) can be expressed as

$$Q(x,y) = \sum_{n=0}^{\infty} \int_{1-x}^{y} \Pr\{X_n \leq y-w,\ S > 1-w - X_n\} \frac{dG(w)}{1-G(1-x)}.$$

Finally, therefore, using the independence of successive S_i, we get

$$Q(x,y) = \sum_{n=0}^{\infty} \int_{1-x}^{y} \int_{0}^{y-w} \frac{1 - G(1-w-s)}{1 - G(1-x)}\, dF_{X_n}(s)\,dG(w), \quad y > 1-x$$

$$= 0, \quad y \leq 1-x. \tag{12}$$

Later, using a uniform distribution for $G(x)$, we shall present closed form results.

Note that in the limit $x \to 0$, $Q(x,y)$ degenerates to a distribution concentrated at $y = 1$. Also, although $Q(x,y)$ for $x > 0$ will generally be a continuous function of y in our application, this is not required for all that follows, and in particular, it will not possess a continuous density.

Now let us continue to suppose $G(x)$ has a density $g(x)$ that is strictly positive on $[0,1]$. Under this assumption, a number of qualitative properties of the chain $\{L_i\}$ can be established.

1. For each x, $Q(x,y)$ must clearly be a monotonically increasing function of y in the interval $[1-x,1]$. Moreover, the n-step transition probability $Q^{(n)}(x,y)$, defined by

$$F_{L_{i+n}}(y) = \int_{0}^{1} Q^{(n)}(x,y)\,dF_{L_i}(x), \tag{13}$$

is easily shown to increase monotonically with y over the entire interval $[0,1]$.

2. Since $L_i \le \frac{1}{2}$ implies $W_{i+1} > \frac{1}{2}$ and hence $L_{i+1} > \frac{1}{2}$, we have $E\left[L_{i+1} | L_i \le \frac{1}{2}\right] > \frac{1}{2}$. Moreover, for each x there exists an $\frac{1}{2} < \alpha < 1$ such that for all $x > \alpha$ we have $E[L_{i+1}|L_i = x] < x$. Thus, the process has a tendency to move from points near the boundaries towards the interior. Also, since $L_i + L_{i+1} > 1$ for all $i \ge 1$, it is easy to see that $E[L_i] > \frac{1}{2}$ for all $i \ge 1$.

It follows directly from the above properties [Tw] that $\{L_i\}$ is an ergodic Markov chain having a strictly positive stationary probability distribution $F_L(y) = \lim_{i\to\infty} F_{L^\cdot}(y)$ satisfying

$$F_L(y) = \int_0^1 Q(x,y)dF_L(x) = \lim_{n\to\infty} Q^{(n)}(x,y). \qquad (14)$$

An even stronger property of $\{L_i\}$ can be verified.

<u>Theorem 2</u>. $\{L_i\}$ converges geometrically fast.

This result is proved [see Lo] by examining the Markov measure $\Delta_n(x,y,z) = |Q^{(n)}(x,z) - Q^{(n)}(y,z)|$, and showing that for some $n \ge 1$, $\Delta_n = \sup_{x,y} \sup_z \Delta_n(x,y,z) < 1$. In the present case it is readily verified that in fact $\Delta_1 = \Delta_1(1,0,1) = 1$, but that $\Delta_n < 1$ for all $n \ge 2$.

Using this result we can characterize the expected efficiency of Next-Fit packings, relative to the best achievable. To this end we consider first the expected cumulative size of the pieces packed in the prefix $B_1,\ldots,B_m$, for arbitrary m. The following result shows that this differs by at most a constant from the value obtained when approximating $E(L_i)$ by $\overline{L} = \lim_{i\to\infty} E(L_i)$.

<u>Theorem 3</u>. There exists a constant γ such that for all m

$$\left| m\bar{L} - \sum_{i=1}^{m} E[L_i] \right| < \gamma. \tag{15}$$

Proof. We begin by writing

$$\sum_{i=1}^{m} E[L_i] = m\bar{L} + \sum_{i=1}^{m} (E[L_i]-\bar{L}).$$

Therefore,

$$\left| m\bar{L} - \sum_{i=1}^{m} E[L_i] \right| \le \sum_{i=1}^{m} |E[L_i]-\bar{L}|. \tag{16}$$

We complete the proof by showing that the sum $\gamma' = \sum_{i=1}^{\infty} |E[L_i]-\bar{L}|$ exists. Since the L_i are non-negative random variables we have

$$E[L_i] = 1 - \int_0^1 F_{L_i}(y)dy.$$

Consequently, γ' can be expressed as

$$\gamma' = \sum_{i=1}^{\infty} \left| \int_0^1 [F_{L_i}(y) - F_L(y)]dy \right|$$

from which, using (13), we derive

$$\gamma' \le \sum_{i=1}^{\infty} \int_0^1 \int_0^1 |Q^{(i-1)}(x,y) - F_L(y)| dF_1(x)dy \tag{17}$$

where $Q^{(0)}(x,y)$ may be written using the Heaviside theta function as $\theta(y-x)$. From the definition of Δ_h it is not difficult to verify (see [Lo] for example) for some fixed $h \ge 1$

$$\left| Q^{(n)}(x,y) - F_L(y) \right| \leq \Delta_h^{\lfloor n/h \rfloor}.$$

Substituting into (17) we have

$$\gamma' \leq \sum_{i=0}^{\infty} \Delta_n^{\lfloor i/h \rfloor}. \tag{18}$$

Convergence of this sum for any $h \geq 2$ follows directly from Theorem 2. □

Extension of the results to the more general case where $g(x)$ is strictly positive with support $[0,a]$, $a \leq 1$, is easily managed. Note that in this case the state space reduces to $[1-a,1]$, so that the earlier characterizations of $\{L_i\}$ must be specialized accordingly.

Important, specific instances of our problem occur when the piece sizes are assumed to be uniformly distributed over the interval $[0,a]$, $a \leq 1$. Although numerical solutions for general a can be worked out in principle, closed-form expressions for measures of interest are not generally possible. Such expressions are available for the case $a = 1$, however, which we shall now present. It is simple to argue that smaller values of a will yield very similar results, except that convergence may be even faster.

Theorem 4. For piece sizes uniformly distributed over $[0,1]$ we have the kernel

$$\begin{aligned} Q(x,y) &= 1 - (1-y)e^{-(1-y)\left(\frac{e^x}{x}\right)}, \quad 1-x < y \leq 1, \\ &= 0, \quad 0 \leq y \leq 1-x, \end{aligned} \tag{19}$$

with the invariant measure (stationary distribution) and expectation

$$F_L(y) = y^3, \quad 0 \leq y \leq 1, \qquad \bar{L} = \frac{3}{4}. \tag{20}$$

Also,

$$m \leq \frac{4}{3} \sum_{i=1}^{m} E[L_i] + 3. \tag{21}$$

Proof. We omit the largely routine, but occasionally heavy calculation producing (19) and (20). As before, (20) is obtained by substituting (19) into (14) and transforming the integral equation into an easily solved differential equation.

To prove (21) we use Theorem 3 with $\bar{L} = \frac{3}{4}$; to verify the additive constant we calculate Δ_2 from

$Q^{(2)}(x,z) = \int_0^1 Q(y,z)Q(x,dy)$. Omitting the details we find $\Delta_2 \leq 3/10$. Next, from (16) and (18) we obtain

$\gamma \geq \sum_{i=0}^{\infty} \Delta_2^{\lfloor n/2 \rfloor} = 2(1-\Delta_2)$. Since $\Delta_2 < 3/10$, $\gamma = 3$ suffices and (21) follows. □

In comparing the performance of NF packings for finite problems with that of optimum packings, Theorem 4 gives rise to a simple approach, avoiding difficulties in perhaps more direct approaches. One uses the fast rate of convergence of $E[L_i]$* to estimate the ratio for even moderately large lists by the asymptotic bound 4/3. Further details on this estimate are given in [CSHY].

REFERENCES

[BC] Baker, B. S. and E. G. Coffman, Jr., "A Tight Asymptotic Bound on Next-Fit-Decreasing Bin-Packing," SIAM J. on Algebraic and Discrete Methods, 2, 2(1981), 147-152.

---

*Numerically, it has been shown that the first five expected bin levels are .71828, .75797, .74846, .75035 and .749935, respectively.

[BCR] Baker, B. S., E. G. Coffman, Jr., and R. L. Rivest, "Orthogonal Packings in Two Dimensions," SIAM J. on Computing, 9, 4(1980), 846-855.

[CHS] Coffman, E. G., Jr., M. Hofri, and K. So, "On the Expected Performance of Schedules of Independent Tasks," in preparation.

[CSHY] Coffman, E. G., Jr., K. So, M. Hofri, and A. C. Yao, "A Stochastic Model of Bin-Packing," Information and Control, 44, 2(1980), 105-115.

[Fe] Feller, W., An Introduction to Probability Theory and Its Applications, Vol. II, John Wiley & Sons, Inc., New York, 1966.

[GJ] Garey, M. R. and Johnson, D. S., "Approximation Algorithms for Bin Packing Problems: A Survey," in Analysis and Design of Algorithms in Combinatorial Optimization (G. Ausiello and M. Lucertini, eds.), Springer-Verlag, New York, 1981, pp. 147-172.

[Gr] Graham, R. L., "Bounds on Performance of Scheduling Algorithms," in Computer and Job-Shop Scheduling Theory, E. G. Coffman, Jr. (ed.), John Wiley & Sons, Inc., New York, 1976.

[JDUGG] Johnson, D.S., Demers, A., Ullman, J. D., Garey, M. R., and Graham, R. L., "Worst-Case Performance Bounds for Simple One-Dimensional Packing Algorithms," SIAM J. on Computing, 3 (1974), 299-325.

[Lo] Loeve, M. Probability Theory, Van Nostrand, Princeton, N.J., 3rd Ed., 1963.

[Tw] Tweedie, R. L., "Sufficient Conditions for Ergodicity and Recurrence of Markov Chains on a General State Space," Stochastic Processes and Their Appl., Vol. 3 (1975), pp. 385-403.

A STOCHASTIC APPROACH TO HIERARCHICAL PLANNING AND SCHEDULING

M.A.H. Dempster

Dalhousie University, Halifax
Balliol College, Oxford

This paper surveys recent results for stochastic discrete programming models of hierarchical planning problems. Practical problems of this nature typically involve a sequence of decisions over time at an increasing level of detail and with increasingly accurate information. These may be modelled by multistage stochastic programmes whose lower levels (later stages) are stochastic versions of familiar NP-hard deterministic combinatorial optimization problems and hence require the use of approximations and heuristics for near-optimal solution. After a brief survey of distributional assumptions on processing times under which SEPT and LEPT policies remain optimal for m-machine scheduling problems, results are presented for various 2-level scheduling problems in which the first stage concerns the *acquisition* (or assignment) of machines. For example, heuristics which are asymptotically optimal in expectation as the number of jobs in the system increases are analyzed for problems whose second stages are either *identical* or *uniform* m-machine scheduling problems. A 3-level location, distribution and routing model in the plane is also discussed.

1. INTRODUCTION

Practical hierarchical planning problems typically involve a sequence of decisions over time at an increasing level of detail and with increasingly accurate information. For example, a 3-level hierarchy of planning decisions in terms of increasingly finer time units is often utilized for manufacturing operations (see Figure 1). The first level concerns medium term planning, which works with projected quarterly or monthly averages and is primarily concerned with the acquisition of certain resources. The next level treats

M. A. H. Dempster et al. (eds.), Deterministic and Stochastic Scheduling, 271–296.

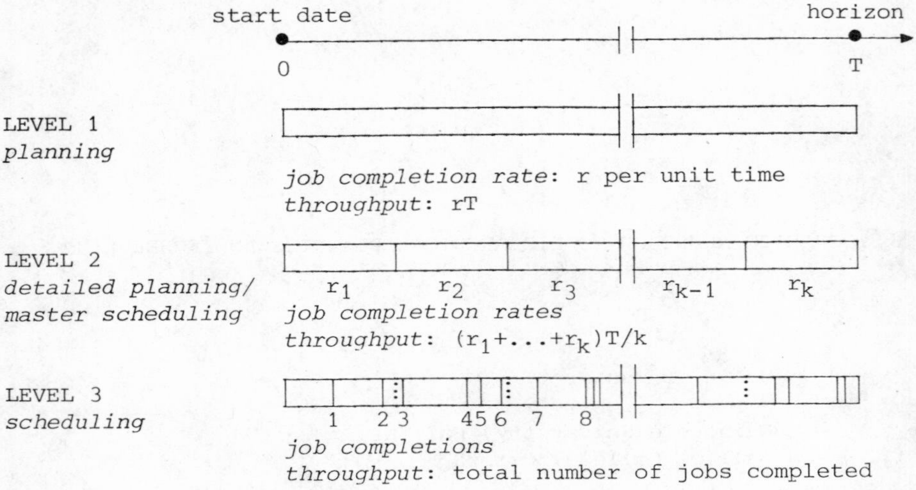

Figure 1. 3-Level scheme for hierarchical production planning/ scheduling.

weekly production scheduling, while the third level is concerned with the real-time sequencing of jobs through various machine centres on the shop floor. The first two levels can currently be handled adequately by respectively deterministic linear programming and combinatorial permutation procedures, but the third realistically involves a network of stochastic m-machine scheduling problems whose natural setting is in continuous time.

More generally, many hierarchical planning problems can be modelled by multistage stochastic programmes whose later stages (lower levels) are stochastic versions of familiar NP-hard deterministic combinatorial optimization problems. Hence they usually require the use of approximations and heuristics for near-optimal solution. For these systems, in which at the higher levels - as in their practical counterparts - details are suppressed and instead replaced by *approximate* aggregates, one would hope to demonstrate that the instances of the data for which these higher level assumptions are *severely* violated occur with increasingly negligible probability as the number of tasks in the system becomes large. Asymptotic probabilistic analysis of heuristics has therefore an important role to play in the analysis of hierarchical stochastic programming models (*cf.* [Dempster *et al.* 1981A]).

Recently, computer-based planning systems have become popular for practical multilevel decision problems in a variety of applications including manufacturing production planning and scheduling, trade training school planning and scheduling, distribution planning and vehicle scheduling, manpower planning, crew routing and scheduling and computer utilization and scheduling (see, for example, [Dempster *et al.* 1981A; Dempster & Whittington 1976; Dirickx & Jennergren 1979; Giannessi & Nicoletti 1979; Kao & Queyranne 1981; Kleinrock 1976]). In principle, the performance of such systems can be evaluated relative to optimality for the appropriate multistage stochastic programming model.

This paper primarily reports on a programme of research conducted jointly with M.L. Fisher, B.J. Lageweg, J.K. Lenstra, A.H.G. Rinnooy Kan and L. Stougie.

Section 2 contains a brief survey of distributional assumptions on processing times under which *shortest expected processing time* (SEPT) and *longest expected processing time* (LEPT) policies remain optimal for m-machine scheduling problems with appropriate expected value criteria. Section 3 sets out various 2-level scheduling problems as 2-stage stochastic programmes with recourse in which the first stage concerns the *acquisition* (or assignment) of machines and the second stage is an m-machine scheduling problem. The difficulty of exact solution of such problems is also discussed in §3 and representative results are quoted. In §4 results are presented which analyze heuristics for the 2-level scheduling problems in §3. For example, heuristics which are almost surely asymptotically optimal as the number of jobs in the system increases are analyzed for problems whose second stages are either *identical* or *uniform* m-machine scheduling problems. A 3-level location, distribution and routing model in the plane is discussed in §5 and some conclusions drawn in §6. Open problems and directions for further research are indicated throughout the paper.

2. RECENT RESULTS IN PARALLEL MACHINE STOCHASTIC SCHEDULING

This section surveys recent results for the following basic m-machine scheduling problem.

Problem 2.1. Schedule n *jobs* $j \in J := \{1,\ldots,n\}$ with independent random *processing requirements* $\underset{\sim}{p}_j$ on m *uniform machines* $i \in \mathcal{M} := \{1,\ldots,m\}$ with *speeds* s_i, subject to the usual constraints that at any moment at most one job can be processed by any machine and at most one machine can process any job.

We may think of the processing requirements as defined relative to standard time units so that if job j is assigned for processing solely to machine i it will be completed in the random (clock) *time* $\underset{\sim}{p}_j/s_i$. The *machine set* $\mathcal{M}$ will be assumed to be ordered in decreasing order of speed so that $s_1 \geq s_2 \geq \ldots \geq s_m \geq 0$. When $s_i \equiv 1$, we speak of m *identical* machines.

In order to complete the definition of Problem 2.1 various alternative assumptions may be made. These concern the nature of the *time* set, the possibility of *preemption* of running jobs and the possibility and nature of *release dates* or *arrivals* of some of the jobs *after* time zero.

The time set for the problem may be either discrete ($\mathbb{N}$) or continuous ($\mathbb{R}_+$). Scheduling problems are most naturally set in *continuous time* (CT); usually a *discrete time* (DT) setting is generated by a discrete time step, for the purposes of approximation or simplification. It is usually not an entirely trivial or even straightforward matter to extend discrete time results to their continuous time analogues.

Although finer classifications are possible (see, *e.g.*, [Pinedo & Schrage 1982]) we shall be interested simply in whether or not any job currently being processed may, at any point, be interrupted and set aside for later processing or immediate assignment to a different machine. In the affirmative situation we say *preemption* is allowed, and otherwise we say the problem allows *no preemption*.

In case all n jobs are available for processing at time zero we say that *no arrivals* are allowed. If some, say ℓ $(0 < \ell \le n)$, of the n jobs are not available for processing at time zero and are released at subsequent random times $\underset{\sim}{r}_j$ according to some (possibly labelled) stochastic point process independent of the processing requirements we speak of a problem with *arrivals*. In this case we condition the problem on the occurrence of exactly ℓ events of the arrival process.

We shall be interested in constructing schedules which are "optimal" in terms of two schedule measures. Let $\underset{\sim}{C}_j$ denote the (random) *completion time* of job $j \in J$ under a given scheduling policy. Then the schedule *makespan* $\underset{\sim}{C}_{max}$ is the earliest time at which all jobs are completed, defined by

$$\underset{\sim}{C}_{max} := \max_{j \in J} \{\underset{\sim}{C}_j\},$$

while the schedule *flowtime* $\Sigma \underset{\sim}{C}_j$ is the sum of the job completion times given by

$$\Sigma \underset{\sim}{C}_j := \Sigma_{j=1}^{n} \underset{\sim}{C}_j .$$

In the deterministic case, minimization of makespan optimizes the completion time of the *last* job, while minimization of flowtime is equivalent to optimizing the completion time of the *average* job. In the stochastic case, a natural schedule minimization criterion is in terms of the *expected value* of makespan or flowtime. For these two measures we shall also be interested in minimality in *distribution*, *i.e.* the probability of achieving a given makespan or flowtime level γ is, uniformly in γ, at least as great for the schedule resulting from the optimal policy π^o as for any other π. For example, for all $\gamma \in \mathbb{R}_+$,

$$P\{\underset{\sim}{C}_{max}^{\pi^o} \le \gamma\} \ge P\{\underset{\sim}{C}_{max}^{\pi} \le \gamma\}, \text{ i.e.}$$

$$\bar{F}_{\underset{\sim}{C}_{max}^{\pi^o}}(\gamma) \le \bar{F}_{\underset{\sim}{C}_{max}^{\pi}}(\gamma), \tag{2.1}$$

where $\bar{F}_{\underset{\sim}{C}_{max}} := 1 - F_{\underset{\sim}{C}_{max}}$ denotes the *survivor function* of $\underset{\sim}{C}_{max}$ under the appropriate policy. If $\tilde{x}$ and $\tilde{y}$ are two random variables, then $\tilde{x}$ is *stochastically dominated* by $\tilde{y}$, written $\tilde{x} <_D \tilde{y}$, if, and only if, $\bar{F}_{\tilde{x}}(t) \leq \bar{F}_{\tilde{y}}(t)$ for all t. It is easy to see that $\tilde{x} <_D \tilde{y}$ implies $E\tilde{x} \leq E\tilde{y}$ (when the expectations exist), but not conversely.

Choosing one of the possibilities discussed above regarding the time set, preemption and job arrivals - and specifying an optimality criterion - generates from the basic Problem 2.1 a stochastic scheduling problem. Since we shall allow scheduling decisions at t = 0 and at the epochs of subsequent job arrivals and completions, the resulting stochastic scheduling problems can be formulated as *semi-Markov* decision problems over an infinite horizon (see, *e.g.*, [Ross 1970, Ch.4]). In this section we are interested in conditions on the processing requirement distributions under which optimal scheduling policies for these problems can be specified in a simple form which utilizes dynamic *priority indices* (*cf.* Gittins' "dynamic allocation" indices [Gittins 1979]). At any moment these policies assign to each unfinished job a number - its priority index - and at decision epochs unfinished jobs are assigned to (speed ordered) free machines in monotonic order of their current indices. (When preemptions are allowed all m machines are considered free at job completion epochs.) Policies of similar form have recently been found applicable to a large class of related *Markov* decision problems in discrete time including 1-machine scheduling, search problems and multiarmed bandit and superbandit processes (see [Presman & Sonin 1979; Gittins 1979; Whittle 1980; Nash 1980]).

More formally, let U_t ($\subset J$) denote the set of *unfinished jobs* at time $t \geq 0$. Then a *priority policy* π defines for each decision epoch t a permutation of the elements of U_t,

$$\pi_t : U_t \xrightarrow{1-1} U_t, \; j \longrightarrow \pi_t(j),$$

and assigns jobs to free machines in (speed) order according to their permutation (priority) order. Let $p_j(t)$ be the amount of processing already received at time t by job $j \in U_t$ with processing requirement $\tilde{p}_j$ and denote by $\mu_j(t) := E\{\tilde{p}_j | \tilde{p}_j > p_j(t)\}$ the expectation of the processing requirement remaining at time t for job j. The *longest expected processing time* (LEPT) policy is the priority policy which at a decision epoch t reorders jobs in *decreasing* order of $\mu_j(t)$. The *shortest expected processing time* (SEPT) policy is the priority policy which at a decision epoch t reorders jobs in *increasing* order of $\mu_j(t)$.

Table 1 sets out currently known results concerning the optimality of LEPT and SEPT policies for the variants of Problem 2.1 resulting from the possible alternatives cited above. Since, as previously mentioned, makespan criteria concern minimizing the completion time of the *last* job, it is intuitively obvious that potentially long jobs should be processed first and hence LEPT is a candidate for optimizing makespan. Similarly, since flowtime criteria require the minimization of the completion time of the *average* job, potentially short jobs should be processed first and

LEPT/Makespan $\underset{\sim}{C}_{max}$		SEPT/Flowtime $\Sigma\underset{\sim}{C}_j$	
Deterministic			
$P\|$*pmtn*$\|C_{max}$ (LPT)	(see text)	$P\|\|\Sigma C_j$ (SPT)	[Conway *et al.* 1967, Ch.4.4]
Exponential			
$P\|\|E\underset{\sim}{C}_{max}$ $Q\|$*pmtn*$\|E\underset{\sim}{C}_{max}$	[Weiss & Pinedo 1980]	$P\|\|E\Sigma\underset{\sim}{C}_j$ $Q\|$*pmtn*$\|E\Sigma\underset{\sim}{C}_j$	[Weiss & Pinedo 1980]
$P\|\|\bar{F}_{\underset{\sim}{C}_{max}}$	?	$P\|\|\bar{F}_{\Sigma\underset{\sim}{C}_j}$	?
$P\|$*pmtn*$,\underset{\sim}{r}_j\|\bar{F}_{\underset{\sim}{C}_{max}}$	?	$P\|$*pmtn*$,\underset{\sim}{r}_j\|\bar{F}_{\Sigma\underset{\sim}{C}_j}$	?
Geometric			
$P\|\|E\underset{\sim}{C}_{max}$	?	$P\|\|E\Sigma\underset{\sim}{C}_j$	[Gittins 1981]
$Q\|$*pmtn*$\|E\underset{\sim}{C}_{max}$	?	$Q\|$*pmtn*$\|E\Sigma\underset{\sim}{C}_j$	?
$P\|\|\bar{F}_{\underset{\sim}{C}_{max}}$	?	$P\|\|\bar{F}_{\Sigma\underset{\sim}{C}_j}$	?
$P\|$*pmtn*$,\underset{\sim}{r}_j\|\bar{F}_{\underset{\sim}{C}_{max}}$	?	$P\|$*pmtn*$,\underset{\sim}{r}_j\|\bar{F}_{\Sigma\underset{\sim}{C}_j}$	?
Log Convex Similar			
$P\|$*pmtn*$\|\bar{F}_{\underset{\sim}{C}_{max}}$ $P\|$*pmtn*$,\underset{\sim}{r}_j\|\bar{F}_{\underset{\sim}{C}_{max}}$	DT [Weber 1979] CT [Weber 1981]	$P\|\|\bar{F}_{\Sigma\underset{\sim}{C}_j}$	DT [Weber 1979] CT [Weber 1981]
ICR *Similar*			
$P\|$*pmtn*$\|E\underset{\sim}{C}_{max}$	DT [Weber 1979] CT [Weber 1981]	$P\|\|E\Sigma\underset{\sim}{C}_j$	DT [Weber 1979] CT [Weber 1981]
Log Concave Similar			
$P\|\|\bar{F}_{\underset{\sim}{C}_{max}}$ $P\|$*pmtn*$,\underset{\sim}{r}_j\|\bar{F}_{\underset{\sim}{C}_{max}}$	DT [Weber 1979] CT [Weber 1981]	$P\|$*pmtn*$\|\bar{F}_{\Sigma\underset{\sim}{C}_j}$	DT [Weber 1979] CT [Weber 1981]
DCR *Similar*			
$P\|\|E\underset{\sim}{C}_{max}$	DT [Weber 1979] CT [Weber 1981]	$P\|$*pmtn*$\|E\Sigma\underset{\sim}{C}_j$	DT [Weber 1979] CT [Weber 1981]

Table 1. Summary of independent processing requirement distributions under which priority policies optimize makespan and flowtime criteria for multimachine scheduling.

SEPT is the obvious candidate for an optimal policy regarding flowtime.

Problems in Table 1 are specified by a natural modification for stochastic problems of the 3-field problem classification $\alpha|\beta|\gamma$ currently in use for deterministic scheduling problems. The fields α, β and γ refer respectively to the *machine environment*, *job characteristics* and *optimality criterion*. (The reader is referred to [Lawler *et al.* 1982] for more details.) As mentioned above, we are interested here only in identical (P) and uniform (Q) parallel machine environments. The job characteristics of interest are whether preemption is permitted (*pmtn*) or not (blank field) and whether random *release dates* corresponding to an arrival process are specified for all jobs ($\underset{\sim}{r}_j$) or all jobs are available at t = 0 (blank field). Results for two stochastic optimality criteria are reported for both makespan and flowtime, *viz.* minimization in expectation (*e.g.* $E\Sigma\underset{\sim}{C}_j$) and in distribution, *i.e.* with respect to the partial ordering of *stochastic dominance* (*e.g.* $F_{\underset{\sim}{C}_{max}}$) (*cf.* (2.1)). Since the families of processing requirement distributions for which results have been obtained have fairly complex specifications for which no acronyms - or even terminology - have been generally agreed, distributional assumptions have not been incorporated in the symbolic problem classifications (for the opposite approach see [Pinedo & Schrage 1982]). Table 1 reports only the best results obtained to date; no attempt has been made to supply complete references on a problem (but in this regard see [Weber 1981; Weiss 1982]). In order to appreciate the information contained in Table 1, some remarks are in order.

First notice that for the deterministic problem $P||\Sigma C_j$ the priority policy *shortest processing time first* (SPT) is actually a *list scheduling* policy - jobs may be placed in order (of increasing processing requirement) at t = 0 and assigned to machines as they become free in this order without subsequent permutation - and hence it may be implemented in O(n log n) running time. Since remaining processing requirements decrease linearly with processing, the SPT order of unfinished jobs never changes and hence reordering of U_t and preemption are never required to optimize flowtime. On the other hand, the *largest processing time first* (LPT) order of unfinished jobs will of course change with processing, and hence the LPT *list scheduling* policy is easily seen to be suboptimal for the *NP-hard* [Karp 1972] nonpreemptive problem $P||C_{max}$.

The preemptive problem $P|pmtn|C_{max}$ is usually solved by McNaughton's *wrap-around rule* (see, *e.g.*, [Baker 1974, Ch.5.2.1]) which yields the optimal value

$$C^{o}_{max} = \max\{P_n/m, p_{max}\}, \tag{2.2}$$

where

$$P_n := \Sigma^{n}_{j=1}\, p_j,\quad p_{max} := \max_{j\in J}\{p_j\}, \tag{2.3}$$

in O(n) time. This algorithm gives only one of many optimal schedules and, although it creates at most m-1 preemptions, makes no attempt to minimize this number. The problem of minimizing the number of preemptions is in fact NP-hard. Alternatively, an optimal schedule can be obtained by a simple preemptive LPT priority policy, which is based on *processor sharing*. The algorithm may be described as follows. Arrange the jobs such that $p_1 \geq \ldots \geq p_n$. At time zero start processing jobs 1,...,m', where $m' = \max\{j | p_j = p_m\}$; if $m' > m$, a number of jobs with processing requirement p_m must equally share a (smaller) number of machines. The next decision epoch occurs when the remaining processing requirement of another job becomes equal to that of a job with initial processing requirement p_m. Then repeat, with remaining rather than original processing requirements. Apply McNaughton's rule in each of the intervals generated to resolve processor sharing. All this requires $O(n^2)$ time.

In the case of deterministic *uniform* machine problems, $Q||\Sigma C_j$ and $Q|pmtn|\Sigma C_j$ can be solved in polynomial time by appropriately modified SPT policies, and $Q|pmtn|C_{max}$ is still solvable in O(n) time given an LPT ordering of the jobs (see [Lawler *et al.* 1982]).

By virtue of the *memoryless* property of the exponential distribution, for exponentially distributed processing requirements the expectation of the remaining processing requirement is always equal to the expectation of the original requirement. Hence preemption may be expected to be irrelevant and, in the parallel machine case, jobs may be initially monotonically ordered in terms of expected processing requirement and both LEPT and SEPT implemented as list scheduling (*i.e.* nondynamic priority) policies in O(n log n) time. This has been established in [Weiss & Pinedo 1980]. They have also given the only treatment to date of (optimal) stochastic scheduling for uniform machine models. For exponential processing requirements, preemption will only be necessary in optimal LEPT and SEPT priority policies for these models to move running jobs to faster machines. Preemption and priority policies are required for *all* problems involving random release dates, since preemption and job reordering may be needed at job arrival epochs. Results involving optimality in distribution for problems with job arrival processes and exponential processing requirements are currently open, as (with the sole exception of the treatment of $P||E\Sigma \underset{\sim}{C}_j$ in [Gittins 1981]) are discrete time - *i.e. geometric* processing requirement distribution - analogues of the Weiss-Pinedo results. Nevertheless, sufficient is known about various stochastic scheduling problems with exponential processing requirements to begin the analysis of their computational complexity, see [Pinedo 1982] where several apparently anomalous results are presented.

Some definitions are in order to continue the discussion of Table 1. The *completion rate* $h_{\underset{\sim}{p}}$ of a job with processing requirement $\underset{\sim}{p}$, density $f_{\underset{\sim}{p}}$ and survivor function $\bar{F}_{\underset{\sim}{p}}$ is given by

$$h_{\underset{\sim}{p}}(t) := \begin{cases} f_{\underset{\sim}{p}}(t)/\overline{F}_{\underset{\sim}{p}}(t) & \text{if } \underset{\sim}{p} \text{ is absolutely continuous,} \\ f_{\underset{\sim}{p}}(t+1)/\overline{F}_{\underset{\sim}{p}}(t) & \text{if } \underset{\sim}{p} \text{ is discrete.} \end{cases} \tag{2.4}$$

The distribution of p is *increasing* (*decreasing*) *completion rate* (ICR, respectively DCR) if, and only if, $h_{\underset{\sim}{p}}$ is a nondecreasing (nonincreasing) function on $\mathbb{R}_+$. If p is absolutely continuous, its distribution is *log(arithmically)* *convex* (*concave*) if, and only if, log $f_{\underset{\sim}{p}}$ is convex (concave). Alternatively, the distribution of p is said to be *increasing* (*decreasing*) *likelihood ratio* (ILR, respectively DLR). If p is discrete, its distribution is ILR (DLR) if, and only if, the function given by

$$h_{\underset{\sim}{p}}(t+1)[1-h_{\underset{\sim}{p}}(t)]/h_{\underset{\sim}{p}}(t) \tag{2.5}$$

is nonincreasing (nondecreasing). This allows a definition of log convexity (concavity) for discrete p. Since their completion rates are constant and the logarithm, respectively (2.5), of its density is linear, both the exponential and geometric distributions are simultaneously ICR, DCR, log convex and log concave. The uniform, hyperexponential, gamma, beta, Gaussian and folded-normal distributions all have either log convex or log concave densities.

Log convex and ICR processing requirement distributions correspond to practical situations (such as are found, for example, in manufacturing) in which processing tends to accelerate job completion. Log concave and DCR distributions, on the other hand, correspond to situations in which *work hardening* of jobs occurs and processing tends to delay job completion (as, for example, with some types of faulty software running on a computer system). It may be shown (*cf.* [Weiss 1982]) that a log convex (concave) processing requirement distribution is necessarily ICR (DCR), but not conversely. (For more details on these concepts see [Barlow & Proschan 1975; Karlin 1968].)

Counterexamples to the optimality of LEPT and SEPT priority policies for multimachine problems with arbitrary processing requirement distributions are easily constructed, see, *e.g.*, [Sevcik 1974; Weber 1979; Weiss 1982]. What is needed to obtain the optimality in *expectation* of these policies is that at any moment current processing requirements can be compared in terms of the stochastic ordering $<_D$ introduced above and hence in terms of expectations (which generate a corresponding order). To obtain optimality in *distribution* (and entertain the possibility of job arrivals) current processing requirements must be comparable in terms of the stronger *likelihood ratio ordering* $<_{LR}$. (If $\underset{\sim}{x}$ and $\underset{\sim}{y}$ are two random variables with densities $f_{\underset{\sim}{x}}$ and $f_{\underset{\sim}{y}}$ respectively, then $\underset{\sim}{x}$ is *likelihood ratio dominated* by $\underset{\sim}{y}$, written $\underset{\sim}{x} <_{LR} \underset{\sim}{y}$, if, and only if, $f_{\underset{\sim}{y}}/f_{\underset{\sim}{x}}$ is a nondecreasing function.) This order will again correspond to expectation (and stochastic) order (both of which it implies). It follows that given the expectation functions of remaining processing requirements $O(n^2 \log n)$ running time is needed to implement LEPT and SEPT as preemptive priority policies.

For a processing requirement $\underset{\sim}{p}$, the processing requirement $\underset{\sim}{p}(s)$ remaining after s units of processing has distribution function $[F_{\underset{\sim}{p}}(\cdot+s)-F_{\underset{\sim}{p}}(s)]/\bar{F}_{\underset{\sim}{p}}(s)$ and completion rate $h_{\underset{\sim}{p}}(\cdot+s)$. The remaining processing requirements for ICR and DCR processing requirement distributions are always comparable in stochastic order, *i.e.* for all s,t either $\underset{\sim}{p}(s) <_D \underset{\sim}{p}(t)$ or $\underset{\sim}{p}(s) >_D \underset{\sim}{p}(t)$ (or both). A similar statement can be made with regard to the likelihood ratio order $<_{LR}$ for the remaining processing requirements generated by log convex (ILR) and log concave (DLR) distributions.

The processing requirements of a set J of jobs are *similar* if, and only if, they are given by an independent collection $\underset{\sim}{p}(s_j)$, $j \in J$, of the remaining processing requirements generated by a processing requirement $\underset{\sim}{p}$. Thus similar jobs have identical processing requirements, but may have received differing amounts of processing prior to the problem.

Weiss [Weiss 1982] has shown that when the completion rates $h_{\underset{\sim}{p}j}$, $j \in J$, are continuous, it is necessary for current processing requirement comparability as discussed above to have either similar processing requirements, or processing requirement distributions whose completion ratio $h_{\underset{\sim}{p}}$ may be ordered in the sense of uniform pointwise order. Since these conditions are easily shown to be sufficient for current processing requirement comparability, in order to obtain a best possible result (subsuming all previous ones and settling affirmatively the open problems in Table 1) a direct proof is needed which is based only on current requirement comparability in the appropriate sense and which is equally applicable *mutatis mutandis* to both continuous and discrete time. The most promising approach is through the Bellman-Hamilton-Jacobi sufficiency condition for optimal stochastic control problems along the lines of the arguments from [Weber 1981] for similar processing requirement problems in continuous time. (In fact, Weber defines current priority orderings in terms of completion rates rather than expectations, but under the assumptions necessary for current processing requirement comparability, as we have seen, the two orderings are identical.)

Notice that with similar log convex and ICR (log concave and DCR) processing requirement distributions and makespan (flowtime) criteria, preemption is necessary for LEPT (SEPT) priority policies to be optimal since - analogous to the situation for the deterministic $P|pmtn|C_{max}$ problem - the remaining processing requirements of running jobs tend to diminish (increase) and at decision epochs LEPT (SEPT) reordering of the set of unfinished jobs may be required. For such problems processor sharing - as discussed above for the deterministic problem $P|pmtn|C_{max}$ - is introduced in [Weber 1981] for remaining processing requirements equal in priority. However, processor sharing may be resolved here - as in the preemptive LPT priority algorithm for the deterministic problem - as a consequence of the fact that unfinished job reordering is only necessary at permitted decision (job arrival and completion) epochs.

Observe also that only the LEPT priority policy remains optimal for problems with random release dates. Intuitively this is so because available jobs should be processed in LEPT order to minimize makespan, regardless of job arrival events in the future, whereas SEPT order may need to be violated to minimize flowtime in order to take advantage of future job arrival events, *cf.* [Weber 1979].

Finally, it should be mentioned that in [Weber 1979] a series of counterexamples is given to show that the discrete time results for similar processing requirement distributions in Table 1 are best possible.

3. STOCHASTIC PROGRAMMING MODELS OF 2-LEVEL SCHEDULING

In this section we shall consider some alternative (multistage) dynamic stochastic programming models of 2-level planning and scheduling in continuous time. In these models, the set of machines to be acquired or assigned must be decided at the first level (stage) *before* any processing begins at t = 0. This decision must be made so as to minimize the sum of machine costs and the expected criterion value of an appropriate variant of the stochastic multimachine scheduling problem (Problem 2.1 treated in §2) which forms the second level (second and subsequent stages) of the problem. We consider second stage (dynamic stochastic) scheduling problems which are variants of $Q|pmtn|E\underset{\sim}{C}_{max}$ and $Q|pmtn|E\Sigma\underset{\sim}{C}_j$. Even under distributional assumptions on processing requirements which guarantee the optimality of LEPT or SEPT preemptive priority policies for these problems, we shall see that a closed form expression of second stage cost - which is required to calculate the optimal first stage decision utilizing the usual dynamic programming method of backwards recursion - is not readily available.

Let $\mathcal{M}$ denote a set of m uniform machines with ordered speeds s_i (as in §2) and costs $c_i \geq 0$, and suppose that the n jobs of the set J of jobs to be processed have independent processing requirements $\underset{\sim}{p}_j$ with means $E\underset{\sim}{p}_j$. Denote by $\underset{\sim}{p} := (\underset{\sim}{p}_1,\ldots,\underset{\sim}{p}_n)$ the *vector* of nonnegative processing requirements and by $c(M) := \Sigma_{i\in M} c_i$ the *cost* of employing the $|M|$ machines in $M \subset \mathcal{M}$. At the second level we shall permit (as before) preemptive scheduling policies π with decision epochs at t = 0 and subsequent job completions. All jobs are assumed available at t = 0.

Consider the following planning decision problems. Choose a subset $M \subset \mathcal{M}$ of the available machines to be applied so as to:

$$\text{(P)}\quad \min_{M\subset\mathcal{M}}\{c(M) + \inf_\pi E\{C_{max}(M,\pi,\underset{\sim}{p})\}\};$$

$$\text{(P')}\quad \min_{M\subset\mathcal{M}}\{c(M) + \inf_\pi E\{\Sigma_{j=1}^{n} C_j(M,\pi,\underset{\sim}{p})\}\};$$

$$\text{(P")}\quad \min_{M\subset\mathcal{M}}\{c(M) + \inf_\pi E\{(C_{max}(M,\pi,\underset{\sim}{p})-T)_+\}\}.$$

Here $C_{max}(M,\pi,p)$ and $\Sigma_{j=1}^{n} C_j(M,\pi,p)$ denote respectively the makespan and flowtime of the jobs in J with processing requirements p performed on the machines in M under scheduling policy π. Without loss of generality, total machine allocation cost c(M) may be assumed to be expressed in terms of schedule delay costs in time units. In stochastic programming terminology (see, *e.g.*, [Dempster 1980]), these problems are multistage *recourse* problems and the second terms in their objective functions are the total expected costs of *recourse* to the first stage decision M through the scheduling policy π. The policy π is a *complete* recourse decision - *i.e.* its choice imposes only considerations of cost on the choice of M - and hence ordinary dynamic programming methods are applicable to the problems at hand. For the problem (P') the *(total) recourse cost* is linear, while for (P) and (P") it is piecewise linear, in job completion times. These costs are clearly monotonically decreasing in $|M| = k$ for (speed ordered) machine sets of the form $\{1,\ldots,k\}$, but depend in a complicated nonlinear manner on the scheduling policy π. The more realistic recourse cost of problem (P") is a piecewise linear function of makespan which represents *overtime cost* incurred when the schedule makespan exceeds the scheduling horizon T (such as occurs, for example, when weekend working is necessary to finish work planned for a given week).

The results from [Weiss & Pinedo 1980] may be interpreted to show that when processing times are independently exponentially distributed the optimal recourse decisions for (P) and (P') are (preemptive) list scheduling policies.

THEOREM 3.1. [Weiss & Pinedo 1980] *Let* $\underset{\sim}{p}_j \sim \lambda_j e^{-\lambda_j t}$, $j \in J$, *and assume* $\lambda_1 \leq \lambda_2 \leq \ldots \leq \lambda_n$. *Then the optimal recourse policies are:*

$\pi^o: j \to j$ (LEPT) *for* (P);

$\pi^o: j \to n-j+1$ (SEPT) *for* (P').

Further, assume without loss of generality that $m' := |M| \geq n$ *(for otherwise machines of cost and speed* 0 *can be added to M). Then the optimal expected recourse cost* $E\underset{\sim}{C}^o_{max} = G_{\pi^o}(J,s)$ *for* (P) *can be computed from the backwards recursion over unfinished job sets* U *(with initial value* $G_{\pi^o}(\emptyset,s) := 0$*) given by*

$$G_{\pi^o}(U,s) = \frac{1_U + \Sigma_{j\in U} \lambda_j s_j G_{\pi^o}(U/\{j\},s)}{\Sigma_{j\in U} \lambda_j s_j} \tag{3.1}$$

where $s := (s_1,\ldots,s_{m'})$ *is the vector of machine speeds and* 1_U *denotes the (binary) indicator function of* U. *Similarly, for* (P'), $E\Sigma\underset{\sim}{C}^o_j = G_{\pi^o}(J,s)$, *with*

$$G_{\pi^o}(U,s) = \frac{|U| + \Sigma_{j\in U} \lambda_j s_j G_{\pi^o}(U/\{j\},s)}{\Sigma_{j\in U} \lambda_j s_j}. \quad \Box \tag{3.2}$$

Thus even with independent exponential processing requirements, although the optimal recourse policies for (P) and (P') are known *explicitly*, the optimal total expected recourse costs must be determined *computationally*.

Algorithmic determination of the optimal expected recourse cost of (P) or (P') by means of (3.1) or (3.2) for a fixed machine set M is of course of exponential (time and space) complexity. Moreover, even if these expected recourse costs were known for each M, an argument similar to that from [Dempster *et al.* 1981B, Lemma 4], shows that the problem (P) of minimizing total expected costs over all $M \subset \mathcal{M}$ is NP-hard. The situation is even worse for the more realistic 2-level planning and scheduling problem (P") in which all jobs have a common due date T and the problem is to minimize expected tardiness of the last job. Hence we turn in the next section to consideration of approximate solutions for these problems through the use of heuristics.

4. PROBABILISTIC ANALYSIS OF HEURISTICS FOR 2-LEVEL SCHEDULING

First notice, for example, that since makespan is nonnegative, problem (P) is *value equivalent* to

$$\min_{M \subset \mathcal{M}} \{c(M) + E(\min_{\pi} \{C_{max}(M,\pi,\underset{\sim}{p})\})\}. \tag{4.1}$$

That is, given M, we may find the expected recourse cost corresponding to an optimal stochastic scheduling policy $\pi^o(M)$ by finding an optimal *deterministic* scheduling policy $\pi^o(M,p)$ for each realization $\underset{\sim}{p} = p$ of the processing requirement data (actually for a set of realizations of the data which occurs with probability one). However, since we are considering stochastic scheduling policies which allow preemptions only at t = 0 and job completions, it follows that we must solve an instance of $Q|pmtn|C_{max}$ in which preemptions are limited to moving running jobs to faster machines for each realization $\underset{\sim}{p} = p$ of the data. But as is well known [Karp 1972] even $P||C_{max}$ is NP-hard for $m \geq 2$.

Therefore let us first consider a version of (4.1) involving *identical* machines with identical assignment cost $c_i \equiv c > 0$, *viz.*

$$E\underset{\sim}{z}^o(m^o) := \min_{m \in \mathbb{N}} \{cm + E(\min_{\pi} \{C_{max}(m,\pi,\underset{\sim}{p})\})\} \tag{4.2}$$

$$=: \min_{m \in \mathbb{N}} \{cm + EC^o_{max}(m,\underset{\sim}{p})\}$$

where $EC^o_{max}(m,\underset{\sim}{p})$ denotes the expectation of the minimum makespan for $P||C_{max}$ (a random variable) for a random data instance $\underset{\sim}{p}$. We have thus reduced our multistage problem (P) to the value equivalent 2-stage recourse problem (4.2) in which at the first level the number m of identical machines to be assigned must be chosen, *before* the processing requirement data $\underset{\sim}{p} = p$ is realized at t = 0, after which, at the second level, an instance of $P||C_{max}$ must be solved. At the second stage we are in the realm of *probabilistic analysis* of algorithms - or, in the terminology of stochastic programming, the *distribution problem* - for $P||C_{max}$.

Similar to the situation in §3, however, manifold difficulties are attendant on solving (4.2) for the optimal m by backwards recursion - *i.e.* by solving first the NP-hard lower level combinatorial optimization problem for each m. Moreover, this is not the natural order of decisions in practice. Therefore, consider applying an idea fundamental to planning at the higher levels of a hierarchy: namely, suppression of detailed lower level structure and its replacement by aggregates. Replace the optimal recourse cost $C^o_{max}(m,p)$ by its obvious lower bound $P/m := \Sigma^n_{j=1} p_j/m$, *cf.* (2.3), and - in order to determine the *approximately* optimal higher level decision - solve the easy *lower bounding* problem to (4.2),

$$E\underset{\sim}{z}^{LB}(m^{LB}) := \min_{m\in \mathbb{N}} \{cm + E\underset{\sim}{P}/m\}. \tag{4.3}$$

The solution m^{LB} of (4.3) minimizes $cm+E\underset{\sim}{P}/m$ subject to $m \in \mathbb{N}\cap\{\lfloor\sqrt{E\underset{\sim}{P}/c}\rfloor,\lceil\sqrt{E\underset{\sim}{P}/c}\rceil\}$.

In practice, the scheduling decisions corresponding to a foreman's dispatching task are also not explicitly optimally planned, but rather are handled *ad hoc* in real time through the use of heuristics. Consider the solution of the (deterministic) second stage problem of (4.2) using the *list scheduling* heuristic - *i.e.* place the n jobs in an *arbitrary* order and at each step assign the next job on the list to the earliest available machine.

Let $z^{LS}(m) := cm+C^{LS}_{max}(m,p)$, where for given m and p $C^{LS}_{max}(m,p)$ denotes the earliest time by which jobs are completed under this heuristic. The *2-stage heuristic* procedure defined for problem (4.2) produces a total expected cost of

$$E\underset{\sim}{z}^{LS}(m^{LB}) := cm^{LB} + EC^{LS}_{max}(m^{LB},\underset{\sim}{p}). \tag{4.4}$$

Notice that in the more realistic dynamic stochastic situation of problem (P) in which the realization $\underset{\sim}{p}_j = p_j$ becomes known only upon the completion of job $j \in J$, list scheduling may be implemented in an *on line* manner. At $t = 0$ a job is assigned to each of the m^{LB} machines in list (*e.g.* LEPT) order, as soon as a job is completed on a machine the next job on the list is assigned to that machine, and so on, until $C^{LS}_{max}(m^{LB},\underset{\sim}{p})$ is realized. Hence our 2-stage heuristic procedure is also applicable to the version of the original multistage recourse problem (P) value equivalent to (4.2), *viz.*

$$\min_{m\in \mathbb{N}} \{cm + \inf_{\pi} E\{C_{max}(m,\pi,\underset{\sim}{p})\}\} = E\underset{\sim}{z}^{o}(m^{o}). \tag{4.5}$$

We are thus in a position to study the stochastic performance of our 2-stage stochastic programming heuristic for this problem through the more familiar problem (4.2). (No notational distinction will be made in the sequel between the common optimal value of the two problems.)

Let us first briefly review known results on the *performance ratio* of list scheduling relative to the minimum makespan $C^o_{max}(m,p)$ for the deterministic problem $P||C_{max}$. An easy demonstration yields, for given m and p,

$$P/m \leq C^o_{max}(m,p) \leq C^{LS}_{max}(m,p) \leq P/m + p_{max}. \tag{4.6}$$

It follows that for the performance ratio we have

$$1 \leq C^{LS}_{max}(m,p)/C^o_{max}(m,p) \leq 1 + \frac{p_{max}}{P/m}. \tag{4.7}$$

Graham [Graham 1966, 1969] has obtained the *data independent* worst case bound

$$C^{LS}_{max}(m,p)/C^o_{max}(m,p) \leq 2 - \frac{1}{m} \tag{4.8}$$

and has shown that for list scheduling in LPT order this bound can be considerably improved to

$$C^{LPT}_{max}(m,p)/C^o_{max}(m,p) \leq \frac{4}{3} - \frac{1}{3m}. \tag{4.9}$$

The bound (4.8) is tight for maximum processing requirement ratios $p_{max}/p_{min} \geq 4$ and in [Achugbue & Chin 1981] tight bounds are given for lower values of this ratio. Observe that (4.9) is of little use in analyzing list scheduling heuristics for the multistage recourse problem (4.5) since LPT order requires full knowledge of the data realization $\underset{\sim}{p} = p$ and cannot be implemented in an on line manner when the data is realized sequentially. Moreover, we shall see below (Proposition 4.7) that data independent bounds such as (4.8) or (4.9) do not produce asymptotically tight bounds for the 2-stage heuristic as the number of jobs in the system becomes large.

Although our proper concern is the distribution of the *relative performance ratio* $\underset{\sim}{z}^{LS}(m^{LB})/\underset{\sim}{z}^o(m^o)$ of the 2-stage heuristic for a random data instance $\underset{\sim}{p}$, an easy consequence of (4.6) and the definitions yields a bound on the ratio of the expected value of this heuristic relative to the expected optimum [Dempster *et al.* 1981B].

THEOREM 4.1. $1 \leq E\underset{\sim}{z}^{LS}(m^{LB})/E\underset{\sim}{z}^o(m^o) \leq 1 + E\underset{\sim}{p}_{max}/(2\sqrt{cE\underset{\sim}{P}})$. □

In order to study the asymptotic performance of the 2-stage heuristic for (4.5) as the number of jobs (and machines) in the system becomes large, the following assumption on the processing requirement data was made in [Dempster *et al.* 1981B].

Assumption A. The processing requirements $\underset{\sim}{p}_j$, $j \in J$, are independent identically distributed random variables with two moments finite: $\mu := E\underset{\sim}{p}$ and $\sigma^2 := V\underset{\sim}{p}$.

Thus, from the modelling point of view, Assumption A only allows consideration of random variations in the processing requirements of identical jobs. Under Assumption A, asymptotic extreme value theory may be invoked to conclude that

$$E\underset{\sim}{p}_{max}/\sqrt{n} \longrightarrow 0, \tag{4.10}$$

and

$$\underset{\sim}{p}_{max}/\sqrt{n} \xrightarrow{a.s.} 0, \tag{4.11}$$

read $\underset{\sim}{p}_{max}/\sqrt{n}$ tends *almost surely* to 0 as the number n of jobs in the system tends to infinity (*i.e.* $P\{\lim_{n\to\infty} \underset{\sim}{p}_{max}/\sqrt{n} = 0\} = 1$). Analogous results to those obtained in [Dempster *et al.* 1981B] under Assumption A follow from the observation that (4.11) and (4.10) continue to hold (by a slight extension of the arguments given in [Dempster *et al.* 1981B, Appendix]) under the following weaker assumption.

Assumption A'. The processing requirements $\underset{\sim}{p}_j$, $j \in J$, are independent random variables with two moments finite: $\mu_j := E\underset{\sim}{p}_j$ and $\sigma_j^2 := V\underset{\sim}{p}_j$; $\mu := \lim_{n\to\infty} \Sigma_{j=1}^n \mu_j/n$ and $\sigma^2 := \lim_{n\to\infty} \Sigma_{j=1}^n \sigma_j^2/n$ are finite.

For fixed n, Assumption A' permits more realistic processing requirement distributions, for example, as considered in §2. The asymptotic requirements on processing requirement means and variances ensure that no large (*i.e.* infinite) set of processing requirements dominate. Put another way, Assumption A' ensures that the contributions of individual jobs to long run processing requirement statistics are negligible - exactly the preconditions for aggregation in higher level planning.

An immediate consequence of Theorem 4.1, (4.10) and the observation that under our assumption

$$E\underset{\sim}{P} \to n\mu \quad \text{as } n \to \infty, \tag{4.12}$$

is the asymptotic optimality in expectation of the 2-stage heuristic.

THEOREM 4.2. *Under Assumption* A', $\lim_{n\to\infty} E\underset{\sim}{z}^{LS}(m^{LB})/E\underset{\sim}{z}^{o}(m^{o}) = 1$. □

To obtain the analogue for the performance ratio, two lemmas will be needed. Minor modifications of the arguments in [Dempster *et al.* 1981B] to accommodate the extra passage to the limit entailed in (4.12) yield the required common asymptotic characterizations of heuristic and optimal first stage decisions and expected recourse costs.

LEMMA 4.3. *Under Assumption* A',

$$\lim_{n\to\infty} m^{LB}/\sqrt{n\mu/c} = \lim_{n\to\infty} m^{o}/\sqrt{n\mu/c} = 1. \; □$$

LEMMA 4.4. *Under Assumption* A',

$$C^{LS}(m^{LB},\underset{\sim}{p})/(n\mu/m^{LB}) \xrightarrow{a.s.} 1,\ C^{o}_{max}(m^{o},\underset{\sim}{p})/(n\mu/m^{o}) \xrightarrow{a.s.} 1.$$

Proof. Use Kolmogorov's strong law of large numbers for *nonidentical* independent random variables (see, *e.g*. [Tucker 1967, Theorem 1, p.124]) to conclude that $(\underset{\sim}{P}-n\mu)/n \xrightarrow{a.s.} 0$ in the argument of [Dempster *et al*. 1981B, pp.8,10]. □

Combining the lemmas yields asymptotic optimality of the 2-stage heuristic (in terms of the performance ratio) with probability one, *i.e*. for almost every instance of the requirements data $\underset{\sim}{p}$.

THEOREM 4.5. *Under Assumption* A', $\underset{\sim}{z}^{LS}(m^{LB})/\underset{\sim}{z}^{o}(m^{o}) \xrightarrow{a.s.} 1$. □

Theorem 4.5 constitutes a justification for hierarchical planning procedures - as represented by our 2-stage heuristic - in the 2-level planning and scheduling situation modelled by (4.5). Indeed, it may be interpreted loosely to state that, for large parallel machine shops and many jobs with arbitrary processing requirements, aggregation and approximation at the higher level and *ad hoc* heuristics at the lower level are approximately optimal.

We note also that for the multistage problem (4.5) with a second stage stochastic scheduling problem involving *identical* machines this result comes about through a common asymptote for the optimal and heuristic value.

COROLLARY 4.6. *Under Assumption* A',

$$\underset{\sim}{z}^{LS}(m^{LB})/(2\sqrt{cn\mu}) \xrightarrow{a.s.} 1,\ \underset{\sim}{z}^{o}(m^{o})/(2\sqrt{cn\mu}) \xrightarrow{a.s.} 1.\ \square$$

Before considering problem (P) and its value equivalent (4.1) which involve *uniform* machine scheduling problems at the second level, it is worth noting that the above results are delicate, in that an attempt to use Graham's *data independent* bound (4.8) results in an asymptotic expectation ratio bound greater than one.

PROPOSITION 4.7. *Using* (4.8) *rather than* (4.6) *to bound* $C^{LS}_{max}(m^{LB},p)$ *yields only* $1 \le E\underset{\sim}{z}^{LS}(m^{LB})/E\underset{\sim}{z}^{o}(m^{o}) \le 3/2$ *as* $n \to \infty$.

Proof. A simple argument using (4.8) and (4.6) shows that

$$1 \le E\underset{\sim}{z}^{LS}(m^{LB})/E\underset{\sim}{z}^{o}(m^{o}) \le 1 + \left(1-\frac{1}{m^{LB}}\right)\frac{\sqrt{E\underset{\sim}{P}/c}}{2m^{LB}} + \left(2-\frac{1}{m^{LB}}\right)\frac{E\underset{\sim}{p}_{max}}{2\sqrt{cE\underset{\sim}{P}}}.$$

Lemma 4.3, (4.10) and (4.12) imply that the right hand side of the second inequality $\to 3/2$ as $n \to \infty$. □

Analysis of 2-stage heuristics for the 2-level uniform machine problem (P) is more difficult than the analysis of the identical machine special case presented above. The first problem lies with the obvious extension of the lower bounding problem (4.3) to the general case, *viz.*

$$E\underset{\sim}{w}^{LB}(M^{LB}) := \min_{M \subset \mathcal{M}}\{c(M) + E\underset{\sim}{P}/s(M)\}, \tag{4.13}$$

where $s(M) := \Sigma_{i \in M} s_i$. By a reduction from the NP-complete *partition* problem it is shown in [Dempster *et al.* 1981B] that the lower bounding problem (4.13) is NP-hard! To ensure polynomial time determination of an appropriate first level decision for (P) we shall therefore employ a heuristic for the approximate solution of (4.13). To this end, reorder $\mathcal{M}$ in increasing order of $q_i := c_i/s_i$ and define $C_i := \Sigma_{h=1}^{i} c_h$, $S_i := \Sigma_{h=1}^{i} s_h$ and $W_i := C_i + E\underset{\sim}{P}/S_i$. The *greedy heuristic* chooses the machine set $M^G := \{1,\ldots,g\} \subset \mathcal{M}$ so that g is the largest index such that $W_{g-1} > W_g$.

Making the obvious definition of extreme machine costs, speeds and q ratios, it may be shown [Dempster *et al.* 1981B] that the greedy decision M^G satisfies

$$W_g = \min_{i \in \mathcal{M}}\{W_i\}, \tag{4.14}$$

$$E\underset{\sim}{w}^{LB}(M^G) \le E\underset{\sim}{w}^{LB}(M^{LB}) + c_{max}, \tag{4.15}$$

and that in the present case the analogue of Theorem 4.1 for the expectation ratio follows from (4.15) by a simple argument (*cf.* (4.6)).

THEOREM 4.8. $1 \le E\underset{\sim}{w}^{LS}(M^G)/E\underset{\sim}{w}^{O}(M^O) \le 1+(c_{max}+E\underset{\sim}{p}_{max}/s_{min})/(2\sqrt{q_{min}E\underset{\sim}{P}})$. □

In order to obtain the analogues of Theorems 4.2 and 4.5 in the uniform machine case, some reasonable assumptions are needed about the growth of the available machine set $\mathcal{M}$ as the number of jobs in the system tends to infinity.

Assumption B. The bounds $c_{min} \le c_i \le c_{max}$ and $s_{min} \le s_i \le s_{max}$ $(i \in \mathcal{M})$ are fixed constants. Moreover, there exist constants $D, D' > 0$ and $\varepsilon' \ge \varepsilon > 0$ such that $Dn^{\frac{1}{2}+\varepsilon} \le |\mathcal{M}| \le D'n^{\frac{1}{2}+\varepsilon'}$.

Assumption B allows an efficient implementation of the greedy heuristic (in O(n log n) time) in that the number of *available* machines remains polynomially bounded in n. We shall see that it ensures that the number of *selected* machines grows as $\sqrt{n}$, as in the identical machine case.

Theorem 4.8, (4.10) and (4.12) yield immediately the asymptotic optimality in expectation of the 2-stage heuristic for (P) defined as the first stage greedy heuristic followed by arbitrary (on line) list scheduling for the uniform machine set M^G chosen. (Recall that in the present context list scheduling is implemented so that at job completion epochs preemption may be applied to move running jobs to faster machines in list order, as described in §2).

THEOREM 4.9. *Under Assumptions* A' *and* B, $\lim_{n\to\infty} E\underset{\sim}{w}^{LS}(M^G)/E\underset{\sim}{w}^{O}(M^O) = 1$. □

Let $g(n) = \Theta(f(n))$ denote the existence of constants $C, C' > 0$ such that $Cf(n) \le |g(n)| \le C'f(n)$ for all n sufficiently large. Then an easy extension of the argument given in [Dempster *et al.* 1981B] (to account for the extra passage to the limit necessitated by Assumption A') yields the analogue of Lemma 4.3 for uniform machines.

LEMMA 4.10. *Under Assumptions* A' *and* B,

$$s(M^{LB}) = \Theta(\sqrt{n}),\ s(M^G) = \Theta(\sqrt{n}),\ s(M^O) = \Theta(\sqrt{n}).\ \square$$

The extension of Theorem 4.5 to the performance ratio of the 2-stage heuristic for the uniform machine case under Assumption A is due to [Stougie 1981]. He has given a direct proof which is (trivially) extended below to accommodate Assumption A'.

THEOREM 4.11. *Under Assumptions* A' *and* B, $\underset{\sim}{w}^{LS}(M^G)/\underset{\sim}{w}^{O}(M^O) \xrightarrow{a.s.} 1$.

Proof. For every realization $\underset{\sim}{p} = p$ of the data it may be shown (*cf.* [Dempster *et al.* 1981B, pp.15-16]) that

$$w^{LS}(M^G) \le w^O(M^O) + |P - \Sigma_{i=1}^n \mu_i| \left(\frac{1}{s(M^G)} + \frac{1}{s(M^O)}\right) + c_{max} + \frac{p_{max}}{s_{min}} \quad (4.16)$$

and

$$w^O(M^O) \ge q_{min}\, s(M^O) + P/s(M^O). \quad (4.17)$$

The existence of an $\alpha > 0$ such that $w^O(M^O) \ge \alpha\sqrt{n}$, for sufficiently large n, follows from (4.17) and Lemma 4.10. Hence, combining (4.16) and (4.17) and observing that $w^O(M^O) \le w^{LS}(M^G)$ for every realization $\underset{\sim}{p} = p$, we have that

$$1 \le \underset{\sim}{w}^{LS}(M^G)/\underset{\sim}{w}^{O}(M^O) \le$$
$$1 + \left|\frac{\underset{\sim}{P} - \Sigma_{i=1}^n \mu_i}{\alpha\sqrt{n}}\right| \left(\frac{1}{s(M^G)} + \frac{1}{s(M^O)}\right) + \frac{c_{max}}{\alpha\sqrt{n}} + \frac{\underset{\sim}{p}_{max}}{s_{min}\alpha\sqrt{n}} \quad (4.18)$$

surely. But Assumptions A' and B imply that $c_{max}/(\alpha\sqrt{n}) \to 0$ and $\underset{\sim}{p}_{max}/(s_{min}\alpha\sqrt{n}) \xrightarrow{a.s.} 0$. Moreover, Lemma 4.10 implies that there are constants α' and α" for which, for sufficiently large n, $1/s(M^G) \le \alpha'/\sqrt{n}$ and $1/s(M^O) \le \alpha''/\sqrt{n}$. It follows that the second term of the right hand side of (4.18) tends almost surely to 0 with $|\underset{\sim}{P}/n - \Sigma_{i=1}^n \mu_i/n|$. Applying Kolmogorov's strong law of large numbers (*op. cit.*) to this expression yields the desired result. □

It is perhaps worth observing that Assumption A' could be weakened to be necessary and sufficient through the use of canonically truncated processing requirements and Kolmogorov's three series theorem (see, *e.g.*, [Tucker 1967, Theorem 4, p.113]). From a modelling point of view however little would be gained but unnecessary mathematical complexity.

Also note that no analogue of Corollary 4.6 appears to be possible in the uniform machine case. The most precise statement about the asymptotic form of the optimal and heuristic values we can give is the existence of constants C,C' > 0 such that, for sufficiently large n,

$$C\sqrt{n} \leq \underset{\sim}{w}^{O}(M^{O}) \leq \underset{\sim}{w}^{LS}(M^{G}) \leq C'\sqrt{n} \tag{4.19}$$

almost surely.

Consider next the problem (P') of §3 involving a flowtime recourse cost. This problem is difficult for reasons which illuminate the intricacies of hierarchical problems. As noticed in §2, the second stage problem $Q|pmtn|\Sigma C_j$ for the value equivalent problem to (P') is easy! In the identical machine case it can be solved (nonpreemptively) in O(n log n) running time by SPT, while in the uniform machine case it can be solved in O(n log n + mn) time by an extension of SPT due to Gonzalez, which only preempts running jobs to move them to faster machines (see [Lawler *et al.* 1982]). An explicit expression for the optimal flowtime of the nonpreemptive problem $Q||\Sigma C_j$ is given [Conway *et al.* 1967, p.97] by

$$\Sigma_{j=1}^{n} C_j^{O}(M,p) = \Sigma_{i\in M} \Sigma_{k=1}^{n_i} (n_i-k+1)p_{i[k]}/s_i, \tag{4.20}$$

where n_i is the number of jobs assigned to machine $i \in M$ by the optimal schedule and $p_{i[1]},\ldots,p_{i[n_i]}$ are their processing requirements in SPT order. This expression could be used to form an upper bound on the optimal flowtime of the preemptive problem $Q|pmtn|C_{max}$. Although the asymptotic expectation of (4.20) could in principle be evaluated (under Assumption A) using the theory of order statistics, it appears to be of little use in developing a lower bounding problem for (P'). More generally (and unlike the situation for the NP-hard problem $Q||C_{max}$) bounds for the criterion value produced by suboptimal schedules for the easy problem $Q|pmtn|\Sigma C_j$ useful in the analysis of heuristics for its stochastic counterpart $Q|pmtn|E\Sigma\underset{\sim}{C}_j$ - which forms the second stage of (P') - are not readily apparent.

The more realistic problem (P") of §3 involving a schedule tardiness recourse cost is also difficult. If, in the identical machine case, one uses the obvious lower bounding problem

$$\min_{m\in \mathbb{N}} \{cm + E(\max\{\underset{\sim}{P}/m-T,0\})\} \tag{4.21}$$

to determine the first stage heuristic decision m^{LB}, then it can only be determined as a nearest integer to the solution of the integral equation

$$m^2 - \int_{mT}^{\infty} P dF_{\underset{\sim}{P}}(P)/c = 0. \tag{4.22}$$

(If T = 0, (4.22) yields m^{LB} as a nearest integer to $\sqrt{E\underset{\sim}{P}/c}$ as before.) Under the realistic assumption that $T = \Theta(\sqrt{n})$ - which models the idea of many small jobs whose processing requirements are small relative to the schedule horizon T - asymptotic analysis of the performance ratio of the above first stage heuristic followed by list scheduling as the number of jobs in the system tends to infinity appears complicated. This is in no small measure because - unlike the above higher level heuristics based on expected values and aggregation - distributional information on the processing requirements must be taken into account at the higher level due to the nonnegativity restriction on the lower bound for the second stage cost. Unfortunately, a complete analysis of (P") is a prerequisite to the analysis of a realistic *multiperiod* planning and scheduling model in which work is allowed either to overflow from one period to the next or to be finished in overtime at a higher recourse cost.

In the presence of random nonnegative job release dates $\underset{\sim}{r}_j$ in the 2-level models (P) and (P'), list scheduling will no longer suffice and priority policies become necessary (as noted in §2). Unfortunately, the list scheduling bounds (4.6) and its uniform machine extension fundamental to our asymptotic analysis then cease to hold and a more careful analysis is required.

5. A 3-LEVEL DISTRIBUTION PLANNING MODEL

To illustrate the complexities involved, this section briefly sets out a realistic 3-level hierarchical spatial planning and scheduling model for which suitable heuristics are currently under investigation. The problem concerns the location of distribution facilities and delivery vehicles in a region in order to ultimately route the vehicles at the facilities through the customers in the region in a cost effective way. As in the 2-level stochastic machine scheduling problems treated in this paper, the random data is realized successively at each level *after* decisions are taken at the previous level.

More precisely, suppose given a random natural number $\underset{\sim}{n}$ of *customers* and a random finite sequence $\underset{\sim}{x}(\underset{\sim}{n})$ of the Cartesian coordinates $(\underset{\sim}{x}_{11},\underset{\sim}{x}_{12}),(\underset{\sim}{x}_{21},\underset{\sim}{x}_{22}),\ldots,(\underset{\sim}{x}_{\underset{\sim}{n}1},\underset{\sim}{x}_{\underset{\sim}{n}2})$ of their *locations* in a planar (simply connected) region Ω of area A. Assume that $\underset{\sim}{n} = n$ will be realized before $\underset{\sim}{x}(n)$ is known and consider the following 3-level hierarchical distribution planning and vehicle routing problem. At level:

1. Choose the *number* k and *locations* $y(k) := ((y_{11},y_{12}),(y_{21},y_{22}),\ldots,(y_{k1},y_{k2}))$ of identical distribution *facilities* to be placed in the region Ω before $\underset{\sim}{n}$ is realized.
2. Observe $\underset{\sim}{n} = n$. At each facility i, choose the *territory* $\Omega_i \subset \Omega$ ($\Omega_i \cap \Omega_j = \emptyset$, $i \neq j$, $\cup_{i=1}^{k} \Omega_i = \Omega$) to be served and the *number* ℓ_i of identical *vehicles* of unlimited capacity to service customers in the territory before $\underset{\sim}{x}(n)$ is realized.

3. Observe $\underset{\sim}{x}(n) = x(n)$. At each facility, *allocate* realized *customers* to vehicles and *route vehicles* so as to minimize the length of the longest vehicle tour through the allocated customer locations in Ω_i.

If C denotes the cost of a distribution facility and c denotes the cost of a vehicle (in transportation cost units), this problem may be given the following 3-stage complete recourse stochastic programming formulation:

$$\inf_{k,y(k)} \{Ck + \Sigma_{i=1}^{k} E(\inf_{\ell_i,\Omega_i} \{c k_i + E(V^o(\ell_i,\Omega_i;\underset{\sim}{n},\underset{\sim}{x})|\underset{\sim}{n})\})\} \quad (5.1)$$

where $V^o(\ell_i,\Omega_i;n,x)$ denotes the minimal longest vehicle tour length (in terms of Euclidean distance) for the ℓ_i vehicles servicing territory Ω_i, $i = 1,\ldots,k$, when the n customers in Ω have locations x.

As in the machine scheduling models, it is prudent first to attempt to analyze very simple special cases of (5.1). Even these raise some intriguing and nontrivial questions. For example, suppose Ω is the unit disk $\{(x_1,x_2) \in \mathbb{R}^2: x_1^2+x_2^2 \leq 1\}$, $\underset{\sim}{n}$ is geometric on $[N,\infty)$ for some large N, *i.e.* $f_{\underset{\sim}{n}}(n) = p(1-p)^{i-N+1}$, $i = N,N+1, N+2,\ldots$ $(0 < p < 1)$, and $\underset{\sim}{x}(n)$ is spatial Poisson on Ω, *i.e.* the n customer locations are distributed uniformly at random in Ω. It is an obvious advantage in analysis to have all second level problems identical. But is it even approximately optimal at the first level to choose and partition Ω into pie shaped territories Ω_i of equal area with the i-th facility located at, say, the centroid of Ω_i, $i = 1,\ldots,k$? More generally, what is the effect of ignoring the partition constraints $\Omega_i \cap \Omega_j = \emptyset$, $i \neq j$, $i,j = 1,\ldots,k$ (while maintaining $\cup_{i=1} \Omega_i = \Omega$) - *cf.* U.S. national oil distribution in 1975 and 1979 - on the optimal choice of territories? Answers to these questions of course depend on the nature of the metric imposed on the higher level problems by the minimal longest vehicle tour cost measure, and results in random graph theory (see, *e.g.*, [Erdös & Spencer 1973]) can be expected to be helpful.

A single second level problem has been analyzed for a fixed circular Ω_i of area π in [Marchetti Spaccamela *et al.* 1982] building on earlier work reported in [Beardwood *et al.* 1959; Karp 1977; Steele 1980]. They observe that the length of the longest of the optimal tours of the vehicles through the customers assigned to them exceeds $1/\ell_i$ times the length of an optimal travelling salesman tour. Using a theorem from [Steele 1980], which gives an almost sure asymptote for this tour involving a constant β, they define the lower bounding problem

$$\min_{\ell_i \in \mathbb{N}} \{c\ell_i + \beta\sqrt{n\pi}/\ell_i\}$$

to yield a second stage heuristic decision $\ell^{LB} = O(n^{\frac{1}{4}})$ for sufficiently large n. Their third level multivehicle routing heuristic is based on a modification appropriate to a circular region of Karp's [Karp 1977] "divide and conquer" polynomial time approximation algorithm for the NP-hard Euclidean travelling salesman problem posed in a rectangle. (Such approximation algorithms - unfortunately sometimes termed *probabilistic* in the literature - have the property of arbitrary ε-optimality for sufficiently large finite n with a probability which has a precisely known lower bound tending to 1 with n tending to ∞ and hence are almost surely asymptotically optimal.) Marchetti Spaccamela *et al.* demonstrate that the expectation ratio of this 2-stage heuristic relative to the optimal value approaches 1 and that the heuristic is optimal in performance ratio almost surely for random data instances as the number n of customers in the system tends to infinity. They also obtain similar results for the case of random $\underset{\sim}{n}$ and the realistic third level *repetitive* vehicle routing situation in which customers in given locations require a (Bernoulli) random delivery with probability p.

The first level problem defined by (5.1) is essentially a *non*-Euclidean *planar* k-*median problem*. Thus there is some hope in extending the analysis of the *Euclidean* k-median problem given in [Fisher & Hochbaum 1980] to the metric defined on the n customers by the sum of the second and third stage costs of (5.1). These authors give an asymptotic probabilistic analysis of a polynomial time approximation algorithm for the NP-hard [Papadimitriou 1980] Euclidean problem in a planar region of area A (including the almost sure asymptote $n\sqrt{A/k}$ for the sum of the minimal Euclidean distances to each point from the k centres) whose extension would provide a suitable first stage heuristic for (5.1). We are currently working in this direction.

Although it retains the essence of practical hierarchical planning in the distribution field, the simplified model set out here could be usefully extended in many directions to improve its realism. It is however already sufficiently difficult and, for example, addition of vehicle capacity constraints (as in deterministic models) would complicate matters even more.

6. CONCLUSIONS

Open problems and directions for further research have been indicated throughout this paper. Rather than collect them here, some remarks on the nature of stochastic models for hierarchical planning and scheduling decisions seem more appropriate.

First, it is worth observing that many of the parallel machine scheduling problems of §2 provide instances of NP-hard deterministic problems for which simple suboptimal heuristics (*e.g.* LEPT) become optimal when the problem data is (more realistically) taken to be suitably random. The implication - a central thesis of this paper - is that in a practical situation suboptimality of relatively simple heuristics can be the erroneous conclusion of the wrong model, which has been taken to be deterministic for analytic convenience rather than stochastic for realism.

More generally, multistage recourse stochastic programming

models appear to provide a realistic representation of hierarchical planning and scheduling decision problems in several fields of application. Heuristics for such problems are necessitated by their analytic and computational complexity and the sequential availability of data and can be made to mirror the top down sequential nature of actual hierarchical decision making based on averaging and aggregation until more refined data becomes available. Analyses which demonstrate the asymptotic optimality of these heuristics with the growth of random instances of the problem data tend to reinforce the long held views of practical persons faced with difficult decisions - in sufficiently complex environments *suitable* rules of thumb can be highly efficient.

Finally, the project described in this paper can be seen as part of a current general trend in mathematical sciences. Driven by the exigencies of numerical computation, approximation methods are moving from applications to functions, equations and other relatively simple deterministic structures to the approximation of more and more complex stochastic problems.

ACKNOWLEDGEMENTS

It is a pleasure to acknowledge the support of IIASA, where much of the preparation of this paper was completed. The wider work of my colleagues (named in §1) and myself on this topic has been partially supported by IIASA, by NSF Grant ENG-7826500 to the University of Pennsylvania, by NATO Special Research Grant 9.2.02 (SRG.7) and by NATO Research Grant 1575. I am indebted to my colleagues for stimulating discussions and for providing me with manuscripts of work in progress which permitted this attempt at a comprehensive survey of our collective effort. In particular, I would like to thank Jan Karel Lenstra and Leen Stougie for their helpful comments on the manuscript.

Many thanks are also due to Marjolein Roquas for very rapidly producing the typescript to her usual exacting standards.

REFERENCES

J.O. ACHUGBUE, F.Y. CHIN (1981) Bounds on schedules for independent tasks with similar execution times. *J. Assoc. Comput. Mach.* 28,81-99.

K.R. BAKER (1974) *Introduction to Sequencing and Scheduling*, Wiley, New York.

R.E. BARLOW, F. PROSCHAN (1975) *Statistical Theory of Reliability and Life Testing: Probability Models*, Holt, Rinehart and Winston, New York.

J. BEARDWOOD, H.J. HALTON, J.M. HAMMERSLEY (1959) The shortest path through many points. *Proc. Cambridge Phil. Soc.* 55, 299-327.

R.W. CONWAY, W.L. MAXWELL, L.W. MILLER (1967) *Theory of Scheduling*, Addison-Wesley, Reading, MA.

M.A.H. DEMPSTER (ed.) (1980) *Stochastic Programming*, Academic, London.

M.A.H. DEMPSTER, M.L. FISHER, L. JANSEN, B.J. LAGEWEG, J.K. LENSTRA, A.H.G. RINNOOY KAN (1981A) Analytical evaluation of hierarchical planning systems. *Oper. Res.* 29,707-716.

M.A.H. DEMPSTER, M.L. FISHER, L. JANSEN, B.J. LAGEWEG, J.K. LENSTRA, A.H.G. RINNOOY KAN (1981B) Analysis of heuristics for stochastic programming: results for hierarchical scheduling problems. Report BW 142, Mathematisch Centrum, Amsterdam.

M.A.H. DEMPSTER, C.H. WHITTINGTON (1976) *Computer Scheduling of REME Training*, FS 3/01 Final Report I & II, Council for Educ. Tech., London.

Y.M.I. DIRICKX, L.P. JENNERGREN (1979) *Systems Analysis by Multilevel Methods: With Applications to Economics and Management*, International Series on Applied Systems Analysis 6, Wiley, Chichester.

P. ERDÖS, J. SPENCER (1973) *Probabilistic Methods in Combinatorics*, Academic, New York.

M.L. FISHER, D.S. HOCHBAUM (1980) Probabilistic analysis of the planar K-median problem. *Math. Oper. Res.* 5,27-34.

F. GIANNESSI, B. NICOLETTI (1979) The crew scheduling problem: a travelling salesman approach. In: N. CHRISTOFIDES *et al.* (eds.) (1979) *Combinatorial Optimization*, Wiley, Chichester, 389-408.

J.C. GITTINS (1979) Bandit processes and dynamic allocation indices. *J. Roy. Statist. Soc. Ser. B* 41,148-177.

J.C. GITTINS (1981) Multiserver scheduling of jobs with increasing completion rates. *J. Appl. Probab.* 18,321-324.

R.L. GRAHAM (1966) Bounds for certain multiprocessing anomalies. *Bell System Tech. J.* 45,1563-1581.

R.L. GRAHAM (1969) Bounds on multiprocessing timing anomalies. *SIAM J. Appl. Math.* 17,263-269.

E.P.C. KAO, M. QUEYRANNE (1981) Aggregation in a two-stage stochastic program for manpower planning in the service sector. Research Report, Department of Quantitative Management Science, University of Houston.

S. KARLIN (1968) *Total Positivity, Vol. I*, Stanford University Press, Stanford.

R.M. KARP (1972) Reducibility among combinatorial problems. In: R.E. MILLER, J.W. THATCHER (eds.) (1972) *Complexity of Computer Computations*, Plenum, New York, 85-103.

R.M. KARP (1977) Probabilistic analysis of partitioning algorithms for the traveling-salesman problem in the plane. *Math. Oper. Res.* 2,209-224.

L. KLEINROCK (1976) *Queueing Systems, Vol. II: Computer Applications*, Wiley, New York.

E.L. LAWLER, J.K. LENSTRA, A.H.G. RINNOOY KAN (1982) Recent developments in deterministic sequencing and scheduling: a survey. This volume.

A. MARCHETTI SPACCAMELA, A.H.G. RINNOOY KAN, L. STOUGIE (1982) Hierarchical vehicle routing. To appear.

P. NASH (1980) A generalized bandit problem. *J. Roy. Statist. Soc. Ser. B* 42,165-169.

C.H. PAPADIMITRIOU (1980) Worst-case analysis of a geometric location problem. Technical Report, Laboratory for Computing Science, Massachusetts Institute of Technology, Cambridge, MA.

M.L. PINEDO (1982) On the computational complexity of stochastic scheduling problems. This volume, 355.

M.L. PINEDO, L. SCHRAGE (1982) Stochastic shop scheduling: a survey. This volume, 181.

E.L. PRESMAN, I.M. SONIN (1979) On the asymptotic value function of the many-armed bandit problem. In: V.I. ARKIN, H.YA. PETRAKOV (eds.) (1979) *Theoretical Probabilistic Methods for Problems of Economic Process Control*, Central Economic Mathematical Institute, USSR Academy of Science, Moscow. (In Russian.)

S.M. ROSS (1970) *Applied Probability Models with Optimization Applications*, Holden-Day, San Francisco.

K.C. SEVCIK (1974) Scheduling for minimum total loss using service time distributions. *J. Assoc. Comput. Mach.* 21,66-75.

J.M. STEELE (1980) Subadditive Euclidean functionals and non-linear growth in geometric probability. Research Report, Department of Statistics, Stanford University.

L. STOUGIE (1981) Private communication.

H.G. TUCKER (1967) *A Graduate Course in Probability*, Academic, New York.

R.R. WEBER (1979) Optimal organization of multi-server systems. Ph.D. Thesis, University of Cambridge.

R.R. WEBER (1981) Scheduling jobs with stochastic processing requirements on parallel machines to minimize makespan or flowtime. *J. Appl. Probab.*, to appear.

G. WEISS (1982) Multiserver stochastic scheduling. This volume.

G. WEISS, M.L. PINEDO (1980) Scheduling tasks with exponential service times on non-identical processors to minimize various cost functions. *J. Appl. Probab.* 17,187-202.

P. WHITTLE (1980) Multi-armed bandits and the Gittins index. *J. Roy. Statist. Soc. Ser. B* 42,143-149.

ON STOCHASTIC ANALYSIS OF PROJECT-NETWORKS

W. Gaul

Universität Karlsruhe

ABSTRACT

If the activity-completion-times of a project-network are random variables the project-completion-time is a random variable the distribution function of which is difficult to obtain. Thus, efforts have been made to determine bounds for the mean and bounding distribution functions for the distribution function of the project-completion-time some results of which are shortly surveyed.
Then, a new approach using stochastic programming for a cost-oriented project scheduling model is presented. Generalizing a well-known Fulkerson-approach planned execution-times for the random activity-completion-times are computed where nonconformity with the actual realizations impose compensation costs (gains). Taking into consideration a prescribed project-completion-time constraint the expected costs for performing the activities according to the planned execution-times are minimized. A solution procedure is described which constructs a sequence of nonstochastic Fulkerson project scheduling models. It is demonstrated by means of an example.

KEYWORDS: Network Programming, Scheduling Theory, Stochastic Programming.

1. INTRODUCTION

A project is described by a set of activities, a relation on this set representing restrictions between the activities and activity-completion-times.
A project-network, as graphtheoretical description of a project,

M. A. H. Dempster et al. (eds.), Deterministic and Stochastic Scheduling, 297–309.

is given by

$$D_{s,t} = (V,X,f,Y_X)$$

where (V,X,f) is a finite, directed, simple, acyclic, weakly connected graph with point set V, arc set X, incidence mapping $f=(f^1,f^2)$ with $f^i:X\to V$, i=1,2 ($f^1(x),f^2(x)$ denote the starting-, end point of $x\in X$) and single-element point basis s, single-element point contrabasis t, see e.g. Harary, Norman and Cartwright [12] for the graphtheoretical notations. X corresponds with the set of activities of the project at least after introducing dummy activities (The case where V corresponds with the activities is handled e.g. by MPM, Metra Potential Method, but not discussed here.). The restrictions between the activities are described by the chosen project-network's arc-adjacency relation.
$Y_X=(Y_x,x\in X)$ is a random vector defined on a probability space $(\Omega,\mathcal{G},Pr)$ the components of which give the activity-completion-times (The more general case where additional stochastic aspects influence the project structure is handled e.g. by GERT, Graphical Evaluation and Review Technique, but not discussed here, see Neumann and Steinhardt [15] for a recent contribution.).
Such project-networks $D_{s,t}$ have proved to be an appropriate tool when a schedule for coordinating and supervising of the single activities of a project is needed. One of the aims of project scheduling is to determine the project-completion-time which is yielded by maximizing over the sums of the completion-times of those activities which form paths from s to t. Even under the assumption of stochastic independence for $Y_x,x\in X$, however, the distribution function of the project-completion-time is difficult to obtain (activities can be used by different paths). Thus, in Van Slyke [20] one of the first attempts to apply Monte-Carlo methods was described. Efforts which have been made to determine bounds for the mean and bounding distribution functions for the distribution function of the project-completion-time are shortly surveyed in section 2. Together with the well-known CPM, Critical Path Method, and PERT, Program Evaluation and Review Technique, approaches the results of Fulkerson [8], Clingen [4], Robillard and Trahan [16] and Devroye [5] concerning bounds for the mean of the project-completion-time are mentioned some of which are shown to be special cases of a more general result of Gaul [10]. The results of Kleindorfer [14] and Shogan [18] on bounding distribution functions for the distribution function of the project-completion-time close section 2. As in a typical planning situation execution-times for the activities have to be planned before the actual realizations of the random activity-completion-times are known, in section 3 a new approach of Cleef and Gaul [3] is presented using stochastic programming for a cost-oriented project scheduling model. Generalizing a well-known approach of Fulkerson [7], see also Ford and Fulkerson [6], planned execution-times for the random activity-completion-times are computed where nonconformity with the actual realizations impose compensation costs (gains). The planned execution-times are determined in such a way that the expected compensation costs together with a nonstochastic cost-term are minimized.

Using discrete random activity-completion times (e.g. as approximation of the actual ones) a solution procedure is described which constructs a finite sequence of nonstochastic Fulkerson project scheduling models. The size of the subproblems in the sequence is independent of the number of realizations of the activity-completion times. In section 4 the new approach is demonstrated by means of an example.

2. BOUNDS, BOUNDING DISTRIBUTION FUNCTIONS

Let m be the number of points of $D_{s,t}$.
For $D_{s,t}$ there exists a bijective labelling $l:V\to\{1,\dots,m\}$ with

$$l(s)=1,\ l(t)=m \text{ and } x\in X \Rightarrow l(f^1(x))<l(f^2(x)).$$

In this section such a labelling is needed for the sequential determination of bounds and bounding distribution functions.
For graphtheoretical considerations sometimes the notation $(V(D),X(D))$ (omitting the incidence mapping and the random vector) is used for a network D with point set $V(D)$, arc set $X(D)$. With these abbreviations

$$D^1 \subset D^2 \text{ iff } V(D^1)\subset V(D^2),\ X(D^1)\subset X(D^2),$$

$$D^1 \overset{\cup}{\cap} D^2 \text{ iff } V(D^1 \overset{\cup}{\cap} D^2) = V(D^1) \overset{\cup}{\cap} V(D^2),$$

$$X(D^1 \overset{\cup}{\cap} D^2) = X(D^1) \overset{\cup}{\cap} X(D^2)$$

describe the subnetwork, union-intersection-network notation. Now,

$$D_{i,j} \subset D_{1,m} \text{ is called subproject-network}$$

if $D_{i,j}$ is a project-network with point basis i and point contrabasis j. One has $V(D_{i,j})\subset\{i,i+1,\dots,j-1,j\}$, $X(D_{i,j})\subset f^{-1}(V(D_{i,j})\times V(D_{i,j})\cap f(X))$, the incidence mapping $f/_{X(D_{i,j})}$ and the random vector $Y_{X(D_{i,j})}=(Y_x,x\in X(D_{i,j}))$ are mostly omitted. If for $i,j\in V$ a subproject-network $D_{i,j}$ exists $\tilde{D}_{i,j}$ denotes the maximal one (notice $\tilde{D}_{1,m}=D_{1,m}$ the underlying project-network). A path $P_{i,j}$ with $V(P_{i,j})=\{i_1,\dots,i_n \mid i_1=i,\ i_n=j\}$, $X(P_{i,j})=\{x_1,\dots,x_{n-1} \mid f(x_\mu)=(i_\mu,i_{\mu+1}),\mu=1,\dots,n-1\}$ is a special subproject-network. $(P_{i,j})_k$ resp. $(P_{i,j})^k$, $k\in V(P_{i,j})$, gives the subpath of $P_{i,j}$ from i to k resp. k to j. Instead of $D_{f^1(x),f^2(x)}$ the arc notation x is used.

(1) $$L(D_{i,j}) = \max_{P_{i,j}\subset D_{i,j}} \sum_{x\in X(P_{i,j})} Y_x$$

is the $D_{i,j}$-completion-time ($L(\tilde{D}_{1,m})$ is the project-completion-time). Next, for $v\in V$, $v>1$, consider subproject-network systems of the form

$$\delta_v = \{D_{i,v} \mid i<v\}.$$

With $B(\delta_v)=\{i \mid i\in V, D_{i,v}\in\delta_v\}$ call δ_v proper (with respect to $\tilde{D}_{1,m}$) if

(2) $$\forall D^1_{i_1,v}, D^2_{i_2,v}\in\delta_v: D^1_{i_1,v}\cap D^2_{i_2,v} = \begin{cases}(\{i,v\},\emptyset) & \text{if } i_1=i_2=i,\\ (\{v\},\emptyset) & \text{otherwise,}\end{cases}$$

(3) $$\forall P_{1,v}\subset \tilde{D}_{1,m}\ \exists\ D_{i,v}\in\delta_v: P_{1,v}=(P_{1,v})_i \cup (P_{1,v})^i$$

$$\text{with } (P_{1,v})_i\cap\delta_v\subset(B(\delta_v),\emptyset),\ (P_{1,v})^i\subset D_{i,v}.$$

Proper δ_v always exist, e.g. $\delta_v = \{\tilde{D}_{1,v}\}$ is proper. A useful property of proper δ_v is, see Gaul [10],

(4) $$L(\tilde{D}_{1,v}) = \max_{D_{i,v}\in\delta_v} \{L(\tilde{D}_{1,i}) + L(D_{i,v})\} \text{ if } \delta_v \text{ is proper.}$$

To define lower bounds for the $L(\tilde{D}_{1,v})$-mean let, for $X^*\subset X$, E_{X^*} denote the integration with respect to $Y^*_X = (Y_x, x\in X^*)$, E (without subscript) the expectation. Assume, for proper $\delta_v = \{D_{i,v}\}$, that $T_i, i\in B(\delta_v)$, are known lower bounds for $E\, L(\tilde{D}_{1,i})$, and that $X_v, \overline{X_v}$ is a partition of X with $X_v \subset X(\delta_v)$, then, under adequate stochastic independence assumptions

(5) $$E\, L(\tilde{D}_{1,v}) \geq E_{X_v} \max_{D_{i,v}\in\delta_v} \{T_i + E_{\overline{X_v}} L(D_{i,v})\} = T(\delta_v, X_v, L(\tilde{D}_{1,v})).$$

For different choices of δ_v and X_v one gets well-known special cases:

$$\delta^1_v = \{D_{iv} \mid D_{iv} \text{ coincides with x with } f(x) = (i,v)\}, X^1_v = \emptyset$$

yields

(6) $$T^1_v = T(\delta^1_v, \emptyset, L(\tilde{D}_{1,v})) = \max_{x\in\delta^1_v}\{T^1_i + E_{\overline{X^1_v}} Y_x\} = \max_{x\in\delta^1_v}\{T_i + EY_x\}.$$

Using recursive arguments, if $T^1_i, i\in B(\delta^1_v)$, are determined in the same way as described by (6) (with $T^1_1 = 0$), T^1_v gives the PERT lower bound of $E\, L(\tilde{D}_{1,v})$. If all $Y_x, x\in X$, have degenerate distributions, (6) describes the CPM-approach.

$$\delta^2_v = \delta^1_v,\ X^2_v = X(\delta^2_v)$$

yields

(7) $$T^2_v = T(\delta^2_v, X^2_v, L(\tilde{D}_{1,v})) = E_{X^2_v} \max_{x\in\delta^2_v}\{T^2_i + E_{\overline{X^2_v}} Y_x\} = E \max_{x\in\delta^2_v}\{T^2_i + Y_x\}.$$

Using recursive arguments, if $T^2_i, i\in B(\delta^2_v)$, are determined in the same way as described by (7) (with $T^2_1 = 0$), T^2_v gives the Fulkerson [8] lower bound, see also Clingen [4], of $E\, L(\tilde{D}_{1,v})$.
Whereas it is easy to see that $\delta^1_v = \delta^2_v$ is proper, now, among the set of paths P_{iv} one has to choose

$$\delta^3_v = \{P_{iv} \mid P_{iv} \text{ is path from i to v}, i<v\} \text{ proper}, X^3_v = X(\delta^3_v)$$

which yields

(8) $$T^3_v = T(\delta^3_v, X^3_v, L(\tilde{D}_{1,v})) = E \max_{P_{iv}\in\delta^3_v} \{T^3_i + L(P_{iv})\}.$$

Again, using recursive arguments one gets a method suggest by Robillard and Trahan [16]. For the exact computation of $E\, L(\tilde{D}_{1,v})$ choose

$$\delta^4_v = \{\tilde{D}_{1,v}\}, X^4_v = X(\tilde{D}_{1,v})$$

which yields

(9) $$T^4_v = T(\delta^4_v, X^4_v, L(\tilde{D}_{1,v})) = E[T^4_1 + L(\tilde{D}_{1,v})] = E\, L(\tilde{D}_{1,v})$$

with $T^4_1 = 0$ as usual.
Under assumptions given in Gaul [10] one can show

$$E\, L(\tilde{D}_{1,v}) = T^4_v \geq T^3_v \geq T^2_v \geq T^1_v$$

and construct improved lower bounds.
An easy method to determine upper bounds is given in Devroye [5]. Knowing EY_x, var Y_x, and using the recursive approach described in (6),

$$U'_v = \max_{x\in\delta_v^1}\{U'_i + EY_x\} + \sqrt{n_v \max_{x\in\delta_v^1}\{\overline{\operatorname{var} L(\tilde{D}_{1,i})} + \operatorname{var} Y_x\}},$$

(10)
$$U''_v = \max_{x\in\delta_v^1}\{U''_i + EY_x\} + \ldots$$
$$\ldots + \sqrt{(n_v-1)\ [\max_{x\in\delta_v^1}\{2\,\overline{\operatorname{var} L(\tilde{D}_{1,i})} + \operatorname{var} Y_x\} + \min_{x\in\delta_v^1}\{2\,\overline{\operatorname{var} L(\tilde{D}_{1,i})} + \operatorname{var} Y_x\}]}$$

are shown to be upper bounds for $E\,L(\tilde{D}_{1,v})$ if $Y_x, x\in X$, are stochastically independent. Here, n_v is the number of elements of $B(\delta_v^1)$ and $\overline{\operatorname{var} L(\tilde{D}_{1,v})}$ is an upper bound for $\operatorname{var} L(\tilde{D}_{1,v})$ recursively defined by

$$\overline{\operatorname{var} L(\tilde{D}_{1,v})} = \sum_{x\in\delta_v^1} [\overline{\operatorname{var} L(\tilde{D}_{1,i})} + \operatorname{var} Y_x] \quad (\text{with } \overline{\operatorname{var} L(\tilde{D}_{1,1})} = 0).$$

Lower and upper bounds for the mean and for higher moments of the project-completion-time can also be determined if one knows bounding distribution functions for the distribution function of the project-completion-time, see Kleindorfer [14] and Shogan [18]. With restriction to discrete random activity-completion-times and the abbreviations

$$p(y(v)) = \Pr(Y_{X(\delta_v^1)} = y(v)), \quad y(v) = (y_x, x\in X(\delta_v^1)),$$

the following recursive definition of bounding distribution functions is possible:

Under the assumption of stochastic independence for $Y_{X(\delta_v^1)} = \{Y_x, x\in X(\delta_v^1)\}$, $v\in\{2,\ldots,m\}$,

(11)
$$\text{(a)}\quad F^u_{\tilde{D}_{1,v}}(t) = \sum_{y(v)} p(y(v))\,[\min_{x\in\delta_v^1} F^u_{\tilde{D}_{1,i}}(t-y_x)],$$

$$\text{(b)}\quad F^l_{\tilde{D}_{1,v}}(t) = \sum_{y(v)} p(y(v)) \max\{0, [\sum_{x\in\delta_v^1} F^l_{\tilde{D}_{1,i}}(t-y_x)] - n_v + 1\}$$

$$\text{with } F^l_{\tilde{D}_{1,1}}(t) = F^u_{\tilde{D}_{1,1}}(t) = \begin{cases} 1, & t\geq 0,\\ 0, & \text{otherwise}\end{cases}$$

fulfill

$$F^l_{\tilde{D}_{1,v}}(t) \leq F_{\tilde{D}_{1,v}}(t) = \Pr(L(\tilde{D}_{1,v}) \leq t) \leq F^u_{\tilde{D}_{1,v}}(t), \qquad t\in\mathbb{R}.$$

Obviously, (11) is based on the well-known Frechet-bounds for $\Pr(\bigcap_{x\in\delta_v^1}\{L(\tilde{D}_{1,i}) \leq t-y_x\})$.

Under the additional stochastic dependence assumption of association for $Y_x, x\in X(\delta_v^1)$, $v\in\{2,\ldots,m\}$ (often used in context with reliability problems), an improved lower bounding distribution function

(12)
$$F^{l(ass)}_{\tilde{D}_{1,v}}(t) = \sum_{y(v)} p(y(v)) \left[\prod_{x\in\delta_v^1} F^{l(ass)}_{\tilde{D}_{1,i}}(t-y_x)\right]$$

$$\text{with } F^{l(ass)}_{\tilde{D}_{1,1}}(t) = \begin{cases} 1, & t\geq 0,\\ 0, & \text{otherwise}\end{cases}$$

can be computed which fulfills

$$F^l_{\tilde{D}_{i,v}}(t) \leq F^{l(ass)}_{\tilde{D}_{1,v}}(t) \leq F_{\tilde{D}_{1,v}}(t), \qquad t\in\mathbb{R}.$$

Of course, having established lower, upper bounding distribution functions the determination of lower, upper bounds for the mean and the variance of $L(\tilde{D}_{1,v})$ is straightforward and, thus, omitted. The Kleindorfer bounding distribution functions are also not explicitly reported because, although they were developped under the stronger assumption of stochastic independence for $Y_x, x \in X$, the Shogan bounding distribution functions (11a), (12) are tighter. In all cases, using recursive arguments and increasing v up to m, the desired results for the project-completion-time are obtained.

3. STOCHASTIC PROGRAMMING PROJECT SCHEDULING

Knowing the difficulties originating from the stochastic description of project scheduling problems as discussed in section 2 the question arises whether a new approach might be more appropriate. As in a typical planning situation one has to plan execution-times for the single activities under cost-viewpoints before the actual realizations of the random activity-completion-times are known a "two-stage stochastic programming with simple recourse" approach was described in Cleef and Gaul [3] which generalizes the non-stochastic Fulkerson [7] project scheduling model. A first attempt to apply stochastic programming to project scheduling was formulated by Charnes, Cooper and Thompson [1] within a "chance-constrained stochastic programming" approach.

For an introduction to stochastic programming Kall [13], for an extensive bibliography on papers dealing with various topics of stochastic programming Stancu-Minasian and Wets [19] are recommended, for considerations where the here described stochastic programming model is used for the general linear case with simple recourse see Cleef [2].

The new stochastic programming project scheduling approach is formulated as follows:

For the arcs of the given project-network $D_{s,t}=(V,X,f,Y_x)$ assume

(13) $\Pr(Y_x \geq y_x^o) = 1, \quad x \in X,$

where $y_x^o \geq 0$ is the lowest possible (crash) completion-time.

In the nonstochastic case, see Fulkerson [7], Y_x have degenerate distributions at $y_x \geq y_x^o$ but if one puts up with additional costs for extra efforts assumed to be describable by linear cost-functions of the form

(14) $c(d_x)=b_x-o_x d_x \quad , \quad d_x \in [y_x^o, y_x], \quad x \in X,$

where $b_x, o_x \geq 0$ are known values allowing for costs of needed resources (machines, material, staff etc.),

planned execution-times d_x can be determined which minimize the total costs $\sum_{x \in X} c(d_x)$ under a project-completion-time constraint $\lambda > 0$.

In the stochastic case assume that X_d, X_r is a partition of X into the sets of arcs with deterministic or random activity-completion-times. X_d contains the dummy activities (with $y_x^o = y_x = 0$, $b_x = 0$), $X_r = \emptyset$ gives the nonstochastic Fulkerson-approach. For $x \in X_r$

additional costs for compensating nonconformity between the actual realizations of the activity-completion-times $Y_x(\omega)$, $\omega\in\Omega$, and the planned execution-times d_x (which have to be determined before the realizations are known)

(15) $$\varphi_{d_x}(Y_x(\omega))=\begin{cases} q_x^+(Y_x(\omega)-d_x) & > \\ 0 & , \; Y_x(\omega) = d_x , \quad \omega\in\Omega,\; x\in X_r, \\ -q_x^-(Y_x(\omega)-d_x) & < \end{cases}$$

have to be taken into consideration where q_x^+, q_x^- are known compensation cost-terms satisfying

(16) $$-q_x^+ < q_x^- \leq o_x \quad , \quad x\in X_r .$$

With

$$\Phi_{d_x}=\begin{cases} E\varphi_{d_x}+c(d_x) & , \quad x\in X_r, \\ c(d_x) & , \quad x\in X_d, \end{cases}$$

the following SPPS, Stochastic Programming Project Scheduling, approach can be formulated:

(17) $$\begin{aligned} &\sum_{x\in X} \Phi_{d_x} = \min \\ &d_x+\pi_{f^1(x)}-\pi_{f^2(x)} \leq 0 \quad , \quad x\in X, \\ &-\pi_s \quad + \pi_t \quad \leq \lambda \\ &y_x^o \leq d_x \leq y_x \quad , \quad x\in X . \end{aligned}$$

Here, π_i, $i\in V$, are upper time-bounds when with respect to the planned execution-times d_x all activities x with $f^2(x)=i$ have to be completed. For an appropriate choice of y_x for $x\in X_r$ see (18). (17) describes a linear program if one assumes that for $x\in X_r$, Y_x are (or are approximated by) discrete random activity-completion-times. With the realizations and probabilities

$$y_x^k \text{ with } \Pr(Y_x=y_x^k)=p_x(k)>0,\; k=1,\dots,r_x,\; \sum_{k=1}^{r_x} p_x(k)=1$$

and, for computational convenience, with the choice of y_x^o, $y_x^{r_x+1}$ with $p_x(0)=p_x(r_x+1)=0$ according to

(18) $$0\leq y_x^o<y_x^1<y_x^2<\dots<y_x^{r_x}<y_x \;(=y_x^{r_x+1}=\max\{\max\{y_x^{r_x}\},\lambda\}+1),$$

the reformulation of (17) gives the following large (dependent on the number of realizations of the random variables) linear program:

(19) $$\begin{aligned} &\sum_{x\in X_r}\sum_{k=1}^{r_x}\Big[p_x(k)\,[q_x^+u_x^+(k)+q_x^-u_x^-(k)]+c(d_x)\Big]+\sum_{x\in X_d} c(d_x)=\min, \\ &d_x+\pi_{f^1(x)}-\pi_{f^2(x)} \leq 0 \quad , \quad x\in X , \\ &-\pi_s \quad + \pi_t \quad \leq \lambda \\ &d_x+u_x^+(k)-u_x^-(k) = y_x^k \quad , \quad x\in X_r, \quad k=1,\dots,r_x, \\ &-d_x \leq -y_x^o \quad , \quad x\in X , \\ &d_x \leq y_x \quad , \quad x\in X , \\ &u_x^+(k),u_x^-(k) \geq 0 \quad , \quad x\in X , \quad k=1,\dots,r_x . \end{aligned}$$

Instead of handling (19) a finite sequence of Fulkerson project scheduling models (independent of the number of realizations of the random variables) is solved. For the n-th subproblem select

(20) $$s_x^n\in\{0,1,\dots,r_x\},\; x\in X_r \text{ (with } s_x^n\equiv o,\; x\in X_d),$$

denote

$$(21)\qquad \alpha_x=\begin{cases} y_x^{s_x^n} \\ y_x^o \end{cases},\quad \beta_x=\begin{cases} y_x^{s_x^n+1} \\ y_x \end{cases},\quad \gamma_x=\begin{cases} O(s_x^n) & , x\in X_r, \\ o_x & , x\in X_d, \end{cases}$$

with $O(s_x^n)=q_x^+-(q_x^++q_x^-)\Pr(Y_x\leq s_x^n)+o_x$,
and consider

$$\text{SUB }(s_x^n)\qquad \begin{aligned} &\sum_{x\in X} \gamma_x d_x = \max, \\ &d_x+\pi_{f^1(x)}-\pi_{f^2(x)} \leq 0, \\ &\quad -\pi_s \quad +\pi_t \leq \lambda, \\ &-d_x \leq -\alpha_x, \\ &d_x \leq \beta_x, \end{aligned}$$

and its dual

$$\text{DSUB }(s_x^n)\qquad \begin{aligned} &\lambda v+\sum_{x\in X}[\beta_x g_x-\alpha_x h_x] = \min, \\ &w_x+g_x-h_x = \gamma_x \qquad\qquad , x\in X, \\ &\sum_{\{x|f^1(x)=i\}} w_x-\sum_{\{x|f^2(x)=i\}} w_x=\begin{cases} v & , i=s \\ 0 & , i\neq s,t \\ -v & , i=t \end{cases} , i\in V, \\ &w_x,\ g_x,\ h_x \geq 0 \qquad\qquad , x\in X, \\ &v \geq 0. \end{aligned}$$

DSUB(s_x^n) has restrictions which remind of flow problems in networks, SUB(s_x^n) has restrictions which coincide with those of SPPS except for the random activity-completion-times where the variation of d_x-values is bounded by subsequent realizations y_x^{sn}, y_x^{sn+1}. Obviously, optimal solutions of SUB(s_x^n) are feasible for SPPS, thus, the question arises under which conditions optimal solutions of SUB(s_x^n) are also optimal for SPPS.
A sufficient optimality condition is the following:

Let d_x^*, $x\in X$, π_i^*, $i\in V$, be optimal for SUB(s_x^n)

and w_x^*, g_x^*, h_x^*, $x\in X$, v^* be optimal for DSUB(s_x^n).

If

$$(22)\qquad \begin{aligned} g_x^* &\leq (q_x^++q_x^-)p_x(s_x^n+1) \quad , \quad x\in X_r, \\ h_x^* &\leq (q_x^++q_x^-)p_x(s_x^n) \quad , \quad x\in X_r \text{ with } s_x^n>0, \end{aligned}$$

then d_x^*, $x\in X$, π_i^*, $i\in V$, is optimal for SPPS.

If (22) fails to be satisfied new problems SUB(s_x^{n+1}), DSUB(s_x^{n+1}) have to be selected which allow improvements. The selection instructions use properties of the out-of-kilter algorithm applied to the following modified graph

$$\begin{aligned} &(\tilde{V},\tilde{X},\tilde{f}) \text{ with} \\ &\tilde{V}=V, \\ &\tilde{X}=\{x_1|x\in X\}\cup\{x_2|x\in X\}\cup\{x_o\}, \\ &\tilde{f}(z)=\begin{cases} f(x) & , z=x_k,\ x\in X,\ k=1,2 \\ (t,s) & , z=x_o \end{cases} , z\in\tilde{X}, \end{aligned}$$

because with

$$c_z = \begin{cases} -\beta_x , & z=x_1 , \quad x\in X \\ \lambda , & z=x_o \\ -\alpha_x , & z=x_2 , \quad x\in X \end{cases} , z\in\tilde{X}$$

and

$$l_z = \begin{cases} \gamma_x , & z=x_1 , \quad x\in X \\ \infty , & \text{otherwise} \end{cases} , z\in\tilde{X}$$

the following circulation problem in $(\tilde{V},\tilde{X},\tilde{f})$:

CIRC(s_x^n)

$$\sum_{z\in\tilde{X}} c_z w_z = \min,$$

$$\sum_{\{z|\tilde{f}^1(z)=i\}} w_z - \sum_{\{z|\tilde{f}^2(z)=i\}} w_z = 0 \quad , \quad i\in\tilde{V},$$

$$0\leqq w_z \leqq l_z \quad , \quad z\in\tilde{X},$$

is equivalent to DSUB(s_x^n).

A well-known solution procedure for CIRC(s_x^n) is the out-of-kilter algorithm which consists of an initial phase, a labeling phase, a circulation-alteration phase and a point value-alteration phase. It is easy to check that having obtained optimal point values $\tau_i^*, i\in\tilde{V}$, and an optimal circulation $w_z^*, z\in\tilde{X}$ (by application of the out-of-kilter algorithm to CIRC(s_x^n))

(23) $$w_x^* = w_{x_1}^* + w_{x_2}^*, \; g_x^* = \gamma_x - w_{x_1}^*, \; h_x^* = w_{x_2}^*, \; x\in X, \; v^* = w_{x_o}^*,$$

is optimal for DSUB(s_x^n),

(24) $$\pi_i^* = -\tau_i^*, \; i\in V, \; d_x^* = \min\{\beta_x, \pi_{f^2(x)}^* - \pi_{f^1(x)}^*\}, \; x\in X,$$

is optimal for SUB(s_x^n).

Using CIRC(s_x^n), (23), (24) the optimality-condition (22) can be reformulated as

(25)
$$w_{x_1}^* \geqq \gamma_x - (q_x^+ + q_x^-) p_x(s_x^n+1) \quad , \quad x\in X_r,$$
$$w_{x_2}^* \quad (q_x^+ + q_x^-) p_x(s_x^n) \quad , \quad x\in X_r \text{ with } s_x^n > 0.$$

If (25) fails to be satisfied,

$$X_r^+ = \{x\in X_r \,|\, 0\leqq w_{x_1}^* < \gamma_x - (q_x^+ + q_x^-) p_x(s_x^n+1)\},$$

$$X_r^- = \{x\in X_r \,|\, w_{x_2}^* > (q_x^+ + q_x^-) p_x(s_x^n), \; s_x^n > 0\},$$

$$X_r^o = X_r \setminus \{X_r^+ \cup X_r^-\}$$

is a partition of X_r from which one gets

$$s_x^{n+1} = \begin{cases} s_x^n + 1 , & x\in X_r^+, \\ s_x^n - 1 , & x\in X_r^- \text{ (with } s_x^{n+1}\equiv 0, \; x\in X_d), \\ s_x^n , & x\in X_r^o, \end{cases}$$

and new problems SUB(s_x^{n+1}), DSUB(s_x^{n+1}), CIRC(s_x^{n+1}) for which improvements are possible if a "modified" out-of-kilter algorithm is used. It can be shown, see Cleef and Gaul [3]:

Let $(d_x^*)^{n+1}$, $x\in X$, $(\pi_i^*)^{n+1}$, $i\in V$, (resp. $(d_x^*)^n$, $x\in X$, $(\pi_i^*)^n$, $i\in V$) be optimal for SUB (s_x^{n+1}) (resp. SUB (s_x^n)) obtained by subsequent applications of the "modified" out-of-kilter algorithm.

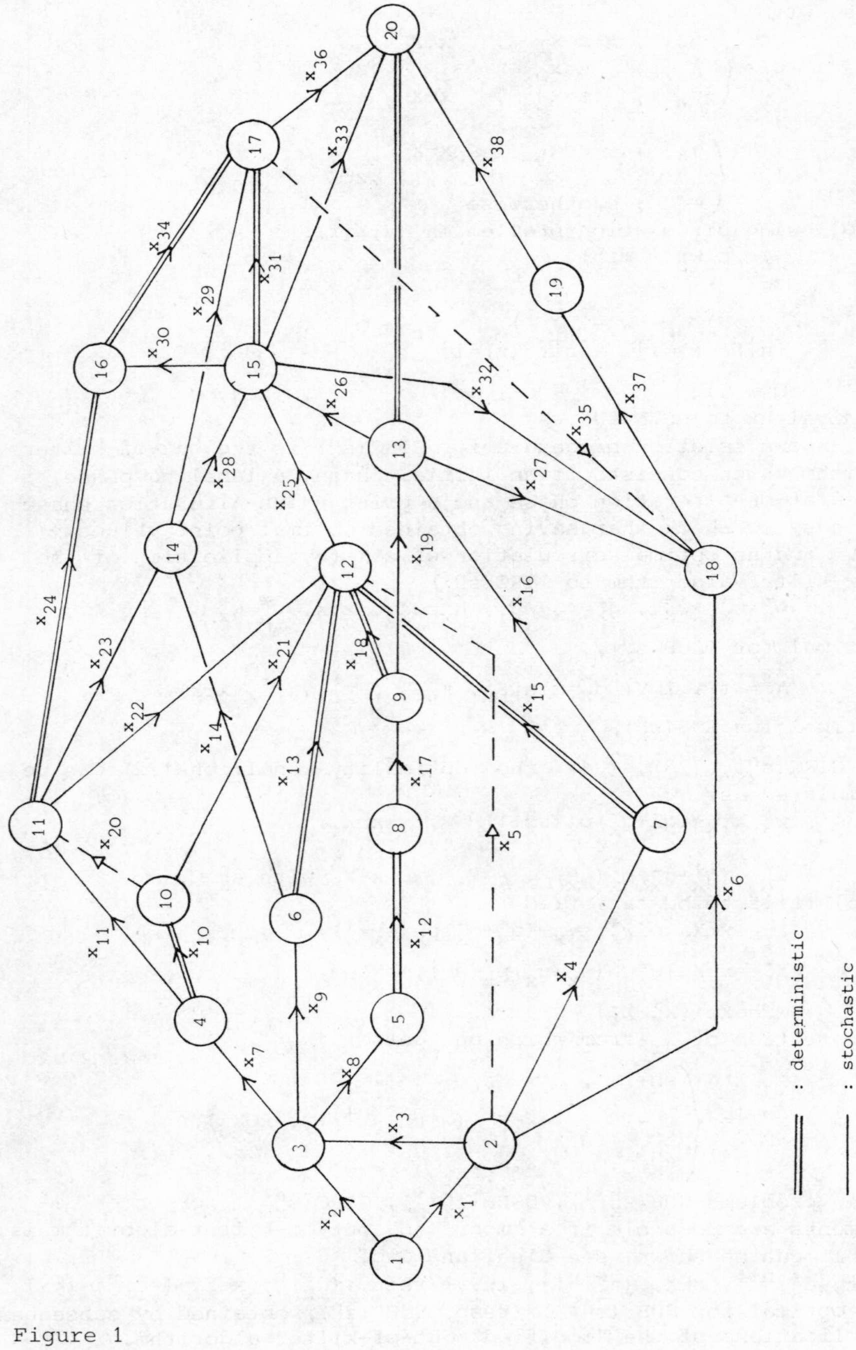

1
2
3
4
5
6
7
8
9
10
11
12
13
14
15
16
17
18
19
20
x_1
x_2
x_3
x_4
x_5
x_6
x_7
x_8
x_9
x_{10}
x_{11}
x_{12}
x_{13}
x_{14}
x_{15}
x_{16}
x_{17}
x_{18}
x_{19}
x_{20}
x_{21}
x_{22}
x_{23}
x_{24}
x_{25}
x_{26}
x_{27}
x_{28}
x_{29}
x_{30}
x_{31}
x_{32}
x_{33}
x_{34}
x_{35}
x_{36}
x_{37}
x_{38}
: deterministic
: stochastic
: dummy

Figure 1

activity x_i	realizations $y^K_{x_i}$	probabilities $p^k_{x_i}$	$y^o_{x_i}$	b_{x_i}	o_{x_i}	$q^+_{x_i}$	$q^-_{x_i}$	$S^n_{x_i}$ n=0,...,7	$d^*_{x_i}$
x_1	3, 5, 10, 13, 20	0.200, 0.300, 0.300, 0.150, 0.050	1	500	30	5	- 3	0 0 0 0 0 0 0 0	1
x_2	3, 5, 10, 13, 20	0.200, 0.300, 0.300, 0.150, 0.050	1	1200	25	4	- 4	0 1 2 2 2 2 2 3	11
x_3	4, 6, 8, 10, 12	0.150, 0.250, 0.250, 0.200, 0.150	2	400	23	10	- 4	0 1 1 1 2 2 2 3	10
x_4	2, 3, 5, 6, 7	0.100, 0.200, 0.500, 0.100, 0.100	1	800	15	10	1	0 1 2 2 2 2 2 2	3
x_5	0, dummy activity	-	0	0	0	-	-	0 ——— 0	0
x_6	2, 4, 8, 16, 22	0.100, 0.200, 0.250, 0.250, 0.200	1	1500	5	8	5	0 1 2 3 4 4 4 4	22
x_7	6, 9, 15, 20, 25	0.175, 0.550, 0.200, 0.025, 0.050	3	3000	20	4	20	0 1 1 1 1 1 0 0	3
x_8	6, 7, 8, 12, 18	0.150, 0.075, 0.300, 0.300, 0.175	2	1550	10	10	0	0 1 2 2 2 1 0 0	4
x_9	6, 9, 15, 20, 25	0.175, 0.550, 0.200, 0.025, 0.050	3	1100	4	6	2	0 1 2 1 1 1 1 0	4
x_{10}	12, deterministic	-	6	1000	8	-	-	0 ——— 0	6
x_{11}	2, 3, 5, 6, 7	0.100, 0.200, 0.500, 0.100, 0.100	1	1100	1	6	0	0 1 2 3 4 4 3 3	6
x_{12}	13, deterministic	-	5	1800	12	-	-	0 ——— 0	5
x_{13}	12, deterministic	-	3	1800	14	-	-	0 ——— 0	12
x_{14}	3, 5, 10, 13, 20	0.200, 0.300, 0.300, 0.150, 0.050	1	600	3	14	2	0 1 2 2 2 2 2 2	10
x_{15}	24, deterministic	-	6	5000	18	-	-	0 ——— 0	23
x_{16}	6, 9, 15, 20, 25	0.175, 0.550, 0.200, 0.025, 0.050	3	300	1	5	- 3	0 1 2 3 3 3 3 3	20
x_{17}	3, 5, 10, 13, 20	0.200, 0.300, 0.300, 0.150, 0.050	1	2000	9	3	1	0 0 0 0 0 0 0 0	1
x_{18}	18, deterministic	-	6	2000	10	-	-	0 ——— 0	6
x_{19}	4, 6, 8, 10, 12	0.150, 0.250, 0.250, 0.200, 0.150	2	500	4	2	0	0 1 1 0 0 0 0 0	3
x_{20}	0, dummy activity	-	0	0	0	-	-	0 ——— 0	0
x_{21}	2, 3, 5, 6, 7	0.100, 0.200, 0.500, 0.100, 0.100	1	700	9	12	2	0 1 2 3 4 5 5 5	7
x_{22}	6, 7, 8, 12, 18	0.175, 0.550, 0.200, 0.025, 0.050	2	2000	11	10	8	0 1 2 2 2 2 2 2	7
x_{23}	6, 9, 15, 20, 25	0.175, 0.550, 0.200, 0.025, 0.050	3	800	3	6	1	0 0 0 0 0 0 0 0	5
x_{24}	9, deterministic	-	3	1000	4	-	-	0 ——— 0	9
x_{25}	2, 3, 5, 6, 7	0.100, 0.200, 0.500, 0.100, 0.100	1	400	5	4	2	0 0 0 0 0 0 0 0	1
x_{26}	4, 6, 8, 10, 12	0.150, 0.250, 0.250, 0.200, 0.150	2	200	1	2	0	0 1 0 1 1 1 1 1	4
x_{27}	6, 7, 14, 15, 30	0.300, 0.100, 0.250, 0.250, 0,100	2	500	2	10	2	0 1 2 2 2 2 2 2	12
x_{28}	6, 9, 15, 20, 25	0.175, 0.550, 0.200, 0.025, 0.050	3	300	15	2	15	0 0 0 0 0 0 0 0	3
x_{29}	2, 3, 5, 6, 7	0.200, 0.300, 0.300, 0.150, 0.050	1	600	3	8	3	0 1 2 3 4 4 4 4	7
x_{30}	2, 4, 8, 16, 22	0.175, 0.550, 0.200, 0.025, 0.050	1	800	2	3	2	0 0 0 0 0 0 0 0	1
x_{31}	15, deterministic	-	3	1200	5	-	-	0 ——— 0	8
x_{32}	6, 9, 15, 20, 25	0.100, 0.200, 0.500, 0.100, 0.100	3	1000	2	8	1	0 1 1 1 1 1 1 1	8
x_{33}	17, deterministic	-	5	2400	1	-	-	0 ——— 0	12
x_{34}	25, deterministic	-	7	1000	2	-	-	0 ——— 0	7
x_{35}	0, dummy activity	-	0	0	0	-	-	0 ——— 0	0
x_{36}	2, 4, 5, 8, 15	0.200, 0.300, 0.300, 0.150, 0.050	0	600	4	8	3	0 1 2 1 1 1 1 2	4
x_{37}	6, 9, 15, 20, 25	0.150, 0.250, 0.250, 0.200, 0.150	3	400	30	16	- 2	0 0 0 0 0 0 0 0	3
x_{38}	2, 3, 5, 6, 7	0.100, 0.200, 0.500, 0.100, 0.100	1	450	25	15	- 3	0 0 0 0 0 0 0 0	1

$\lambda = 40$

Table 1

If during the application of the "modified" out-of-kilter algorithm to CIRC(s_x^{n+1})

(a) a point value-alteration phase was performed then $(d_x^*)^{n+1}$, $x \in X$, $(\pi_i^*)^{n+1}$, $i \in V$, gives a better solution of SPPS than $(d_x^*)^n$, $x \in X$, $(\pi_i^*)^n$, $i \in V$,

(b) no point value-alteration phase was performed then $(d_x^*)^n$, $x \in X$, $(\pi_i^*)^n$, $i \in V$, is optimal for SPPS.

To ensure finiteness a restriction to rational data for the description of SPPS has to be made.

4. EXAMPLE

The dimension of the following example indicates that the suggested approach can handle problems of application-relevant size. For the project-digraph of Fig. 1 the random activity-completion-times are assumed to have five realizations which together with its associated probabilities and the crash completion-times y_x^o, the cost-terms b_x, o_x, the compensation cost-terms q_x^+, q_x^- are given in Tab. 1. Taking into consideration a project-completion-time constraint $\lambda=40$ the optimal solution d_x^* was optained in seven iterations starting with $s_x^o \equiv 0$, $x \in X$. For the computation about 15 seconds CPU-time on UNIVAC 1108 were needed. For other examples see Cleef and Gaul [3] or Gaul [11].

REFERENCES

[1] Charnes, A., Cooper, W.W., Thompson, G.L. (1964) Critical Path Analysis via Chance-Constrained and Stochastic Programming. Operations Research 12, pp. 460 - 470.

[2] Cleef, H.J. (1981) A Solution Procedure for the Two-Stage Stochastic Program with Simple Recourse. Zeitschrift für Operations Research 25, pp. 1 - 13.

[3] Cleef, H.J., Gaul, W. (1980) Project Scheduling via Stochastic Programming. SFB-72-Preprint 367, University of Bonn.

[4] Clingen, C.T. (1964) A Modification of Fulkerson's PERT Algorithm. Operations Research 12, pp. 629 - 632.

[5] Devroye, L.P. (1979) Inequalities for the Completion Times of Stochastic PERT Networks. Mathematics of Operations Research 4, pp. 441 - 447.

[6] Ford, L.R., Fulkerson, D.R. (1962) Flows in Networks. Princeton University Press.

[7] Fulkerson, D.R. (1961) A Network Flow Computation for Project Cost Curves. Management Sc. 7, pp. 167 - 178.

[8] Fulkerson, D.R. (1962) Expected Critical Path Length in PERT Type Networks. Operations Research 10, pp. 808 - 817.

[9] Fulkerson, D.R. (1964) Scheduling in Project Networks. Proc. IBM Scient. Comp. Symp. Comb. Problems, pp. 73 - 92.

[10] Gaul, W. (1981 a) Bounds for the Expected Duration of a Stochastic Project Planning Model. J. Infor. & Optimiz. Sc. 2, pp. 45 - 63.

[11] Gaul, W. (1981 b) Stochastische Projektplanung und Marketingprobleme. Paper presented at: 6. Symposium Operations Research, Augsburg.

[12] Harary, F., Norman, F.Z., Cartwright, D. (1965) Structural Models: An Introduction to the Theory of Directed Graphs. Wiley.

[13] Kall, P. (1976) Stochastic Linear Programming. Springer Verlag.

[14] Kleindorfer, G.B. (1971) Bounding Distributions for a Stochastic Acyclic Network. Operations Research 19, pp. 1586 - 1601.

[15] Neumann, K., Steinhardt, U. (1975) GERT-Networks and the Time-Oriented Evaluation of Projects. Lecture Notes in Economics and Mathematical Systems 172. Springer Verlag.

[16] Robillard, P., Trahan, M. (1976) Expected Completion Time in PERT Networks. Operations Research 24, pp. 177 - 182.

[17] Robillard, P., Trahan, M. (1977) The Completion Time of PERT Networks. Operations Research 25, pp. 15 - 29.

[18] Shogan, A.W. (1977) Bounding Distributions for a Stochastic PERT Network. Networks 7, pp. 359 - 381.

[19] Stancu-Minasian, I.M., Wets, M.J. (1976) A Research Bibliography in Stochastic Programming. Operations Research 24, pp. 1078 - 1119.

[20] Van Slyke, R.M. (1963) Monte Carlo Methods and the PERT Problem. Operations Research 11, pp. 839 - 860.

Part II. Advanced Research Institute Proceedings

INTRODUCTION

The nine contributions to the second part of these proceedings are representative of current interests in scheduling theory - all of them involve probability theory in some essential way. During the Institute, a formal discussant was assigned to each paper, and following its presentation detailed discussion of its contents and related questions usually ensued. In this introduction we shall attempt to sketch the contents of the papers and to present any important questions raised in the discussion concerning future directions for research.

In the first paper, Coffman, Frederickson and Lueker give a probabilistic analysis of the average case performance of the *longest processing time* (LPT) heuristic for the problem of scheduling n independent jobs on m identical parallel machines so as to minimize makespan when processing times are independent random variables with uniform distributions on the interval $(0,1]$. They show that the ratio of the expected length of an LPT schedule to the expected length of an optimal (preemptive) schedule is $1+O(m^2/n^2)$ and that in the case $m = 2$ this can be improved to $1+\Omega(1/n^2)$.

Due to the authors' remark on the specificity of the proof techniques to the uniform distribution, a suggestion was made in the discussion concerning the possibility of conducting probabilistic analysis of algorithms by building approximate results for more complex data distributions from the convolution of results for elementary distributions such as the uniform or triangular.

It was also pointed out that a bound on the ratio of the expected heuristic criterion value to the expected optimum as derived in this paper - although usually not itself easy to establish - is only an approximation to the ideal of obtaining results for the expectation of the performance ratio, or better for its distribution.

M. A. H. Dempster et al. (eds.), Deterministic and Stochastic Scheduling, 313–318.

A probabilistic analysis of the performance ratio distribution of the *next fit* bin packing heuristic was in fact presented at the Institute [Hofri 1981].

Discussion also centered on the efficacy of a stochastic upper bound on the performance ratio which is known to converge to one in a suitable sense (*e.g.* almost surely) as the number of jobs tends to infinity. The consensus was that in this difficult area even such asymptotic results are useful, particularly if the *speed* of convergence can be established or estimated.

The next four contributions concern *optimality* results for various stochastic scheduling problems. In the first of these, Whittle gives a simple proof of the optimality of the *Gittins index policy* for the *multi-armed bandit* problem (preemptive single machine scheduling) in discrete time. The proof is based on an analysis of the dynamic programming recursion for a modified problem, in which an option to *retire* for a specific reward is introduced. It is shown that the index policy is optimal for this modified problem and that for a small enough value of the reward the option to retire is never optimal. In the setting of single machine scheduling, this approach is extended to the case of Poisson job arrivals, deadlines and no preemption, which generalizes some work of Harrison on multiclass customer scheduling in a single server queueing system.

From the scheduling viewpoint, the optimality criterion employed in the multi-armed bandit problem (*discounted* reward) is somewhat unusual. Discussion raised the question of rigorously establishing the optimality of index policies in the limit as the discount factor tends to one (*i.e.* the more usual *undiscounted* case).

It was pointed out that, as indicated by the original proofs of the Gittins school, there is a close relation between the optimality of index policies and *interchange* arguments in which strict discounting plays a major role. The alternative approach taken by Whittle heralds an increasing tendency to apply the Bellman-Carathéodory-Hilbert-Hamilton-Jacobi theory of optimal control (and its variants) to stochastic sequencing and scheduling problems.

This is the general approach taken by Nash and Weber in the next paper. They attempt to characterize Markov decision processes with costs depending only on the state of the process in which there is a *unique* optimal strategy minimizing total costs in expectation or distribution up to *every* finite time horizon. Such a strategy is termed respectively *expectation* or *stochastically dominant*. Although the problems treated are posed in continuous time, the authors reduce the provision of a sufficient condition for the existence of an expectation dominant strategy for such a process to an algebraic problem. This is accomplished through the *uniformization* device [Keilson 1979] of constructing an equivalent semi-Markov process with Poisson jump epochs and studying the dynamics of the value function for the embedded discrete time Markov decision process. Applications are given to assigning arriving customers to parallel servers in order to maximize throughput, optimally allocating dynamic computer memory, optimizing the stake in an advantageous repeated gamble, and optimally repairing a series system. In each of these cases a particular matrix figuring in the sufficient conditions for the existence of an expectation dominant strategy can be constructed, and the authors raise the question of whether, when such a strategy exists, the matrix in question must also exist.

The discussant of this paper noted that the preemptive single machine scheduling problem with negative exponential processing times, deterministic release dates and total weighted completion time criterion has an expectation and stochastically dominant strategy which is similar to Smith's rule, but that in the case of deterministic processing times and zero release dates this rule, while optimal, does not necessarily minimize the costs up to every time horizon.

Pinedo considers the computational complexity of determining policies that are optimal in expectation for a variety of stochastic scheduling problems, in which all the relevant data (processing times, release dates, due dates, *etc.*) are independent negative exponentially distributed random variables. Examples are given of models in which the optimal policy can be determined in polynomial time while the deterministic version is NP-hard, as well as of NP-hard stochastic models.

The discussion centered around the question: what is it that often makes stochastic scheduling models easier than their deterministic counterparts? First, in some stochastic models no distinction between preemptive and nonpreemptive scheduling need be made, due to the memoryless property of the negative exponential distribution. Secondly, stochastic data tend to smooth the hard discontinuities encountered in deterministic models. A third reason is that these stochastic and deterministic problems have quite different optimality requirements: a scheduling policy minimizing an expected value over (almost) all possible realizations of the data versus a schedule minimizing a single value for a specific realization of the data. On the other hand, the paper cites the example of minimizing makespan in a two-machine open shop: there is a linear-time algorithm for the deterministic version, while the polynomial-time algorithm for the stochastic version is not as fast. More investigation is needed before these phenomena are completely understood.

The final paper concerning optimal schedules provides more data for this investigation. Bruno reviews some of the known results for the problem of scheduling identical jobs on identical parallel machines subject to *tree*-like precedence constraints so as to minimize makespan. In the deterministic case, the job processing times are all equal to one; in the stochastic case, they are independent random variables with unit mean negative exponential distributions. While *highest level first* (HLF) scheduling is optimal for deterministic problems with either *intree* or *outtree* constraints, in the stochastic case the situation is more complicated and not so favorable. For intree constraints, HLF is optimal only for two machines; for outtree constraints, HLF is not optimal for two or more machines.

In the discussion, the complexity of stochastic multiprocessor problems with other types of precedence constraints, such as *series-parallel* constraints, was noted to be open. Simple modifications of HLF were considered worthy of study.

It was observed that the optimality of HLF for the two-machine problem with intree constraints can be proved using the methods of Nash and Weber discussed above, even in the case of independent decreasing completion rate processing time distributions. The conjecture was made that these methods could also establish the optimality of HLF for the m-machine problem with *chain*-like constraints and decreasing completion rate jobs.

The next two contributions concern the quality of some *approximative* stochastic scheduling strategies. In the first of these, Glazebrook considers a variety of preemptive single machine problems in both discrete and continuous time. He first gives sufficient conditions on processing time distributions for the optimality of (nonpreemptive) *permutation* schedules for a precedence constrained discrete time model with discounted setup, switching and teardown costs. He next examines the quality of such schedules as approximate solutions to, *e.g.*, a more standard continuous time model with independent nonmonotone completion rate processing time distributions and (discounted or undiscounted) total completion costs criterion.

Performance guarantees are obtained for the heuristic, but, as the discussant pointed out, more information on their quality would be useful.

Nash and Weber consider the efficacy of a generalization of Gittins' *forwards induction* policies - termed *sequential open loop* strategies in engineering parlance - for a variety of single and multiple machine stochastic scheduling problems. Their general approach involves detailed study of necessary (Pontryagin-Hamilton maximum principle) and sufficient (Bellman-Hamilton-Jacobi partial differential inequality) optimality conditions for a deterministic control formulation applicable to many stochastic scheduling problems. This point of view suggests sequential open loop approximative policies as more analytically and computationally tractable than optimal policies with regard to standard techniques of control theory. The intricacies are illustrated, for example, on a problem with identical machines and identical negative exponential due dates.

Although the methods are not directly based on dynamic programming, the results - particularly on asymptotic optimality - have the flavor of *value iteration* in dynamic programming. More investigation is needed before sequential open loop scheduling can be recommended in practice, as the authors themselves suggest.

In the penultimate paper of this volume, Mitrani studies the delay functions achievable for a scheduled single server queueing system in equilibrium with Poisson arrivals and general service time distribution. A *delay function* gives the expected time W(x) spent in the equilibrium queue as a function of the job length x which is drawn from the service time distribution but *announced on arrival*. The delay function depends of course on the scheduling policy, which may make use of job length information for jobs in the current queue. Mitrani addresses the problem of characterizing the delay functions generated by nonpreemptive scheduling strategies. For the case of a *finite discrete* service time distribution, the set of achievable delay functions is characterized in terms of a system of linear equations and inequalities, which defines a convex polytope whose extreme points are known explicitly. The analogous integral system is conjectured to be a valid characterization for the *continuous* case.

The discussant pointed out that the set of achievable functions could be characterized by a linear program with parametric objective function, given by an arbitrarily weighted sum of the delay function values, and that it would therefore be interesting to study the dual program. In the continuous case, progress with regard to synthesizing specific delay functions in terms of convex combinations of extreme points of the achievable set might benefit from *Choquet theory* (see *e.g.* [Asimow & Ellis 1980]). It was also mentioned that the Gittins index policy solves the optimization problem of minimizing a specific weighted sum of the delay function values in the cases of both preemptive and nonpreemptive scheduling.

Other interesting scheduling problems arise for single server queueing systems. One such problem, discussed at the Institute by Gelenbe and typical of a much larger class, is scheduling a set of *mutually destructive* users on a single resource. If one of the tasks accesses the resource by itself, it uses it successfully for a finite time. However, if several users request it simultaneously, then all the requests fail and they have to try their luck again. The problem is then to choose the trial and retrial instants optimally: the optimality criterion may be, *e.g.*, the rate of successful attempts. This leads to a class of problems related to random access techniques, whose applications range from satellite communications to local area networks and to conflict of access to common data in data bases.

The final paper, contributed by Herzog, describes two non-von Neumann architecture multiprocessor projects at Erlangen from the point of view of conception and performance modelling. Both systems - a three-level hierarchical array and a distributed reconfigurable multiprocessor kit - are *synchronous* (*i.e.* clock-based) systems. *Asynchronous* systems also appear promising and offer many advantages over clock-based machines (see *e.g.* [Backus 1978; Davis 1980]).

In discussion, it was pointed out that program tearing, assignment and scheduling, and in particular combinatorial algorithm design, so as to take optimum advantage of parallelism for non-von Neumann computer systems is an important area for research. This area is currently being actively pursued in both East (Kiev Cybernetics Institute) and West (see *e.g.* [Burton *et al.* 1981]).

REFERENCES

L. ASIMOW, A.J. ELLIS (1980) *Convexity Theory and Its Applications in Functional Analysis*, London Mathematical Society Monograph 16, Academic Press, London.

J. BACKUS (1978) Can programming be liberated from the von Neumann style? A functional style and its algebra of programs. *Comm. ACM* 21,613-614.

F.W. BURTON, G.P. McKEOWN, V.J. RAYWARD-SMITH, M.R. SLEEP (1981) Parallel processing and combinatorial optimization. School of Computing Studies, University of East Anglia.

A.L. DAVIS (1980) A data flow evaluation system based on the concept of recursive locality. *Proc. National Computer Conf. 1979*, USA, 1079-1086.

M.R. HOFRI (1981) On the number of bins required to pack N pieces using the next fit algorithm. Unpublished manuscript.

J. KEILSON (1979) *Markov Chain Models: Rarity and Exponentiality*, Springer, New York.

PROBABILISTIC ANALYSIS OF THE LPT PROCESSOR SCHEDULING HEURISTIC

E.G. Coffman, Jr.[1], G.N. Frederickson[2], G.S. Lueker[3]

[1]Bell Laboratories, Murray Hill, NJ
[2]Pennsylvania State University
[3]University of California, Irvine

ABSTRACT

The average performance of the LPT processor scheduling algorithm is analyzed, under the assumption that task times are drawn from a uniform distribution on (0,1]. The ratio of the expected length of the schedule to the expected length of a preemptive schedule is shown to be bounded by $1 + O(m^2/n^2)$, where m is the number of processors, and n is the number of tasks.

1. INTRODUCTION

We consider the following fundamental processor scheduling problem: Given n tasks with known execution times, schedule the tasks on m identical processors, with no preemptions, so as to minimize the finish time. The problem has been shown to be NP-hard (8), indicating that there is a very great likelihood that no polynomial time algorithm exists that solves it. As a consequence, a number of authors have considered heuristics that generate near optimal optimal solutions in polynomial time.

The performance of a heuristic may be characterized in terms of a worst-case bound on the ratio of the cost of a heuristic solution to the cost of an optimal solution. One of the simplest heuristics for the processor scheduling problem is the LPT heuristic, or largest processing time first. Tasks are ordered in a list from largest to smallest and scheduled in the order of this list, with the next task always placed on a processor that is currently least used. The LPT heuristic has been shown in (7) to have

M. A. H. Dempster et al. (eds.), Deterministic and Stochastic Scheduling, 319–331.

a worst-case ratio of $4/3 - 1/(3m)$. A better algorithm in terms of worst-case behavior is MULTIFIT (2), with a bound of which has recently been improved in (6). Heuristics with constant worst-case bounds arbitrarily close to 1 are known (7, 12), but these require computation time that is exponential in m.

It has been observed by several authors that the worst-case bounds for these heuristics are not indicative of average performance, which simulation results suggest is considerably better. Until now, relatively little probabilistic analysis has been applied to heuristics for any of the NP-hard problems. Examples are the traveling salesman's problem (9, 10) and several bin packing problems (5, 11, 13).

In this paper, we analyze the average performance of the LPT heuristic, under the assumption that task times are drawn from a uniform distribution on $(0,1]$. This distribution is chosen for the sake of tractability, but we note that simulations of various heuristics have been made using this distribution (2). We bound the ratio of the expected finish time for the heuristic to the expected finish time of an optimal preemptive schedule. This type of ratio was employed in (5) to analyze the average performance of two simple bin packing heuristics. We show that this ratio is $1 + O(m^2/n^2)$ for the LPT heuristic, confirming analytically that the heuristic does do well on average. For the case $m = 2$, we demonstrate that this ratio is $1 + \Omega(1/n^2)$. We also analyze a related heuristic for $m = 2$.

2. ANALYSIS OF LPT FOR 2 PROCESSORS

We first analyze the average performance of LPT in the case in which there are two processors. The analysis determines a bound on the probability that the r-th smallest task is the last task inserted on the later-finishing processor, and a bound on the expected value of the mismatch given that r-th task is the last. To facilitate the analysis, tasks smaller than the r-th will be considered to come in any order at the end. Let X be a set of n task times, and let the tasks be assigned to the processors in the order $x_n, x_{n-1}, \ldots, x_1$. The scheduling algorithm will induce a partition of X into Q_1 and Q_2, the subsets of tasks on each processor. Assume without loss of generality that $c(Q_1) \geq c(Q_2)$. Let A_r be the condition that x_r is the last element inserted into Q_1. Let $g_i = \Sigma_{j=1}^{i} x_j$. Let B_r be the condition that $x_r \geq g_{r-1}$, given that $x_i \leq x_r$ for $i=1,\ldots,r-1$, and $x_r \leq x_i$ for $i=r+1,\ldots,n$. We note that in any case in which A_r holds,

B_r also holds. Furthermore, we have $c(Q_1) - c(Q_2) \le x_r - g_{r-1}$. Thus,

$$\begin{aligned} E(\hat{F}) &= \Sigma_{r=1}^{n} P\{A_r\} E(c(Q_1)|A_r) \\ &= \Sigma_{r=1}^{n} P\{A_r\} E((g_n + c(Q_1) - c(Q_2))/2|A_r) \\ &= E(g_n)/2 + \Sigma_{r=1}^{n} P\{A_r\} E((c(Q_1) - c(Q_2))/2|A_r) \\ &< E(g_n)/2 + \Sigma_{r=1}^{n} P\{B_r\} E((x_r - g_{r-1})/2|B_r). \end{aligned}$$

We first determine the probability that condition B_r holds. An expression that is useful in determining this probability is evaluated in the following lemma.

Lemma 2.1.

Let $r \ge 1$, and $0 \le x_i \le x$ for $i=1,\dots,r-1$. Let

$$a(r, x) = \int_0^x \int_0^{x-g_1} \dots \int_0^{x-g_{r-2}} dx_{r-1} \dots dx_2 dx_1.$$

Then $a(r, x) = x^{r-1}/(r-1)!$.

Proof.

The proof proceeds by induction on r. For $r = 1$, $a(r, x)$ is trivially 1. For $r > 1$, let $y = x - x_1$. Then

$$a(r, x) = \int_0^x a(r-1, y)\, dx_1 = \int_0^x (x-x_1)^{r-2}/(r-2)! \; dx_1,$$

which yields the desired result. □

Lemma 2.2.

Let $r-1$ elements, $x_1,\dots,x_{r-1}$, be drawn from a uniform distribution on $(0, x]$. Then $P\{x \ge g_{r-1}\} = 1/(r-1)!$.

Proof.

The condition $x \ge g_{r-1}$ implies that

$$0 \le x_i \le x-g_{i-1}, \quad i-1,\dots,r-1.$$

Thus $P\{x \ge g_{r-1}\} = a(r, x)/x^r$. Substituting from Lemma 2.1 gives the claimed result. □

Lemma 2.3. Let $r-1$ elements, $x_1,\dots,x_{r-1}$, be drawn from a uniform distribution on $(0, x]$. Then

$$E(x - g_{r-1} | x \geq g_{r-1}) = x/r.$$

Proof. Using the same set of constraints as in Lemma 2.2, we have

$$E(x-g_{r-1}|x\geq g_{r-1}) = \frac{1}{a(r,x)} \int_0^x \int_0^{x-g_1} \dots \int_0^{x-g_{r-2}} (x-g_{r-1})dx_{r-1}\dots dx_2 dx_1$$

$$= \frac{a(r+1, x)}{a(r, x)} = \frac{x^r}{r!} \frac{(r-1)!}{x^{r-1}} = \frac{x}{r}. \quad \square$$

Lemma 2.4. Let n elements be drawn from a uniform distribution on $(0, 1]$. Let X_r be the order statistic of the r-th smallest element, and g_{r-1} the sum of the $r-1$ smaller elements. Then

$$E(X_r - g_{r-1} | X_r \geq g_{r-1}) = 1/(n+1).$$

Proof. We represent the desired value by Z. Thus

$$Z = \int_0^x f_r(x)\, E(x - g_{r-1} | x \geq g_{r-1})\, dx,$$

where $f_r(x)$ is the probability density function of X_r for a uniform distribution on $(0, 1]$. For $1 \leq r \leq n$, we have from (3):

$$f_r(x) = \frac{n!\, x^{r-1}\, (1-x)^{n-r}}{(r-1)!\,(n-r)!}.$$

The $r-1$ elements smaller than X_r may be viewed as drawn from a uniform distribution on $(0, x]$. Substituting from Lemma 2.3, we get

$$Z = \int_0^1 \frac{n!\, x^{r-1}\, (1-x)^{n-r}}{(r-1)!\,(n-r)!} \frac{x}{r}\, dx,$$

which yields the desired result. $\square$

Theorem 2.1. Let n elements drawn from a uniform distribution on $(0, 1]$ be scheduled on two processors according to the LPT heuristic Let $\hat{F}$ be the length of the schedule generated. Then

$$E(\hat{F}) < \frac{n}{4} + \frac{e}{2(n+1)}.$$

Proof. From a preceding discussion, we have

$$E(\hat{F}) < E(g_n)/2 + \frac{1}{2}\Sigma_{r=1}^{n} P\{x_r \geq g_{r-1}\} \; E(x_r - g_{r-1} | x_r \geq g_{r-1}).$$

Noting that $E(g_n) = n/2$, and using the results from Lemmas 2.2 and 2.4, we get:

$$E(\hat{F}) < \frac{n}{4} + \frac{1}{2}\Sigma_{r=1}^{n} \frac{1}{(r-1)!} \; \frac{1}{n+1} < \frac{n}{4} + \frac{1}{2(n+1)} \; \Sigma_{r=0}^{\infty} \frac{1}{r!}$$

$$= \frac{n}{4} + \frac{e}{2(n+1)}. \quad \square$$

Corollary 2.1. Let n elements drawn from a uniform distribution on $(0, 1]$ be scheduled on two processors according to the LPT heuristic. Let $\hat{F}$ be the length of the schedule generated, and $\tilde{F}$ be the sum of the elements divided by 2. Then

$$E(\hat{F})/E(\tilde{F}) < 1 + O(1/n^2).$$

Proof. We note that $E(\tilde{F}) = E(g_n/2) = n/4$. The result then follows immediately. $\square$

3. A LOWER BOUND FOR TWO PROCESSORS

We have shown in the previous section that the expected length of a schedule generated by the LPT algorithm is $n/4 + O(1/n)$. In this section, we show that the bound is tight. That is, the expected length of a schedule is $n/4 + \Omega(1/n)$.

Theorem 3.1. Let n elements drawn from a uniform distribution on $(0,1]$ be scheduled on two processors according to the LPT heuristic. Let $\hat{F}$ be the length of the schedule generated. Then

$$E(\tilde{F}) \geq \frac{n}{4} + \frac{1}{4(n+1)}.$$

Proof. Let $y = |c(Q_1 - \{x_1\}) - c(Q_2 - \{x_1\})|$.

$$E(c(Q_1) - c(Q_2))$$

$$= \int_0^1 (P\{y \geq x\} \int_0^x (y-x_1)\, dx_1$$

$$+ P\{y < x\} (\int_0^y (y-x_1)\, dx_1 + \int_y^x (x_1-y)\, dx_1)) \quad f_2(x) \quad dx$$

$$= \int_0^1 (P\{y \geq x\}(yx-x^2/2) + P\{y < x\}(x^2/2 - xy + y^2)) \; f_2(x)\, dx$$

$$\geq \int_0^1 (P\{y \geq x\} \; x^2/2 + P\{y < x\} \; x^2/4) \; f_2(x) dx$$

$$\geq \int_0^1 x^2/4\ f_2(x)dx = \int_0^1 x^2/4\ \ n!/(n-2)!\ \ (1-x)^{n-2}\ dx$$

$$= 1/(2(n+1)).$$

The result follows, since $E(\hat{F}) = E(g_n/2) + E((c(Q_1) - c(Q_2))/2)$. □

4. AN UPPER BOUND FOR MORE THAN TWO PROCESSORS

In this section we repeat the analysis of section 2, but here generalized to the problem of scheduling $m \geq 2$ processors. Let X be as before. The scheduling algorithm now induces a partition of X into subsets Q_i, $i=1,2,\ldots,m$. We assume that $c(Q_1) = \max\{c(Q_1)\}_{i=1}^m$. Let A_r and g_i be as defined in section 2.

Let $h_j = \Sigma_{k=1}^j x_{(m-1)(k-1)+1}$. Let C_j be the condition that $x_{j(m-1)} \geq h_{j-1}$, given that $x_{j(m-1)}$ is the $j(m-1)$-th smallest element and $x_{(m-1)(k-1)+1}$ is the $((m-1)(k-1)+1)$-th smallest element, $k=1,\ldots,j-1$. We relax condition C_j to yield condition D_j as follows. Let D_j be the condition that $x_{j(m-1)} \geq h_{j-1}$, given that $x_{j(m-1)}$ is the $j(m-1)$-th smallest element, and $x_i \leq x_{j(m-1)} - h_k$ for $i=(m-1)(k-1)+2,\ldots,(m-1)k$ and $k=1,\ldots,j-2$. Thus,

$$\begin{aligned}
E(\hat{F}) &= \Sigma_{r=1}^n\ P\{A_r\}\ E(c(Q_1)|A_r)\\
&= \Sigma_{r=1}^n\ P\{A_r\}\ E((g_n + (m-1)\ c(Q_1) - \Sigma_{i=2}^m\ c(Q_i))/m|A_r)\\
&= E(g_n)/m + \Sigma_{r=1}^n\ P\{A_r\}\ E(((m-1)\ c(Q_1) - \Sigma_{i=2}^m\ c(Q_i))/m|A_r)\\
&< E(g_n)/m + \Sigma_{j=1}^{\left\lceil \frac{n}{m-1} \right\rceil}\ P\{C_j\}\ E((x_{j(m-1)} - h_{j-1})(m-1)/m|C_j)\\
&< E(g_n)/m + \Sigma_{j=1}^{\left\lceil \frac{n}{m-1} \right\rceil}\ P\{D_j\}\ E((x_{j(m-1)} - h_{j-1})(m-1)/m|D_j).
\end{aligned}$$

For $r=1,\ldots,m-1$, the conditions A_r are mutually exclusive, but each one implies that C_1 holds. Also x_{m-1} is no smaller than $((m-1)\ c(Q_1) - \Sigma_{i=2}^m\ c(Q_i))/m$ when A_r holds for $r=1,\ldots,m-1$. Thus the substitution with $j = 1$ for $r=1,\ldots,m-1$ yields the proper inequality. For $j > 1$, the argument is similar. The conditions A_r, for $r=(j-1)(m-1)+1,\ldots,j(m-1)$, are mutually exclusive, but each implies that C_j holds. Also $x_{j(m-1)} - h_{j-1}$ is no smaller than $((m-1)\,c(Q_1) - \Sigma_{i=2}^{m} c(Q_i))/m$ when A_r holds for $r=(j-1)(m-1)+1,\ldots,j(m-1)$. Again the substitution yields the claimed inequality.

We note that a set of values satisfying C_j will also satisfy D_j, since $x_{j(m-1)} \geq h_{j-1}$ implies $x_{(m-1)k+1} \leq x_{j(m-1)} - h_k$, which together with $x_i \leq x_{(m-1)k+1}$ implies $x_i \leq x_{j(m-1)} - h_k$. Thus substitution again yields the claimed inequality. We now proceed to establish a set of lemmas similar to those in section 2.

Lemma 4.1. Let $j \geq 1$, and $0 \leq x_i \leq x$ for $i=1,\ldots,j(m-1)-1$. Let $b(1,x) = x^{m-2}$ and, for $j > 1$,

$$b(j,x) = \int_0^x \underbrace{\int_0^{x-h_1}\cdots\int_0^{x-h_1}}_{m-1}\cdots\underbrace{\int_0^{x-h_{j-2}}\cdots\int_0^{x-h_{j-2}}}_{m-1}\underbrace{\int_0^x\cdots\int_0^x}_{2(m-1)-2} dx_{j(m-1)-1}\cdots dx_1 .$$

Then $b(j,x) = x^{j(m-1)-1}/\Pi_{k=0}^{j-2}\,(k(m-1)+1)$.

Proof. The proof proceeds by induction on j. □

Lemma 4.2. Let $j(m-1)-1$ elements, $x_1,\ldots,x_{j(m-1)-1}$, be drawn from a uniform distribution on $(0,x]$. Let $x_i \leq x-h_k$, for $i=(m-1)(k-1)+2,\ldots,(m-1)k+1$ and $k=1,\ldots,j-2$. Then

$$P\{D_j\} = 1/\Pi_{k=0}^{j-2}\,(k(m-1)+1).$$

Proof. The condition $x \geq h_{j-1}$ implies that

$$0 \leq x_{(m-1)i+1} \leq x - h_i \quad \text{for } i=0,\ldots,j-2.$$

Thus $P\{D_j\} = b(j,x)/x^{j(m-1)-1}$. □

Lemma 4.3. Let $j \geq 1$, and $0 \leq x_i \leq x$ for $i=1,\ldots,j(m-1)-1$. Let $d(1,x) = x^{m-1}$ and for $j > 1$,

$$d(j,x) = \int_0^x \underbrace{\int_0^{x-h_1}\cdots\int_0^{x-h_1}}_{m-1}\cdots\underbrace{\int_0^{x-h_{j-2}}\cdots\int_0^{x-h_{j-2}}}_{m-1}\;\underbrace{\int_0^x\cdots\int_0^x}_{2(m-1)-2}$$

$(x-h_{j-1})\ dx_{j(m-1)-1}\ldots dx_1$.

Then $d(j,x) = x^{j(m-1)}/\Pi_{k=0}^{j-2}\ (k(m-1)+2)$. □

Let $p(j,m) = \Pi_{k=0}^{j-2} \frac{k(m-1)+1}{k(m-1)+2}$.

Lemma 4.4. Let $j(m-1)-1$ elements, $x_1,\ldots,x_{j(m-1)-1}$, be drawn from a uniform distribution on $(0,x]$. Let $x_i \le x_{j(m-1)} - h_k$ for $i=(m-1)(k-1)+2,\ldots,(m-1)k$ and $k=1,\ldots,j-2$. Then $E(x-h_{j-1}|D_j) = x\ p(j,m)$. □

Lemma 4.5. Let n elements be drawn from a uniform distribution on $(0,1]$. Let $X_{j(m-1)}$ be the order statistic of the $j(m-1)$-th smallest element. Then

$$E(X_j - h_{j-1}|D_j) = \frac{j(m-1)}{n+1}\ p(j,m).\ \square$$

Theorem 4.1. Let n elements drawn from a uniform distribution on $(0,1]$ be scheduled on $m \ge 2$ processors according to the LPT heuristic. Let $\hat{F}$ be the length of the schedule generated. Then

$$E(\hat{F}) < \frac{n}{2m} + \frac{(m-1)^2 e}{m(n+1)}\ .\ \square$$

Corollary 4.1. Let n elements drawn from a uniform distribution on $(0,1]$ be scheduled on m processors according to the LPT heuristic. Let $\hat{F}$ be the length of the schedule generated, and $\tilde{F}$ be the sum of the elements divided by m. Then

$$E(\hat{F})/E(\tilde{F}) < 1 + O(m^2/n^2)\ .\ \square$$

5. ANOTHER ANALYSIS FOR 2 PROCESSORS

We return to the case of 2 processors, and derive the same result in a somewhat different fashion. Let X_i be the ith smallest of the n task times. Let V_i be the mismatch between total times on the two processors after $X_n, X_{n-1},\ldots,X_i$ have been assigned. Then one easily obtains the recurrence

$$\begin{aligned} V_{n+1} &= 0, \\ V_i &= |V_{i+1} - X_i|, \qquad 0<i\le n. \end{aligned} \tag{5.1}$$

We wish to investigate V_1, the mismatch of total times when all tasks are assigned. Before introducing any probabilistic ideas, we prove two simple lemmas. For the first, some additional notation is useful. Let E_i denote the following proposition:

$$X_j \geq \frac{1}{2}X_{j+1} \quad \text{for all} \quad i \leq j < n. \tag{5.2}$$

<u>Lemma 5.1</u>. If E_i holds, then $V_i \leq X_i$.

<u>Proof</u>. We proceed by induction on i, in reverse order from $i=n$ to $i=1$. The basis, $i=n$, is trivial. For the inductive step, suppose the lemma is true for $i = j + 1$; We seek to prove it for $i=j$. Thus, we suppose E_j is true and seek to show that $V_j \leq X_j$. Now E_j implies E_{j+1}, by inspection of (5.2), so by the inductive hypothesis $V_{j+1} \leq X_{j+1}$. Now consider two cases. If $V_{j+1} < X_j$, we have

$$V_j = |V_{j+1} - X_j| = X_j - V_{j+1} \leq X_j,$$

as desired. On the other hand, if $V_{j+1} \geq X_j$, then

$$V_j = |V_{j+1} - X_j| = V_{j+1} - X_j$$
$$\leq X_{j+1} - X_j \leq 2X_j - X_j = X_j. \quad \square$$

<u>Lemma 5.2</u>. If $V_i \leq X_i$, then $V_j \leq X_i$ for $j \leq i$. $\square$

Now we assume that X_i is the i-th smallest of n uniform independent draws from $[0,n]$. (This is a rescaling of the unit interval.) Note that then $E[X_i] = in/(n+1)$. Two lemmas about such X_i are useful.

<u>Lemma 5.3</u>. $P\{X_i \leq \frac{1}{2}X_{i+1}\} \leq 2^{-i}$.

<u>Proof</u>. The distribution of the i smallest of n independent uniform draws from $[0,n]$, given that $i+1^{\text{th}}$ smallest is equal to some value a, is just that of i uniform draws from $[0,a]$. The largest of these is less than $a/2$ if and only if all of them are less than $a/2$, which occurs with probability 2^{-i}. $\square$

<u>Lemma 5.4</u>. Assume $z \geq i$. Then

$$P\{X_i > z\} \leq \exp\left(-\frac{(z-i)^2}{2z}\right).$$

<u>Proof</u>. Using the well-known formula for the distribution of the

i-th order statistic (3, formula 2.1.3) we have

$$P\{X_i > z\} = \Sigma_{j=0}^{i-1} \binom{n}{j} (\frac{z}{n})^j (1-\frac{z}{n})^{n-j} .$$

Now letting $p = z/n$ and $\beta = 1-i/z$, and applying (1, Proposition 2.4(a)), yields the stated bound. □

Theorem 5.1. For any positive real z and positive integer i,

$$P\{V_1 > z\} \le 2^{-i+1} + \exp(-\frac{(z-i)^2}{2z}).$$

Proof. Let $\bar{E}_i$ denote the complement of E_i. Note that

$$P\{V_1 > z\} \le P\{\bar{E}_i\} + P\{V_1 > z \wedge E_i\}.$$

Now by Lemma 5.3,

$$P\{\bar{E}_i\} \le \Sigma_{j=1}^{n-1} 2^{-j} < \Sigma_{j=i}^{\infty} 2^{-j} = 2^{-i+1} .$$

Also,

$$P\{V_1 > z \wedge E_i\} = P\{V_1 > z \wedge V_i \le X_i\}$$

$$\le P\{V_1 > z \wedge V_1 \le X_i\} \le P\{X_i > z\}$$

$$\le \exp(-\frac{(z-i)^2}{2z}). \quad \text{(by Lemma 5.4)} \ \square$$

Note that the bound in the Theorem is independent of n. This easily yields the following.

Corollary 5.1. $E[V_1] = O(1)$.

Proof sketch. If we choose $i = \lfloor z/2 \rfloor$ in Theorem 5.1, the bound decreases exponentially as z grows, so the expectation is clearly finite. □

This is comparable to the result in section 2, since the interval [0,n] is being used.

6. A RESTRICTED GREEDY METHOD

It is interesting to note that a related process, which assigns tasks in pairs, can be analyzed exactly. This process can be described most readily if we sort the task times into decreasing length, so X_i is the i-th largest time. We will analyze a process which always assigns X_{2i-1} and X_{2i} to distinct processors, using the natural greedy method to assign the processes in order of decreasing size. (If n is odd, X_n is the last task assigned, and is assigned to the less busy processor.) One easily obtains the following recurrence for the imbalance V_i after processes $X_1,\dots,X_i$ have been assigned.

$$\begin{aligned} V_0 &= 0, \\ V_{2i} &= |V_{2i-2} - (X_{2i} - X_{2i-1})|, \quad i=1,\dots,\lfloor n/2 \rfloor, \\ V_n &= |V_{n-1} - X_n| \quad \text{if } n \text{ is odd.} \end{aligned} \tag{6.1}$$

For the remainder of this section let the X_i be the decreasing order statistics of n draws from a uniform distribution on $[0,n]$. The analysis of this process is complicated by the fact that $(X_{2i} - X_{2i-1})$ and $(X_{2j} - X_{2j-1})$ are not independent. This problem can be remedied by first analyzing a closely related and much simpler process.

Define new variables $\hat{X}_i$, $i=0,\dots,n$, by letting Z_i, $i=1,\dots,n+1$, be independent draws from the unit exponential distribution, and

$$\begin{aligned} \hat{X}_{n+1} &= 0, \\ \hat{X}_i &= \hat{X}_{i+1} + Z_{i+1}, \qquad 0 \le i \le n. \end{aligned} \tag{6.2}$$

Define $\hat{V}_i$ by putting hats on all X_i and V_i in (6.1). Then (6.1) can be re-expressed as

$$\begin{aligned} \hat{V}_0 &= 0, \\ \hat{V}_{2i} &= |\hat{V}_{2i-2} - Z_{2i}|, \qquad i=1,\dots,\lfloor n/2 \rfloor, \\ \hat{V}_n &= |\hat{V}_{n-1} - Z_{n+1}| \quad \text{if } n \text{ is odd.} \end{aligned}$$

At this point, it is convenient to relabel the variables. We let T_i, $i=1,2,\dots,$ denote independent random variables with pdf e^{-x}, and define

$$\begin{aligned} S_0 &= 0, \\ S_i &= |S_{i-1} - T_i|. \end{aligned}$$

It is not too hard to establish that the pdf f_i of S_i is (see 4, I. 13, problem 1(iv))

$$f_i(x) = e^{-x}, \qquad i=1,2,\dots .$$

<u>Theorem 6.1</u>. If the $\hat{X}_i$ are generated according to the distribution described in (6.2), the density function of $\hat{V}_n$ is e^{-x}.

In particular, $E[\hat{V}_n] = 1$. □

We can now analyze the case in which the X_i are the (decreasing) order statistics of n draws from $[0,n]$. The following fact is very helpful.

Fact 6.1. (4, III. 3 Example (d)). The joint distribution of the $\hat{X}_i$, $i=1,\ldots,n$, given that $\hat{X}_o = s$, is identical to the joint distribution of $\frac{s}{n}X_i$, $i=1,\ldots,n$.

Thus the distribution of the $\hat{X}_i$ is what we would obtain by letting $X_1,\ldots,X_n$ be the order statistics of n uniform independent draws from $[0,n]$, letting Z^{*n+1} be the sum of $n+1$ independent unit exponential draws, and then setting

$$\hat{X}_i = Z^{*n+1} X_1/n.$$

Assuming we generate the $\hat{X}_i$ this way, we see that Z^{*n+1} and V_n are independent, and

$$\hat{V}_n = Z^{*n+1} V_n/n. \qquad (6.3)$$

Theorem 6.2. $E[V_n] = n/(n+1)$.

Proof. Using Theorem 3.1 and taking expectations of both sides of (6.3) we obtain

$$1 = (n+1)\ E[V_n]/n,$$

from which the theorem follows immediately. □

Note that the schedule, when reversed, is a mean flow time minimizing schedule.

REFERENCES

1. D. Angluin and L.G. Valiant, "Fast probabilistic algorithms for Hamiltonian circuits and matchings", J. Comp. Sys. Sci. 18 (1979) 155-193.

2. E.G. Coffman, Jr., M.R. Garey and D.S. Johnson, "An application of bin packing to multiprocessor scheduling", SIAM J. Comput. 7,1 (Feb. 1978) 1-17.

3. H.A. David, Order Statistics , Wiley, New York (1970).

4. W. Feller, An introduction to probability theory and its applications, Vol. I , Wiley, New York (1968).

5. G.N. Frederickson, "Probabilistic analysis for simple one- and two-dimensional bin packing algorithms", Inf. Proc. Lett. 11,4,5 (Dec. 1980) 156-161.

6. D.K. Friesen, "Tighter bounds for the MULTIFIT processor scheduling algorithm", to appear in SIAM J. Comput.

7. R.L. Graham, "Bounds on multiprocessing timing anomalies", SIAM J. Appl. Math. 17,2 (1969) 416-429.

8. R.M. Karp, "Reducibility among combinatorial problems", in Complexity of Computer Computations, R.E. Miller and J.W. Thatcher, eds., Plenum Press, New York (1972) 85-103.

9. R.M. Karp,"Probabilistic analysis of partitioning algorithms for the traveling salesman problem in the plane", Math. Oper. Res. 2,3 (1977) 209-224.

10. G.S. Lueker, "Optimization problems on graphs with independent random edge weights", SIAM J. Comput. 10, 2 (May 1981) 338-351.

11. G.S. Lueker, "An average-case analysis of bin packing", manuscript (1981).

12. S. Sahni, "Algorithms for scheduling independent tasks", J. ACM 23 (1976) 116-127.

13. E.G. Coffman, Jr., M. Hofri, K. So, and A.C. Yao, "A stochastic model of bin-packing", Computer Systems Lab. technical report TR-CSL-7811, University of California, Santa Barbara (1978).

SEQUENTIAL PROJECT SELECTION (MULTI-ARMED BANDITS) AND THE GITTINS INDEX

P. Whittle

University of Cambridge

A direct proof is given of the optimality of the Gittins index policy, and a related identity demonstrated for the loss function. Especial attention is paid to the case when new projects also arrive in a statistically homogeneous stream. A number of general results are obtained, of which those derived by J.M. Harrison, for example, are shown to be a special case.

1. INTRODUCTION

The multi-armed bandit problem (as it has become known) is important as one of the simplest non-trivial problems in which one must face the conflict between taking actions which yield immediate reward and taking actions (such as acquiring information, or preparing the ground) whose benefit will come only later. It has proved difficult enough to become a classic, and has now a large literature.

In its basic version one has N parallel projects or investigations, indexed $i=1,2,\ldots,N$ and at each instant of discrete time can work on only a single project. Let the state of project i at time t be denoted $x_i(t)$. If one works on project i at t then one acquires an immediate expected reward of $R_i(x_i(t))$. Rewards are additive, and discounted in time by factor β. The state value $x_i(t)$ changes to $x_i(t+1)$ by a Markov transition rule (which may depend on i, but not on t), while the state of the projects one has not touched remain unchanged: $x_j(t+1)=x_j(t)$ for $j\neq i$. The problem is how to allocate one's effort over projects sequentially in time so as to maximize expected total discounted reward over $t=0,1,2,\ldots$.

In the gambling context of the actual "multi-armed bandit",

M. A. H. Dempster et al. (eds.), Deterministic and Stochastic Scheduling, 333–341.

projects correspond to arms of the machine, and the state of a project to one's state of knowledge about the parameters of that arm.

The problem resisted essential solution until Gittins and his co-workers made fundamental progress in a series of papers which have yet to be recognized generally (see especially Gittins and Jones, 1974, Gittins and Glazebrook, 1977 and Gittins, 1979, 1980, but also the other papers listed for these authors in the references).

They proved that to each project i one could attach an index $v_i = v_i(x_i(t))$, this being a function of the project i and its current state $x_i(t)$ alone, such that the optimal action at time t is to work on that project for which the current index is greatest. The index v_i is calculated by solving the problem of allocating one's effort optimally between project i and a *standard project* which yields a constant reward (and so effectively has also a single state). In his 1979 paper Gittins gives a second illuminating interpretation of the index v_i in terms of a "forwards induction" rule, maximizing average yield up to an optimally chosen stopping time.

In his deduction of the index rule and his two characterizations of the index Gittins showed a rare intuition. Unfortunately, his proofs of the optimality of the index rule have been difficult to follow, and this has doubtless been a reason why the full merits and point of this work have not yet been generally appreciated.

In this paper I give a simple proof of the optimality of the Gittins index procedure. The treatment begins from a conjecture, which implies an identity (12) between the maximal rewards for the cases of one and several projects. In the course of verifying this identity one both proves optimality of Gittins' rule, and extends the result in a number of directions.

Gittins refers to the index v_i as the "dynamic allocation index", abbreviated DAI. The term has an obvious rationale, but is clumsy to use. I find it both convenient and proper to refer to v_i as the "Gittins index".

2. PRELIMINARIES

Let us denote the combined state vector $(x_1, x_2, \ldots, x_N)$ by x, or by $x(t)$ if we need to refer to its value at a particular time t. We shall assume the rewards uniformly bounded,

$$k(1-\beta) \leq R_i(x_i) \leq K(1-\beta), \tag{1}$$

where k, K are constants (which may be negative) and β is the discount factor. We assume strict discounting, so that $0 \leq \beta < 1$.

The reward at time t will be $R_{i(t)}(x_{i(t)})$ if $i(t)$ is the project engaged at time t; for simplicity we shall write this simply as $R(t)$. The total discounted reward $\Sigma \beta^t R(t)$ is then well defined, and bounded between k and K. Denote the maximal expected value of this reward, conditional on $x(0)=x$, by $F(x)$. Then F is the unique bounded solution of the dynamic programming equation

$$F=\max_i L_i F, \tag{2}$$

where $L_i F(x)$ is the expected reward if, in state x, one works on project i for one step and then, in the new state x' (differing from x only in the ith component) receives reward $F(x')$. That is

$$L_i F(x)=R_i(x_i)+\beta E[F(x(t+1))|x(t)=x,i(t)=i]. \tag{3}$$

Consider now the process when modified by addition of the option that one can "retire" (i.e. abandon all projects) at any time, for an immediate reward of M. Term this the "M-process", the original process the "continuing process", and let $\Phi(x,M)$ be the analogue of $F(x)$ for the M-process. Φ is then the unique bounded solution of

$$\Phi=\max(M,\max_i L_i \Phi). \tag{4}$$

Lemma 1. <u>$\Phi(x,M)$ is a non-decreasing convex function of M with</u>

$$\Phi(x,M) = \begin{cases} F(x), & M\le k, \\ M, & M\ge K. \end{cases} \tag{5}$$

<u>For $M<k$ the optimal policies of the M-process and the continuing process are identical.</u>

Proof. To increase M can only increase reward, so the non-decreasing character is evident. To retire is always optimal for $M\ge K$ and to continue is always optimal for $M\le k$ whence (5) follows. To retire is never optimal if $M<k$, whence the last assertion.

Let V be the expected return from a policy whose prescription is independent of M. Then

$$V=V_c+ME(\beta^T), \tag{6}$$

where V_c is the expected reward before retirement (independent of M), and T is the time of retirement. All expectations in (6) are those determined by the policy, and are conditional on the observational history at $t=0$. The event of non-retirement can be identified with the event $T=+\infty$ since because $|\beta|<1$, either convention will cause this contingency to yield zero contribution to $E(\beta^T)$.

Since Φ is the supremum over policies of expression (6), linear in M, it is convex in M. □

Lemma 2. <u>For almost all M</u>

$$\frac{\partial\Phi(x,M)}{\partial M} = E_M(\beta^T|x(0)=x), \tag{7}$$

where E_M denotes expectation under a policy optimal for the M-process.

Proof. Since in each state there are at most $N+1$ possible actions, optimal policies always exist. If one applies an M-optimal policy to the $(M+\delta)$-process one achieves an expected reward not exceeding $\Phi(x,M+\delta)$; this assertion and relation (6) imply that for any M,δ

$$\Phi(x,M+\delta)-\Phi(x,M)\geq\delta E_M(\beta^T|x(0)=x). \tag{8}$$

(The expectation is conditional on initial history, but for an M-optimal policy the only effective conditioning variable is initial state.)

It follows from (8) that expression (7) is a sub-gradient to Φ (as a function of M), and so coincides with the gradient of Φ wherever this exists. But Φ, being convex, will have a gradient almost everywhere; hence we deduce the assertion of the Lemma.□

If the optimal policy at M is non-unique then $E_M(\beta^T)$ may take more than one value. However, it follows from (8) that *all* such determinations of $E_M(\beta^T)$ are sub-gradients of Φ, and that the determination is in fact then unique at those values for which $\partial\Phi/\partial M$ is defined.

Typically, a value of M at which $\partial\Phi/\partial M$ does not exist, in that there is a discontinuity in its value, corresponds to a value of M at which there is a discontinuous change in the optimal policy. It is then understandable that there are at least two policies optimal at such a point.

3. WRITE-OFF POLICIES AND A CONJECTURE

Let us define a *write-off policy* as a policy in which project i is written-off (i.e. abandoned) when first its state x_i enters a *write-off set* S_i $(i=1,2,\ldots,N)$; one continues as long as there are projects which have not been written-off, working only on those projects; one retires as soon as all projects are written-off. That is, one continues until all projects have been driven into their write-off sets.

Note that a write-off policy is not fully specified by prescription of the write-off sets S_i, because no rule has been given for the order in which active projects are to be engaged. This is something the write-off policy does not specify; moreover, a write-off policy need be neither Markov nor stationary. Note that the Gittins index policy is a write-off policy, with S_i the set of x_i for which $M_i(x_i)\leq M$.

Lemma 3. Let τ be the time to retirement for an N-project policy, and τ_i the time to retirement when only project i is available. If the policy is a write-off policy, then

$$E[\beta^{\tau}|W_0]=\prod_i E[\beta^{\tau_i}|W_0]. \tag{9}$$

Proof. We have $\tau=\Sigma\tau_i'$ where τ_i' is the total *process time* for project i, the total time for which project i is engaged before it is written-off. Equation (9) states effectively that the τ_i' are independently distributed, and that τ_i, τ_i' have the same distribution (conditional on W_0). But since evolution of the projects (in process time) is independent, the process time needed to take x_i into S_i is independent of the states of other projects; hence the assertion.□

We now make the

Conjecture C. <u>There is an M-optimal policy which is a write-off policy</u>.

If this is true, then the set S_i must be exactly the set of x_i for which one would retire if one only had the single project i, and so the only options were those of continuation or retirement. That is, if we define $\phi_i(x_i,M)$ as the analogue of $F(x,M)$ for this single-project case, so that the analogue of (4) is

$$\phi_i=\max(M,L_i\phi_i), \tag{10}$$

then S_i is the set of x_i for which $\phi_i=M$.

Now, if the conjecture were justified, then relations (7), (9) would lead to the conclusion

$$\frac{\partial\Phi(x,M)}{\partial M}=\prod\frac{\partial\phi_i(x_i,M)}{\partial m} \tag{11}$$

and hence to the conclusion

$$\Phi(x,M)=K-\int_M^K\prod_i\frac{\partial\phi_i(x_i,m)}{\partial m}\,dm. \tag{12}$$

In fact, having conjectured the truth of identity (12) in this way, one establishes easily

Theorem 1. (i) <u>Identity (12) holds</u>
(ii) <u>The Gittins index policy is optimal</u>
(iii) <u>Conjecture C is valid</u>

The proof (see Whittle 1980) follows easily if one uses the right-hand member of (12) to define a quantity $\hat{\Phi}(x,M)$, and then demonstrates that $\hat{\Phi}$ satisfies the dynamic programming equation (4), with the maximising option being exactly that recommended by the Gittins policy. This establishes $\Phi=\hat{\Phi}$ and the optimality of the Gittins policy, and so the truth of (i),(ii).

Since the Gittins policy is a write-off policy, assertion (iii) also follows.

4. GENERALISATIONS; THE OPEN PROCESS

The conclusions of Th.1 continue to hold if a finite horizon is imposed upon the time allowable to each project, since this horizon, s_i , can be incorporated in the state variable, x_i . They also hold under some conditions for the case of a *superprocess* (see Whittle 1980), for which, once a project is chosen, there are secondary decisions concerning its running which must be optimised. Further, the method of proof generalises directly to the case of *variable stages*, when a project, once engaged, must be continued with for a time which may be deterministic or random.

A radical but rather necessary generalisation is to allow the process to be *open*, in that new projects become available in a statistically homogeneous Poisson stream. With appropriate modification of definitions relation (11) remains valid (Whittle, 1981), and the Gittins index policy remains optimal (Nash, 1973).

A project was before described by its individual label i and its state x_i . With indefinitely many projects it is better to combine these two indicators into a single label x , the state of a project. If $n(x)$ is the number of projects of state x available at a given time, then the vector $n=\{n(x)\}$ describes the state of the decision process, under appropriate statistical assumptions. The quantities previously denoted $F(x)$ and $\Phi(x,M)$ will now be denoted $F(n)$ and $\Phi(n,M)$.

Let $e(u)=\{e(x,u)\}$ denote the value of the vector n which is zero except for a unit in the uth place:

$$e(x,u) = \begin{cases} 1 & x = u, \\ 0 & x \neq u. \end{cases}$$

The quantity

$$\phi(u,M)=\phi(e(u),M)$$

is then the analogue of $\phi_i(x_i,M)$; the maximal expected discounted reward conditional on starting with a single project, of state u . However, in contrast to the closed problem previously considered, a one-project situation does not now remain so, and the analogue of the one-project optimality equation (10) is bound to be more complicated. In fact, if we also allow random stage length, then the optimality equation will be

$$\Phi(n,M)=\max[M, \max_{u\in U(n)} L_u\Phi(n,M)], \tag{13}$$

where now

$$L_u\Phi(n,M)=R(u)+E[\beta^z\Phi(n-e(u)+e(u')+w,M)\,|\,u], \tag{14}$$

where the conditioning event u indicates that a project of stage u is engaged, so that expression (14) is defined only if $n\geq e(u)$. The set $U(n)$ in (13) is the set of x for which $n(x)>0$; i.e. the set of states occupied by some currently available project. $R(u)$ is the expected stage reward if a project of state u is engaged. The random variables covered by the expectation in (14) are z , the stage length ; u' , the state to which the project of initial state u is brought by the end of the stage; and $w=\{w(x)\}$, where $w(x)$ is the number of projects of state x which enter the system during the stage.

Suppose that the vector number of projects $q=\{q(x)\}$ entering the system in unit time is independent of previous entries and of the state of the system (which justifies our assumption that n is indeed a state variable for the decision process). It is then shown in Whittle (1981) that relation (12) still holds, in the form

$$\Phi(n,M)=K - \int_M^K \prod\left[\frac{\partial\phi(x,m)}{\partial m}\right]^{n(x)} dm, \tag{15}$$

and that the Gittins index policy is again optimal, with the index $M(x)$ for a project of state x being determined as the infimal value of M for which

$$\phi(x,M)=M . \tag{16}$$

We see from (14), (15) that the relations determining the one-project rewards $\phi(x,M)$ are now

$$\phi(u,M)=\max[M,R(u)+E[\beta^z K-\int_M^K\frac{\partial\phi(u',m)}{\partial m}[\beta A\left(\frac{\partial\phi}{\partial m}\right)]^z\, dm|u]], \tag{17}$$

where

$$A(\zeta)=E[\prod_x\zeta(x)^{q(x)}] \tag{18}$$

and $A\left(\frac{\partial\phi}{\partial m}\right)$ is obtained by setting $\zeta(x)$ equal to $\frac{\partial\phi(x,m)}{\partial m}$.

Relation (17) is the open equivalent of (10), a good deal more complicated, even in the case of fixed unit stage-length: $z=1$. It can be solved for decreasing M : one has $\phi(u,M)=M$, certainly for $M\geq K$ (in fact for $M\geq M(u)$) , and

$$\frac{\partial\phi(u,M)}{\partial M} = E\left[\frac{\partial\phi(u',M)}{\partial M}\left[\beta A\left(\frac{\partial\phi}{\partial M}\right)\right]^{z}\middle| u\right] \tag{19}$$

The right-hand member of (19) changes only when M passes through the index value $M(x)$ for some x ; $\phi(u,M)$ is piecewise linear in M .

5. SCHEDULING OF PRIORITIES IN A QUEUE

Suppose that customers of different types $i=1,2,\ldots$ arrive in a single-server queue in independent Poisson streams. They may be served in any order, but service, once begun, is completed. One seeks for an optimal order in which to serve customers, under cost and service-time assumptions to be specified below.

This can be regarded as an open project selection problem, a customer of type i constituting a "project" of type i. The simplifying feature of this particular problem is that service is always completed. That is, a project, once begun, is continued until completion. Since the project is never left in an intermediate state, there is no need to specify state of project development. On the other hand, the time needed to complete the project will in general be random. This problem was solved by J.M. Harrison (1975,1976); his results, obtained by direct methods, can be regarded as special cases of those of §4. Harrison's results are derived in this way in Whittle (1982).

REFERENCES

J.C. GITTINS (1975) The two-armed bandit problem: variations on a conjecture by H. Chernoff. *Sankhyā Ser. A* 37,287-291.

J.C. GITTINS (1979) Bandit processes and dynamic allocation indices. *J. Roy. Statist. Soc. Ser. B* 41,148-164.

J.C. GITTINS (1980) Sequential resource allocation (a progress report). Mathematical Institute, University of Oxford.

J.C. GITTINS, K.D. GLAZEBROOK (1977) On Bayesian models in stochastic scheduling. *J. Appl. Probab.* 14,556-565.

J.C. GITTINS, D.M. JONES (1974) A dynamic allocation index for the sequential design of experiments. In: J. Gani (ed.) (1974) *Progress in Statistics*, North-Holland, Amsterdam, 241-266.

J.C. GITTINS, P. NASH (1977) Scheduling, queues and dynamic allocation indices. In: *Proc. EMS, Prague*, Czechoslovak Academy of Sciences, Prague, 191-202.

K.D. GLAZEBROOK (1976a) A profitability index for alternative research projects. *Omega* 4,79-83.

K.D. GLAZEBROOK (1976b) Stochastic scheduling with order constraints. *Internat. J. Systems Sci.* 7,657-666.

K.D. GLAZEBROOK (1978a) On a class of non-Markov decision processes. *J. Appl. Probab.* 15,689-698.

K.D. GLAZEBROOK (1978b) On the optimal allocation of two or more treatments in a controlled clinical trial. *Biometrika* 65, 335-340.

J.M. HARRISON (1975a) A priority queue with discounted linear costs. *Oper. Res.* 23,260-269.

J.M. HARRISON (1975b) Dynamic scheduling of a multi-class queue: discount optimality. *Oper. Res.* 23,270-282.

J.M. HARRISON (1976) Dynamic scheduling of a two-class queue: small interest rates. *SIAM J. Appl. Math.* 31,51-61.

P. NASH (1973) Optimal allocation of resources between research projects. Ph.D. thesis, University of Cambridge.

P. NASH, J.C. GITTINS (1977) A Hamiltonian approach to optimal stochastic resource allocation. *Adv. in Appl. Probab.* 9,55-68.

P. WHITTLE (1980) Multi-armed bandits and the Gittins index. *J. Roy. Statist. Soc. Ser. B* 42,143-149.

P. WHITTLE (1981) Arm-acquiring bandits. *Ann. Probab.* 9,284-292.

P. WHITTLE (1982) *Optimisation over Time*, Vol. 1, Wiley, Chichester.

DOMINANT STRATEGIES IN STOCHASTIC ALLOCATION AND SCHEDULING PROBLEMS

P. Nash, R.R. Weber

Cambridge University

ABSTRACT

Some problems of stochastic allocation and scheduling have the property that there is a single strategy which minimizes the expected value of the costs incurred up to every finite time horizon. We present a sufficient condition for this to occur in the case where the problem can be modelled by a Markov decision process with costs depending only on the state of the process. The condition is used to establish the nature of the optimal strategies for problems of customer assignment, dynamic memory allocation, optimal gambling, maintenance and scheduling.

1. DOMINANT STRATEGIES

The aim in many stochastic allocation and scheduling problems is to attain a desired state at the least possible cost. In job scheduling problems one may want to minimize the expected value of the makespan or flowtime. In a reliability problem one may want to minimize the expected repair costs. Some problems of this type have the special property that there exists a single strategy which not only minimizes the expected cost of reaching the desired state, but also minimizes the expected value of the cost incurred at every time. A strategy with this property shall be called expectation dominant (ED). A strategy will be called stochastic dominant (SD) if for all times t it minimizes in distribution the cost incurred at time t. Weeks and Wingler [14] have argued that stochastic dominance is the property most desired in a scheduling strategy.

To give an example of a problem in which the optimal strategy

M. A. H. Dempster et al. (eds.), Deterministic and Stochastic Scheduling, 343–353.

is stochastic dominant we shall describe the problem of assigning customers to parallel servers which has been discussed by Winston [16]. Identical jobs with exponentially distributed processing requirements arrive at a service facility in a Poisson stream. As each job arrives it must be assigned to one of a number of identical servers which operate in parallel. Each server serves jobs in its queue in a first come first served order. No jockeying amongst the queues is allowed. We have shown [11] that the strategy of assigning each arrival to the shortest queue minimizes the expected average waiting time of the first k jobs to arrive. In fact the optimal strategy is SD since it minimizes for all times t the distribution of the number of jobs in the system at time t. Before proceeding with a proof of this and other results, we present a general method for investigating expectation and stochastic dominant strategies.

2. A SUFFICIENT CONDITION FOR DOMINANCE

We will conduct most of our analysis in the setting of a continuous time Markov decision process on a finite state space $(i=1,\dots,N)$. Suppose that the cost of residence in state i is c_i per unit time. Starting from state i the rate of transition to state j is given by Q_{ij}, where Q is an NxN matrix in the set of feasible matrices Ω. The set Ω is bounded, closed and convex with the property that when a collection of 1st to Nth rows from different matrices in Ω is assembled together then this is itself a matrix in Ω. An ED strategy was defined as one which minimizes for all t the expected cost incurred at time t. It corresponds to a choice $\bar{Q}\varepsilon\Omega$ minimizing every component of V(t) where,

$$V(0) = c, \quad \text{and} \quad \dot{V}_i(t) = \min_{Q\varepsilon\Omega} \sum_j Q_{ij}V_j(t) = \sum_j \bar{Q}_{ij}V_j(t).$$

If the states have been indexed from greatest to least costly $(c_1>c_2>\cdots>c_N)$, then a SD strategy must minimize for all t and k $(1<k\leqslant N)$ the probability that the process is in a state less than k at time t. Thus a SD strategy is one which is simultaneously ED for (N-1) cost vectors c(k) (k=1,...,N-1) of the form

$$c_j(k) \begin{array}{l} = 1, \ j=1,\dots,k, \\ = 0, \ j=k+1,\dots,N. \end{array}$$

Stochastic scheduling and allocation problems may be studied in either discrete or continuous time formulations. The continuous time formulation is often simpler for the reason that only one event (such as a completion of a job) can occur at any instant of time. Any convenience which is lost in setting the problem in continuous rather than discrete time can be recovered by treating the Markov process as a jumping process in which the residence times between jumps do not depend on the current state. We choose a θ, with $\theta \geqslant \max\{-Q_{ii}: 1\leqslant i\leqslant N, \ Q\varepsilon\Omega\}$, and let $\Pi=\{P: P=(Q+\theta I)/\theta, \ Q\varepsilon\Omega\}$. Starting from state i the original Markov decision process is realized by waiting a time which is exponentially distributed with parameter θ, choosing a matrix $P \in \Pi$, and then jumping to state j with probability P_{ij}. This process does not rule out the possibility of jumps from a state to itself. The strategy whose transition matrix is $\bar{Q}$ is ED for the continuous time Markov decision process if and only if $\bar{P}$ is the transition matrix of an ED strategy for the discrete time Markov decision process with the same cost vector and set of feasible transition matrices Π. The truth of this statement may be verified by considering the equation

$$V_i(t) = (1-e^{-\theta t})c_i + \int_0^t \sum_j P_{ij} V_j(t-s)\theta e^{-\theta s}\, ds.$$

This construction, which Keilson [5] calls uniformizing, has been discussed in number of papers, including [7] and [9].

We shall now investigate conditions under which a matrix $\bar{P}$ is ED for the discrete time jumping chain. To be minimizing, $\bar{P}$ must satisfy

$$(P-\bar{P})\bar{P}^n c \geq 0, \text{ for all } n \text{ and } P \in \Pi.$$

Let R be a matrix whose rows consist of all possible rows of $(P-\bar{P})$ with $P \in \Pi$.

Lemma 1. $\bar{P}$ is ED if and only if there exist sequences of matrices $\{H_n\}$, $\{A_n\}$, $\{B_n\}$, $(n=0,1,2,\ldots)$ such that

$$\begin{aligned} H_0 Rc &\geq 0, \\ R &= A_n H_n, \quad A_n \geq 0, \text{ and} \\ H_{n+1}\bar{P} &= B_n H_n, \quad B_n \geq 0. \end{aligned}$$

Proof. The sufficiency of the conditions follows immediately from

$$R\bar{P}^n c = A_n H_n \bar{P}^n c = \cdots = A_n B_{n-1} \cdots B_0 H_0 c \geq 0.$$

The necessity comes from considering the cone K_n whose generators are $\{c, \bar{P}c, \cdots, \bar{P}^n c\}$. Because K_n is a convex polyhedral cone it may be written as $\{x: H_n x \geq 0\}$. Now if $\bar{P}$ is ED then $H_n x \geq 0$ implies $Rx \geq 0$. Thus $\{x: Rx \geq 0\}$ is contained in $\{x: H_n x \geq 0\}$. When one cone is contained in another we can write $R = A_n H_n$ for some $A_n \geq 0$. Since $H_n x \geq 0$ implies $H_{n+1}\bar{P} \geq 0$ we must also have $H_{n+1}\bar{P} = B_n H_n$ for some $B_n \geq 0$.

Although this condition is not directly helpful, it does suggest that we may easily be able to establish the optimality of an SD strategy if the sequence H_n can be replaced by a single finite-dimensional matrix H. Certainly the existence of such a matrix is a sufficient condition, and it is this that we shall use in the rest of the paper to show that strategies are ED or SD.

Lemma 2. $\bar{P}$ is ED if there exist matrices H, $A \geq 0$ and $B \geq 0$ such that

$$Hc \geq 0,$$
$$R = AH, \text{ and}$$
$$H\bar{P} = BH.$$

3. PROCESSES WHICH PARTIALLY ORDER THE STATES

Let e_i be the ith row of the $N \times N$ identity matrix. When the rows of $(P-\bar{P})$ are proportional to rows of the form (e_j-e_i) the problem may be particularly simple to analyse. $\bar{P}$ defines a partial order on the states. We say that i is better than j (and denote this by $i \downarrow j$) if there is a row of $(P-\bar{P})$ proportional to (e_j-e_i). We also have $i \downarrow i$. It is natural to consider the matrix H whose rows include

e_j-e_i iff $i \downarrow j$, and there is no k with $i \downarrow k$ and $k \downarrow j$, and
e_i iff there is no $j \neq i$ such that $i \downarrow j$.

From lemma 2 we deduce that $\bar{P}$ is ED if there exists a $B \geq 0$ with $H\bar{P} = BH$. This happens if for $i \downarrow j$ the ith row of $\bar{P}$ may be obtained from the jth row of $\bar{P}$ by operations on row j which shift probability from one state to others which are better in the partial order. The following lemma gives a condition guaranteeing that this shifting is possible (the proof is straightforward).

Lemma 3. $\bar{P}$ is ED if for all states $i \downarrow j$ and states k,

$$c_i \leq c_j, \text{ and } \sum_{k \downarrow h} \bar{P}_{ih} \leq \sum_{k \downarrow h} \bar{P}_{jh}. \tag{1}$$

There is a dual sufficient condition in which (1) is replaced by

$$c_i \leq c_j, \text{ and } \sum_{h \downarrow k} \bar{P}_{ih} \geq \sum_{h \downarrow k} \bar{P}_{jh}. \tag{2}$$

When the states can be totally ordered with $1 \downarrow 2, 2 \downarrow 3, \ldots$, then both conditions are equivalent to the statement that for $i<j$ and all k we must have

$$c_i \leq c_j, \text{ and } \sum_{h \geq k} \bar{P}_{ih} \leq \sum_{h \geq k} \bar{P}_{jh}. \tag{3}$$

$\bar{P}$ is then called a monotone matrix. Some properties of monotone matrices are discussed by Keilson and Kester [5].

Example 1. Assignment of Customers to Parallel Servers

We consider the problem of customer assignment that was described at the start of the paper. Suppose that there are m servers, and that each has a large but finite waiting room. Let x(i) be a vector whose kth component is equal to the number of jobs in the kth longest queue of state i. We partially order the states of the system, writing $i \downarrow j$ if

$$\sum_{h=1}^{k} x_h(i) \leqslant \sum_{h=1}^{k} x_h(j), \quad \text{for all } k=1,\ldots,m. \tag{4}$$

The condition of lemma 3 simply says that for all k and $i \downarrow j$ the probability of a transition to a state in the partial ordering which is as bad as k must be no more starting from i than from j. This is obvious from the nature of the shortest queue strategy and the partial order defined by (4). If we let c_i be equal to 1 or 0 as the number of jobs in the system in state i is or is not greater than k, then $i \downarrow j$ implies $c_i \leqslant c_j$, and thus assignment to the shortest queue minimizes for all k and t the probability that there are more than k jobs in the system at time t.

Example 2. Dynamic Memory Allocation

Benes [1] proves a number results for problems of stochastic memory allocation. In one simple example he describes a linear computer memory having room for exactly three units of program. Computer jobs, which require just one or two contiguous units of memory, arrive according to Poisson processes of rates λ_1 and λ_2. The execution times of the programs in memory are exponentially distributed with parameters μ_1 and μ_2. If a program arrives to find insufficient contiguous room in the memory, then the computer crashes. We let (-—) denote the state in which there is one program of each of the lengths in the memory. In an obvious fashion, we can write the states of the system as

1=(   ) 2=(-) 3=(-) 4=(—)
5=(- -) 6=(--) 7=(-—) 8=(---) 9=(crashed).

A decision must be made as to where to allocate a program of length 1 when it arrives to find that the system is in states 1 or 2. The following transitions are possible.

1=(   ) → 2=(-) or 3=(-)   and   2=(-) → 5=(- -) or 6=(--)

We can use lemma 3 to show that in each case the first choice is the one which minimizes the probability that the system crashes by time t. The acorrect partial order is $1 \downarrow 2$, $2 \downarrow 3$, $2 \downarrow 5$, $3 \downarrow 6$, $5 \downarrow 6$, $6 \downarrow 8$, $4 \downarrow 7$, $7 \downarrow 9$, $8 \downarrow 9$. It is easy to write down a transition matrix $\bar{P}$ and verify lemma 3 with $c=(0,\ldots,0,1)$. Thus the probability of a blockage occurring by time t is minimized by the indicated strategy. A similar analysis can be carried out when the memory is 4 cells long. However, we have computed the optimal strategies for some examples with longer memories and found that the optimal strategy is not generally SD.

The optimality of the strategies in examples 1 and 2 may also be established using simple arguments based on realizations of the sample paths. Sample path arguments also show that the optimal strategies remain SD when the arrival processes are general rather than Poisson. This seems to be a general feature of SD strategies. If a strategy is SD optimal for a problem in which there are Poisson arrivals then it is SD optimal for arbitrary arrivals.

4. APPLICATIONS OF THE SUFFICIENT CONDITION

We consider some examples in which ED strategies exist but neither lemma 3 nor sample path arguments are sufficient to prove their optimality. In these cases we work with lemma 2 directly. We try a H consisting of all rows of R (all rows of $P-\bar{P}$) and any other rows which are in some way obvious from the nature of the problem. We then test to see whether we can write $Hc \geq 0$ and $H\bar{P} = BH$ for some nonnegative B. If not, then we construct a new H whose rows are those of the old H and $H\bar{P}$. The process is continued until we have an H satisfying $Hc \geq 0$ and $H\bar{P} = BH$. In practice, we have found that every problem with an ED optimal strategy has a finite-dimensional H satisfying the conditions of lemma 2. Although we have not been able to show that this must necessarily happen, we do have the following result.

Lemma 4. If $\bar{P}$ defines an ED optimal strategy and all the eigenvalues of $\bar{P}$ are real, then there exists a finite dimensional H satisfying the conditions of lemma 2.

Proof. The proof is not particularly illuminating and will be given in outline only. Although the proof can be adapted to include the cases in which eigenvalues are not simple or some have equal modulus, we suppose for simplicity that $|\lambda_1| > \cdots > |\lambda_N|$. We can then write

$$\bar{P}^n = \sum_i \lambda_i^n (z_i y_i'),$$

where y_i and z_i are the left and right eigenvectors of $\bar{P}$ corresponding to the eigenvalue λ_i (y_i' denotes the transpose of y_i). Consider a row of R, say r'. Since $r'\bar{P}^n c \geq 0$ for all n, we deduce that $\lambda_i > 0$ for the least i such that $(r'z_i)(y_i'c) \neq 0$. We repeat this analysis for all rows of R and let $I = \{i$: i is the least index for which $(r'z_i)(y_i'c) \neq 0$, r' some row of R$\}$. Let G be a matrix consisting of all rows of the form,

$$\sum_{i=1}^{j} \beta_i (r'z_i) y_i',$$

where r' is a row of R, $1 \leq j \leq N$, $\beta_i = 1$ for $i \in I$, and $\beta_i = \pm 1$ for $i \notin I$. We can now check that

$Gx \geq 0$ implies $G\bar{P}x \geq 0$,
$G\bar{P}^n c \geq 0$ for all $n >$ some n_0, and
$Gx \geq 0$ implies $R\bar{P}^n x \geq 0$ for all $n >$ some n_1.

Let $n_2 = n_0 + n_1$. We now take H to consist of all the rows appearing in the matrices $R, R\bar{P}, \ldots, R\bar{P}^{n_2}, G\bar{P}^{n_0}$ and check that this H satisfies the conditions of lemma 2.

Note that, $\bar{P}$ has real eigenvalues if it is similar to a symmetric matrix [6]. This happens if $\bar{P}$ is the transition matrix of a time reversible Markov chain or if it can be represented as the limit of transition matrices of time reversible chains.

Example 3. Optimal Gambling

Ross [8] considers a problem of choosing the optimal stake in an advantageous gamble. He supposes that a gambler who has an amount of money £i can gamble any portion of that money in a bet. He wins an amount equal to his stake with probability p ($p \geq 0.5$) and loses his stake with probability $q = (1-p)$. The gambler wants to chose the size of his bets (which must be integral) so as to maximize the probability of eventually increasing his capital to £(N-1). Ross shows that the gambler achieves this by always making the minimum bet of £1, and that this strategy minimizes for all t the probability that he is ruined by time t. We show how this may be proved using lemma 2.

The problem is analyzed in discrete time. Let state i be the state in which the gambler has £(i-1). States 1 and N are absorbing, and $\bar{P}$ is the transition matrix when bets of 1 are made at each turn. Let I be the NxN identity matrix and let J be a NxN matrix with 1's above the diagonal and 0's elsewhere. Let $G = (qI - pJ)$. It is simple to check that $G\bar{P} = CG$ for some $C \geq 0$. Letting $F = (I-J)$ we can also check that $FG\bar{P} = DFG$ for some $D \geq 0$. Rows of R are positive linear combinations of rows of FG, so we take H as all rows of G and FG. Then $R = AH$ some $A \geq 0$, $H\bar{P} = BH$ some $B \geq 0$, and $Hc \geq 0$ for $c = (1, 0, \ldots, 0)$. Thus the strategy of betting the minimum amount minimizes the probability of being ruined by time t.

Example 4. Repair of the Series System

Derman, Lieberman and Ross [3], Katehatis [4], Smith [10] and Weber [12] have considered the problem of optimally maintaining a series system of n components with a single repairman. When functioning, component i fails with constant hazard rate μ_i. When failed, component i takes a time to repair which is exponentially distributed with parameter λ_i. The repairman desires to allocate his repair effort to maximize the probability that all component are functioning at time t. He may allocate his repair effort amongst the failed components in any way he likes, and he may stop the repair of one component in order to begin repair of another. Smith showed that amongst strategies which repair the components according to a fixed precedence list order the least failure rate strategy of always repairing the component with the least value of μ_i is optimal. He conjectured that the strategy should be optimal in the class of all strategies. Derman and others proved this for the case when all the λ_i are equal, and Weber and Katehakis proved it more generally.

We can establish the result using the sufficient condition of lemma 2. Suppose $\mu_1 < \cdots < \mu_n$. Suppose the states of the system are indexed by α $(1 \leqslant \alpha \leqslant N)$. Let $L(\alpha)$ be the set of components which are functioning in the αth state. Let $E\alpha$ be a row vector of N components having 1 in the αth position and 0 in other positions. For $i,j \notin L(\alpha)$ define the row vectors

$$S(i)\alpha = \lambda_i(E\alpha - E\alpha^i), \text{ and}$$
$$T(ij)\alpha = S(i)\alpha - S(j)\alpha,$$

where α^i is the index of the state in which the functioning components are $L(\alpha)+\{i\}$. Similarly, we shall let α_h denote the state in which the functioning components are $L(\alpha)-\{h\}$. If i is the component of least μ amongst those which are failed, k is the component of next largest μ, and $i<j$ then,

$$S(i)\alpha\{\bar{Q}+(\mu_i+\lambda_i+\sum_{h\in L}\mu_h)I\} = \lambda_i S(k)\alpha^i + \sum_{h\in L}\mu_h S(i)\alpha_h.$$

$$T(ij)\alpha\{\bar{Q}+(\mu_j+\lambda_i+\sum_{h\in L}\mu_h)I\} = (\mu_j-\mu_i)S(i) + \lambda_i T(kj)\alpha^i + \sum_{h\in L}\mu_h T(ij)\alpha_h.$$

If component $k \neq i < j$ is the one of least μ amongst those that are failed then,

$$S(i)\alpha\{\bar{Q}+(\mu_i+\lambda_k+\sum_{h\in L}\mu_h)I\} = \lambda_k S(i)\alpha^k + \sum_{h\in L}\mu_h S(i)\alpha_h.$$

$$T(ij)\alpha\{\bar{Q}+(\mu_j+\lambda_k+\sum_{h\in L}\mu_h)I\} = (\mu_j-\mu_i)S(i) + \lambda_k T(ij)\alpha^k + \sum_{h\in L}\mu_h T(ij)\alpha_h.$$

The matrix H which is appropriate for lemma 2 consists of all rows of the form $S(i)\alpha$ and $T(ij)\alpha$, for all α with $(i<j)\notin L(\alpha)$. The matrix R has rows which are positive linear combinations of the rows $T(ij)\alpha$. The statements above and $(\mu_j-\mu_i) \geqslant 0$ are sufficient for there to exist a θ (equal to the sum of all the λ's and μ's) such that $H(\bar{Q}+\theta I) = \theta BH$ for some nonnegative matrix B. Taking c as a vector which is equal to 1 in all components except the one corresponding to the state in which all components are functioning, we deduce that the least failure rate repair strategy maximizes for all t the probability that the all components are functioning at time t.

Example 5. Job Scheduling

A number of authors have proved that the longest expected processing time order strategy (LEPT) minimizes the expected value of the makespan when n jobs with exponentially distributed processing requirements are to be processed on m identical parallel machines (see [2] and [15]). In [13] we have shown that LEPT is stochastic dominant in that it minimizes the makespan in distribution. We shall indicate how the proof can be accomplished using lemma 2. Suppose the failure rates are $\lambda_1 < \cdots < \lambda_n$. We use the same notation as in example 4 and define $S(i)\alpha$ and $T(ij)\alpha$ as before. Suppose $L(\alpha)$ is the set of jobs which are not yet completed in the αth state. Let $K(\alpha)$ be the set of m jobs of least hazard rate amongst those which are uncompleted. For $i,j\in K(\alpha)$, $k\notin K(\alpha)$ and $k\in K(\alpha^i)$ we also define

$$U(ijk)\alpha = \lambda_i T(kj)\alpha^i + \lambda_j T(ik)\alpha^j.$$

Let H be a matrix whose rows are all the rows of $S(i)\alpha$, $T(ij)\alpha$ with $i<j$, and $U(ijk)\alpha$ with $i<j\in K(\alpha)$. All rows of R and $R\bar{P}$ are positive linear combinations of the rows of H. The $S(i)\alpha$ rows state that it is better to start from a state in which a given job is already complete rather than in a state which differs by that job being uncomplete. With $(\Sigma\lambda_i)\bar{P}=\bar{Q}+(\Sigma\lambda_i)I$ we find

$$S(i)\alpha\bar{P} = \sum_{h\in K}\lambda_h S(i)\alpha^h + \sum_{h\notin K}\lambda_h S(i)\alpha + \lambda_i S(k)\alpha^i \quad \text{for } i\in K, \tag{5}$$

$$S(i)\alpha\bar{P} = \sum_{h\in K}\lambda_h S(i)\alpha^h + \sum_{h\notin K}\lambda_h S(i)\alpha + \lambda_i S(k)\alpha \quad \text{for } i\notin K, \tag{6}$$

$$T(ij)\alpha\bar{P} = \sum_{h\in K}\lambda_h T(ij)\alpha^h + \sum_{h\notin K}\lambda_h T(ij)\alpha + \lambda_i T(kj)\alpha^i + \lambda_j T(ik)\alpha^j \quad \text{for } i,j\in K, \tag{7}$$

$$T(ij)\alpha\bar{P} = \sum_{h\in K}\lambda_h T(ij)\alpha^h + \sum_{h\notin K}\lambda_h T(ij)\alpha + \lambda_i T(kj)\alpha^i + \lambda_j T(ij)\alpha \ i\in K, \quad \text{for } j\notin K, \tag{8}$$

$$T(ij)\alpha\bar{P} = \sum_{h\in K}\lambda_h T(ij)\alpha^h + \sum_{h\notin K}\lambda_h T(ij)\alpha + \lambda_i T(ij)\alpha + \lambda_j T(ij)\alpha \quad \text{for } i,j\notin K, \text{ and} \tag{9}$$

$$U(ijk)\alpha\bar{P} = \sum_{h\in K}\lambda_h U(ijk)\alpha^h + \sum_{h\notin K}\lambda_h U(ijk)\alpha + \lambda_i U(ijk)\alpha + (\lambda_j-\lambda_i)S(ik)\alpha^j + \lambda_k U(ij\ell)\alpha^k. \tag{10}$$

In each of (5)-(10) the sums on the right hand side should be read to exclude any of the indices i,j,k which appear as arguments on the left hand side. In (10) ℓ is a job such that $\ell\notin K(\alpha^k)$ and $\ell\in L(\alpha^{ik})$. If there is no such ℓ then the last term of (10) is replaced by $\lambda_k\{\lambda_j S(i)\alpha^k - \lambda_i S(j)\alpha^k\}$. These equations establish $H\bar{P}=BH$ for some

$B \geqslant 0$, and the result follows by taking c as a vector which is 0 in the component corresponding to all jobs complete and 1 in all other components.

5. CONCLUDING REMARKS

The examples of the previous sections have shown that lemma 2 may be useful in establishing the optimality of a dominant strategy for a stochastic allocation or scheduling problem. In fact, for every problem which we know to have an expectation dominant strategy the optimality of that strategy can be established by this method. Winston's problem [17] of customer assignment to heterogeneous servers is another example where the method is useful.

Example 5 illustrates that it can sometimes be quite difficult to construct the appropriate matrix H. Indeed the H required to prove the results of [13] is even more complicated, though the difficulty is one of notation and not methodology. At the beginning of section 4 we described a method of building the appropriate matrix H from rows of of $R, R\bar{P}, R\bar{P}^2, \ldots$, along with rows which were somehow obvious from the nature of the problem. In fact, the obvious rows were simply ones which stated that a transition was preferred or not preferred to remaining in the current state. It would be interesting to know whether the constructive approach will always succeed, or whether there is a problem with a dominant optimal strategy whose optimality cannot be established by this method. One might also consider whether other sufficient conditions, perhaps more easily verified, would ensure that a strategy were ε-expectation dominant.

REFERENCES

[1] Benes, V.: 1981, Models and problems of dynamic memory allocation. "Proceedings of the ORSA-TIMS meeting, Applied Probability-Computer Science: the Interface", (to appear).

[2] Bruno J., Downey P., and Frederickson, G.N.: 1981, Sequencing tasks with exponential service times to minimize the expected flowtime or makespan. JACM 28, pp. 100-113.

[3] Derman, C., Lieberman, G.U. and Ross, S.M.: 1980, On the optimal assignment of servers and a repairman. J. Appl. Prob. 17, pp. 245-257.

[4] Katehakis, M.: 1980, On the repair of a series system. (private correspondence).

[5] Keilson, J. and Kester, A.: 1977, Monotone matrices and monotone Markov processes, Stoc. Proc. and Appl. 5, pp. 231-241.

[6] Keilson, J.: 1979, "Markov Chain Models-Rarity and Exponentiality", Springer-Verlag, New York.

[7] Lippman, S.: 1975, Applying a new device in the optimization of exponential queueing systems. Opns. Res. 23, pp. 687-710.

[8] Ross, S.M.: 1974, Dynamic programming and gambling models. J. Appl. Prob. 6, pp. 593-606.
[9] Serfozo, R.F.: 1979, An equivalence between continuous and discrete time Markov decision processes. Opns. Res. 27, pp. 616-620.
[10] Smith, D.R.: 1968, Optimal repair of the series system. Opns. Res. 4, pp. 653-662.
[11] Weber, R.R.: 1978, On the optimal assignment of customers to parallel servers. J. Appl. Prob. 15, pp. 406-413.
[12] Weber, R.R.: 1980, "Optimal Organization of Multi-Server Systems", Ph.D. Thesis, University of Cambridge.
[13] Weber, R.R.: 1982, Scheduling jobs with unknown processing requirements on parallel machines to minimize makespan or flowtime. J. Appl. Prob. 19, (to appear).
[14] Weeks, J.K. and Wingler, T.R.: 1979, A stochastic dominance ordering of scheduling rules. Decision Sci. 10, pp. 245-257.
[15] Weiss, G. and Pinedo, M.: 1979, Scheduling tasks with exponential service times on non-identical processors to minimize various cost functions. J. Appl. Prob. 17, pp. 187-202.
[16] Winston, W.L.: 1977, Optimality of the shortest line discipline. J. Appl. Prob. 14, pp. 181-189.
[17] Winston, W.L.: 1977, Assignment of customers to servers in a heterogeneous queueing system with switching. Opns. Res. 25, pp. 469-483.

ON THE COMPUTATIONAL COMPLEXITY OF STOCHASTIC SCHEDULING PROBLEMS

Michael Pinedo

Georgia Institute of Technology

In this paper we consider stochastic scheduling models where all relevant data (like processing times, release dates, due dates, etc.) are independent random variables, exponentially distributed. We are interested in the computational complexity of determining optimal policies for these stochastic scheduling models. We give a number of examples of models in which the optimal policies can be determined by polynomial time algorithms while the deterministic counterparts of these models are NP-complete. We also give some examples of stochastic scheduling models for which there exists no polynomial time algorithm if P ≠ NP.

1. INTRODUCTION

In the last decade deterministic scheduling problems have received substantial attention. Two directions have always been very important in the investigation of a deterministic scheduling problem, namely:

(1) The search for algorithms that determine the optimal sequence efficiently, preferably in polynomial time.

(ii) The investigation of the computational complexity, i.e. determining whether or not the problem is NP-complete.

This research has yielded many results. An excellent survey can be found in Graham, Lawler, Lenstra, Rinnooy Kan (3). Noting that the deterministic distribution can be considered as a special case of an arbitrary stochastic distribution, many of these results can provide a key to the structure of the more general stochastic scheduling problems. Another special distribution is the *exponential* distribution. Stochastic scheduling models with exponentially distributed data (like processing times, release dates, due dates,

M. A. H. Dempster et al. (eds.), Deterministic and Stochastic Scheduling, 355–365.

etc.) usually have a very nice structure, too.

In this paper whenever we refer to a specific stochastic scheduling problem we assume, unless otherwise specified, a problem with all data independent *exponentially* distributed. Of these scheduling models the objective functions usually have to be minimized in expectation. Our goal is either to obtain good (i.e. polynomial time) algorithms for determining the optimal policies within specific classes of policies or to show that no fast algorithms exist (assuming $P \neq NP$). We believe that it may be of interest to compare the computational complexity of these stochastic scheduling problems with the computational complexity of their deterministic counterparts. For a number of models the algorithms that determime the optimal policies in the deterministic and the stochastic versions are equally good. Examples of such models are $F2||E(C_{max})$ and $J2||(C_{max})$ where in the deterministic as well as in the stochastic version the algorithms are $O(n \log n)$. For some models there is a better algorithm for the deterministic version; an example is $O2||E(C_{max})$ where the algorithm for the deterministic version is $O(n)$, while the algorithm for the stochastic version is not as fast (see (11)). In Section 3 we give some examples of models of where the deterministic version is NP-complete and the stochastic version is very easy.

This paper is organized as follows: In Section 2 we give a short description of various classes of policies, formulations of stochastic scheduling problems as Markov Decision Processes, levels of complexity and how to determine the complexity of these stochastic scheduling problems. In Section 3 we present six models, exhibiting NP-completeness in the deterministic version, but having polynomial time algorithms in the stochastic setting. In Section 4 we present two examples of stochastic models for which we can show that there exists no polynomial time algorithm to determine the optimal policies (assuming $P \neq NP$). In Section 5 we discuss in which direction we intend to continue this research.

2. COMPLEXITY OF STOCHASTIC SCHEDULING PROBLEMS

In this section we first discuss three classes of policies. Afterwards we explain how we intend to investigate the computational complexity of determining optimal policies in any of these classes.

2.1. Classes of Policies

The first class of policies is the class of *static* policies. Here the decision-maker decides at $t = 0$ what his policy will be during the process; he is not allowed to make any subsequent changes. The optimal policy in this class often (but not always) prescribes a "list" schedule, i.e. whenever a machine becomes idle, the decision-maker starts processing the next job on the list independent of the state of the system at that moment. Consider the following example: $1|\underset{\sim}{p}_j \sim \exp(1), \underset{\sim}{d}_j \sim \exp(\mu_j)|E(\Sigma w_j U_j)$. As the processing times of all the jobs are exponentially distributed with mean one, the expected penalty caused by a particular job not meeting its due date only depends on the number of jobs that are to be processed before this one. Determining the optimal policy in the class of static policies reduces easily to a (deterministic) assignment problem. By solving this assignment problem the decision-maker obtains a list in which to process the n jobs. Determining the optimal policy in the class of static policies usually is equivalent to a deterministic problem. Hence, in determining whether or not this can be done in polynomial time, techniques similar to those used for deterministic scheduling problems can be used. Examples in Section 4 illustrate this point. Determining the optimal static policy is important because this optimal policy may be related to the optimal policy in another class of policies.

The second class of policies is the class of *nonpreemptive dynamic* policies. These can be described as follows: Under such a policy the decision-maker is allowed to determine his action at specified decision moments (e.g. when a job has finished its processing on a machine) making use of any newly available information. However, he is not allowed to interrupt the processing of any job already started.

The third class of policies is the class of *preemptive dynamic* policies. These are similar to our second class of policies, but now the decision-maker *is* allowed to interrupt the processing of any job at any time. Experience seems to show that determining the optimal policy in the class of preemptive dynamic policies is usually easier than determining the optimal policy in the class of nonpreemptive dynamic policies.

2.2. Markov Decision Processes and Levels of Complexity

A stochastic scheduling problem with all data exponentially distributed and in which the decision-maker is allowed at every decision moment to take all available information into account can be described as a Markov Decision Process in continuous time with a finite state space, finite action space and one absorbing state. It is a well known fact that for these MDP's there exists an optimal *stationary* policy. This formulation as an MDP can be done for the preemptive as well as for the nonpreemptive case.

For these MDP's we will make a distinction between three levels of complexity. Firstly, the ideal situation would be to have an algorithm that determines the optimal actions for *all* states simultaneously in polynomial time. Secondly, for some problems there may be an algorithm that determimes the optimal action in any arbitrary state in polynomial time, but that cannot determine the optimal actions in *all* states in polynomial time (the number of states often grows exponentially with the size of the problem). However, one has to keep in mind now that during its realization the process, from beginning to end, usually visits only a limited number of states, a number that usually is polynomial in the size of the problem. So, if the decision-maker has an algorithm that determines the optimal action in an arbitrary state in polynomial time and the number of times he uses this algorithm is polynomial in the size of the problem, the total number of computations the decision-maker has to perform during the process is still polynomial in the size of the problem. Gifford (2) has developed an algorithm for a stochastic scheduling problem where this is the case. Thirdly, for some problems it may not even be possible to develop an algorithm that determines the optimal action in an arbitrary state in polynomial time. The examples of Section 4 are of this kind.

2.3. Determining the Computational Complexity

We present here two approaches that may be useful when trying to determine the computational complexity of stochastic scheduling problems.

(i) *Formulation of the Markov Decision Process as a Linear Program*. It is a well known fact that determining the optimal actions in an MDP can be done via Linear Programming. It has been shown recently that there exists a polynomial time algorithm to solve an LP (5). This, of course, does not imply that there exists a polynomial time algorithm to determine the optimal actions in all states of an MDP, as the transformation from the MDP into an LP may not be possible in polynomial time. For most scheduling problems the size of the LP (the number of variables and constraints) grows exponentially in the size of the problem. However, this does *not* imply that there does not exist a polynomial time algorithm for the scheduling problem. So this approach appears to be useful only when the size of the LP grows polynomially in the size of the scheduling problem, because then we know we have a polynomial time algorithm. The following example illustrates this: Consider m machines and n jobs with precedence constraints. These precedence constraints have the form of $c(c > m)$ chains. The jobs have arbitrary independent exponential distributions, not necessarily identical. This MDP can be formulated as an LP for which it can be shown that, when c is fixed, the number of variables and constraints increase polynomially in n. So there exists an algorithm for determining the optimal actions that is polynomial in n. For the deterministic counterpart of this problem one can find easily an algorithm, based on dynamic programming, that is polynomial in n.

(ii) *Relating Optimal Dynamic Policies with Optimal Static Policies*. Determining whether the optimal static policy can be found in polynomial time is relatively easy, as this is basically a deterministic problem that can be approached with conventional techniques. Determining whether the optimal actions of an MDP can be found in polynomial time tends to be harder. However, it is possible to find special circumstances in which the optimal actions in a subset of the states have a specific one-to-one correspondence with the optimal list in the static problem. In case the optimal static policy cannot be determined in polynomial time we know that, if we indeed can establish a certain relationship between the actions of an optimal dynamic policy and the list given by the optimal static policy, the optimal dynamic policy cannot be determined in polynomial time either. The two examples of Section 4 illustrate this.

3. EXAMPLES OF STOCHASTIC SCHEDULING PROBLEMS THAT ARE EASY

In this section we discuss six models; the deterministic versions of five of these models are known to be NP-complete. Of the remaining one the complexity has not yet been determined, but the problem appears to be not too easy. The stochastic versions, with processing times exponentially distributed, of all of these models turn out to be very easy to analyze.

3.1. $P||C_{max}$ and its Stochastic Counterpart

In this model n jobs have to be scheduled on m identical parallel machines in order to minimize the makespan. Karp (4) showed that the deterministic version of this problem is NP-complete. Several researchers showed independently that when the processing times are exponentially distributed the Longest Expected Processing Time first (LEPT) policy minimizes the expected makespan. This LEPT policy is optimal in all three classes of policies discussed in Section 2.

3.2. $P2|res\ 1|C_{max}$ and its Stochastic Counterpart

In this model we have two machines and n jobs. There is one resource of which there is an amount s. Job j needs when being processed an amount s_j of this resource. At any time during the process the total amount of resource needed by the two jobs being processed, may not be larger than s. It is clear that there is no polynomial time algorithm for the deterministic version of this problem, as there exists no polynomial time algorithm for $P2||C_{max}$. Pinedo (13) considered the problem with exponentially distributed processing times. He assumed the following agreeability condition: Whenever for any two jobs $1/\lambda_i > 1/\lambda_j$ then $s_i \geq s_j$. A nonpreemptive policy was shown to be optimal in all three classes of policies. This nonpreemptive policy can be determined in O(n log n) time.

3.3. $1|d_j = d|\Sigma w_j U_j$ and its Stochastic Counterparts

In the deterministic version of this model we have n jobs, all with the same deadline. If job j does not meet this deadline the decision-maker incurs a penalty w_j. This problem is equivalent to the knapsack problem and therefore NP-complete (see Karp (4)). Of the stochastic version there are two variants: Firstly, all deadlines are i.i.d. random variables with arbitrary distribution (not necessarily exponential). Secondly, all jobs have the same deadline and this deadline is an arbitrarily distributed random variable.

Derman et al.(1) have shown that in order to minimize the expected objective in all three classes of policies for the second variant the decision-maker has to order the jobs in decreasing order of $w_j \lambda_j$ where $1/\lambda_j$ is the mean of the exponentially distributed processing time of job j. Pinedo (12) showed that for the first variant the same policy minimizes the expected objective in the class of static policies.

3.4. $P|d_j = d|\Sigma U_j$ and its Stochastic Counterparts

This model is similar to the model discussed in 3.3. The only difference lies in the fact that instead of one machine we now have m machines in parallel. It is clear that there exists no polynomial time algorithm for the deterministic version of this problem either. Of the stochastic version with exponentially distributed processing times there are again two variants: Firstly all deadlines are i.i.d. random variables with arbitrary distribution (not necessarily exponential). Secondly, all jobs have the same deadline and this deadline is an arbitrarily distributed random variable. Pinedo (12) has shown that for the first variant the policy that schedules the task in increasing order of their expected processing times minimizes the expected objective in the class of static policies provided the distribution function of the deadline is concave. He also showed that for the second variant the same policy is optimal in all three classes of policies, again provided the distribution function being concave.

3.5. $1|d_j = d|\Sigma w_j T_j$ and its Stochastic Counterparts

In the deterministic version of this model all jobs have again the same deadline. If job j does not meet the deadline the decision-maker incurs a penalty $w_j T_j$ (= $w_j(C_j - d_j)$). The computational complexity of this problem has not yet been determined, but Lawler and Moore (8) developed a pseudo-polynomial algorithm for this problem. Again two variants have been considered for the stochastic version. Firstly, all deadlines are i.i.d. random variables, with arbitrary distribution. Secondly, all jobs have the same deadline with arbitrary distribution. Pinedo (12) showed that for both variants the optimal policy in the class of static policies instructs the decision-maker to schedule the jobs in decreasing order of $w_j\lambda_j$. This policy is also optimal in the class of dynamic policies for the second variant. In (12) more general models are discussed; models where the distributions of the deadlines are not identical but "agreeable" in the following sense: whenever $w_j\lambda_j > w_i\lambda_i$ the distribution of the deadline of job j is stochastically smaller than the distribution of the deadline of job i.

3.6. $1|pmtn, r_j|\Sigma w_jC_j$ and its Stochastic Counterpart

In the deterministic version there are n jobs released at given release dates and the sum of the weighted completion times has to be minimized. Labetoulle et al.(6) have shown that this problem is NP-complete. Pinedo (12) considered the stochastic version where the release dates have an arbitrary joint distribution. The class of preemptive dynamic policies was considered. In this class the following policy was shown to be optimal: At any point in time, of the jobs that already have been released the one with the highest value of $w_j\lambda_j$ is released, the machine has to be preempted and this new job has to be put on the machine.

3.7. Additional Remarks

(i) In the stochastic models of 3.3 and 3.5 it is possible to include release dates in such a way that the optimal preemptive dynamic policy still can be determined easily. Consider for example the following extension: All due dates are identical and the release dates may be any combination of time epochs prior to this due date.

(ii) Gifford (2) considered the models of 3.2 and 3.3 with due dates nonidentically exponentially distributed. He developed algorithms which in some cases are polynomial time, in other cases not.

(iii) There are of course more models of which the deterministic version is NP-complete and for which in the stochastic setting there exists a polynomial time algorithm. Pinedo (10) and Weiss (14) considered a problem in reliability theory where this is the case.

4. EXAMPLES OF STOCHASTIC SCHEDULING PROBLEMS THAT ARE HARD

In this section we present two examples of stochastic scheduling problems for which we can show that, if $P \neq NP$, no polynomial time algorithm exists to determine the optimal policies in any of the three classes of policies mentioned in Section 2. In our approach we follow the method described in Section 2, i.e. we first prove that the static problem cannot be solved in polynomial time, after that we show that there exists a one-to-one correspondence between the optimal list of the static problem and the optimal actions of an appropriately chosen subset of states in the preemptive or nonpreemptive dynamic problems. We will not give rigorous proofs in this section, as the technical details are rather lengthy; the complete proofs will appear elsewhere. In both examples precedence constraints play an important role; without these precedence constraints the problems are easy. The first example is a rather straightforward extension of a deterministic result. The second example is more complicated.

4.1. $1|prec, \underset{\sim}{p}_j \sim \exp(1)|E(\Sigma\, w_j C_j)$

In this model we have one machine and n jobs, each of these jobs having a processing time exponentially distributed with mean one. There are precedence constraints which may have an arbitrary form. The expected sum of the weighted completion times has to be minimized. Lawler (7) considered the case where all the jobs have identical deterministic processing times, i.e. $1|prec, p_j = 1| \Sigma\, w_j C_j$. He showed through a reduction from LINEAR ARRANGEMENT that this problem is NP-complete even if $w_j \varepsilon \{1,2,3\}$ for all j. One can show that changing the processing times of the jobs from deterministic with mean one to exponential with mean one does not change the optimal sequence of the jobs and that this sequence is an optimal policy in all three classes of policies. This implies that, if $P \neq NP$, there exists no polynomial time algorithm to determine the optimal policy for this stochastic scheduling problem.

4.2. $1|prec, \underset{\sim}{p}_j \sim \exp(p), \underset{\sim}{d}_j \sim \exp(\mu_j)|E(\Sigma\, U_j)$

We have again one machine and n jobs. Job j has an exponentially distributed processing time with mean $1/p$ and an exponentially distributed due date with mean $1/\mu_j$. There are precedence constraints and the expected number of jobs overdue has to be minimized. We first will show that finding the optimal static policy cannot be done in polynomial time. Suppose job j is the kth job to be processed. The probability that job j will not meet its deadline is $1 - (p/(p+\mu_j))^k$. Now let $1/\mu_j \gg n/p$ for all j. The probability that job j will not meet its deadline is then approximately $\mu_j \cdot k/p$. The objective to be minimized is the expected number of jobs that do not meet their deadline; this may be approximated by $\Sigma\ \mu_j \cdot E(C_j)$ where $E(C_j)$ is the expected completion time of job j. This objective, however, after replacing μ_j with w_j, is identical to Lawler's objective in the problem $1|prec, p_j \sim exp(p)|E(\Sigma\ U_j)$, when $1/\mu_j >> n/p$ for all j, results in a sequence that is identical to an optimal sequence in Lawler's $1|prec, p_j = 1|\Sigma\ w_j C_j$. This implies that the optimal static policy cannot be determined in polynomial time.

Before investigating the optimal dynamic policies for $1|prec, p_j \sim exp(p)|E(\Sigma\ U_j)$, we consider the following model, which at first sight may seem unrelated: We have one machine and n jobs with identical deterministic processing times of one time unit. Of only one of these jobs a due date will occur during [0,n]; of n-1 jobs no due date will occur. The due date will be job j's with probability q_j. This due date is uniformly distributed over [0,n]. There are arbitrary precedence constraints. The objective is to maximize the probability that no job will be overdue. It can be shown that the optimal static policy in this model is equivalent to an optimal sequence in Lawler's deterministic problem $1|prec, p_j = 1|\Sigma\ w_j C_j$ (the q_j's play the role of the w_j's). It can also be shown that this optimal static policy is also optimal in the classes of preemptive and nonpreemptive dynamic policies. In the following paragraph we will refer to the above model as the One Due Date Model.

Now we consider the dynamic versions of $1|prec, p_j \sim exp(p), \tilde{d}_j \sim exp(\mu_j)|E(\Sigma\ U_j)$. Let $\Sigma\ \mu_i = \alpha$, where α is small and $\alpha n/p$ is very small. Making the rates $\mu_i, i=1,\ldots,n$ small has the following two effects: Firstly, the event of any due date occurring during the time needed to process the n jobs has a very low probability. Secondly, if a due date does occur, the time it occurs is approximately uniformly distributed over [0,T], where T is the random time needed to process all jobs. It can be shown that, for n sufficiently large, the probability of no due date occurring during [0,T] is $1 - \alpha n/p + O((\alpha n/p)^2)$, the probability of one due date occurring is $\alpha n/p + O((\alpha n/p)^2)$ and the probability of two or more due dates occurring is $O((\alpha n/p)^2)$. The problem can be formulated as an MDP in continuous time; optimal actions have to be determined in all possible states, a state being characterized by the set of jobs already processed and the set of due dates already occurred. It can be shown through some rather technical bounding arguments that in order to determine the optimal action in a state one may base one's decision solely on the event of exactly one due date occurring in the remaining time needed to finish the process. This is intuitive as in the case of no due dates occurring any sequence would be optimal, and the probability of two or more due dates occurring is very small in comparison with the probability of one due date occurring. But the problem conditional on one due date occurring is approximately equivalent to the One Due Date Model. Suppose the optimal job sequence for the One Due Date Model is $j_1, j_2, \ldots, j_n$ (which represents simultaneously the optimal static policy as well as the optimal preemptive and nonpreemptive dynamic policy) then it can be shown, through bounding arguments, that when the MDP is in a state with jobs $j_1, j_2, \ldots, j_k$ finished and all due dates still to come, the optimal action is to start job j_{k+1}. Now one can reason that if there exists a polynomial time algorithm to determine the optimal actions in the states of the MDP there also exists a polynomial time algorithm for model M. As, if $P \neq NP$, there exists no polynomial time algorithm for model M, there does not exist a polynomial time algorithm for the MDP either.

5. CONCLUSION

This paper should provide an indication of the direction in which we are currently working. To this extent, it constitutes a sort of progress report. Presently, we are dealing with stochastic scheduling problems with due dates; this research is far from complete. However, Gifford (2), as suggested before, has obtained some interesting results with regard to algorithms for these problems. The following aspects of these due date problems are currently under investigation: The influence of processing time distributions other than exponential (we are thinking here mainly of distributions with a Decreasing Failure Rate (DFR)). In addition, the influence of special classes of precedence constraints, namely series-parallel (see (7)) or chains (see (8)) is also worthy of investigation as well as the development of bounds on the expected values of the objective functions when processing times are exponential.

ACKNOWLEDGMENT

I am grateful to Ted Gifford for his many helpful comments and Betty Plummer for her excellent typing.

REFERENCES

(1) Derman, C., Lieberman, G. and Ross, S., "A Renewal Decision Problem": 1978, Management Sci. 24, pp. 554-561.
(2) Gifford, T., "Algorithms for Stochastic Scheduling Problems with Due Dates": in preparation.
(3) Graham, R., Lawler, E.L., Lenstra, J.K, and Rinnooy Kan, A.H.G., "Optimization and Approximation in Deterministic Sequencing and Scheduling: A Survey": 1979, Ann. Discrete Math. 5, pp. 287-326.

(4) Karp, R.M., "Reducibility Among Combinatorial Problems": 1972, R.E. Miller and J.W. Thatcher, eds., Complexity of Computer Computations (Plenum Press, New York), pp. 85-103.
(5) Khachian, L.G., "A Polynomial Algorithm in Linear Programming": 1979, Doklady, Akademii Nauk SSSR 224, pp. 191-194.
(6) Labetoulle, T., Lawler, E.L., Lenstra, J.K. and Rinnooy Kan, A.H.G., "Preemptive Scheduling of Uniform Machines Subject to Release Dates": 1979, Technical Report Mathematisch Centrum.
(7) Lawler, E.L., "Sequencing Jobs to Minimize Total Weighted Completion Time Subject to Precedence Constraints": 1978, Ann. Discrete Math. 2, pp. 75-90.
(8) Lawler, E.L. and J.M. Moore, "A Functional Equation and its Application to Resource Allocation and Sequencing Problems": 1969, Management Science, 16, pp. 77-84.
(9) Lenstra, J.K. and Rinnooy Kan, A.H.G., "Complexity Results for Scheduling Chains on a Single Machine": 1980, Eur. J. of Operational Research 4, pp. 270-275.
(10) Pinedo, M.L., "Scheduling Spares with Exponential Lifetimes in a Two Component Parallel System": 1980, J. of Appl. Prob. 17, pp. 1025-1032.
(11) Pinedo, M.L. and Ross, S.M., "Minimizing the Expected Makespan in Stochastic Open Shops": 1980, in editorial process, J. of Appl. Prob.
(12) Pinedo, M.L., "Comparisons between Deterministic and Stochastic Scheduling Problems with Release Dates and Due Dates": 1981, in editorial process, Operations Research.
(13) Pinedo, M.L., "On Stochastic Scheduling with Resource Constraints": 1981, in editorial process, Management Science.
(14) Weiss, G., "Scheduling Spares with Exponential Lifetimes in a Two Component Parallel System": 1981, in editorial process, Nav. Res. Logistics Quart.

DETERMINISTIC AND STOCHASTIC SCHEDULING PROBLEMS WITH TREELIKE PRECEDENCE CONSTRAINTS

John Bruno

University of California, Santa Barbara

In this paper we present some of the known results for sequencing identical jobs on parallel machines subject to tree-like precedence constraints. In-trees and out-trees are discussed. Highest-level-first policies are shown to minimize the makespan in the deterministic case. These policies are not necessarily optimal for stochastic processing times, except for the case of two machines and in-tree precedence constraints.

1. INTRODUCTION

In this paper we survey the known results for sequencing jobs on m parallel machines subject to tree-like precedence constraints. We shall discuss in-tree and out-tree constraints and for each we will consider deterministic and stochastic processing times. In the stochastic case we shall take the processing times to be independent and identically distributed random variables with distribution function $F(t)=1-e^{-t}$ for $t\geq 0$. In the deterministic case all the processing times will be identical and equal to one.

We shall see that the intuitive scheduling policy, highest-level-first, is an optimal policy for minimizing the makespan for deterministic scheduling problems with in-tree and out-tree precedence constraints and any number m of parallel machines. By way of contrast, in the stochastic case the highest-level-first policies are known to minimize the expected makespan for two machines and in-tree precedence constraints and are not necessarily optimal for three or more machines. They are not necessarily optimal for two or more machines and out-tree prece-

M. A. H. Dempster et al. (eds.), Deterministic and Stochastic Scheduling, 367–374.

dence constraints. In the remainder of this section we present some terminology and notation.

Since the processing times of all the jobs are identical (i.e., either unity in the deterministic case or i.i.d. exponential random variables in the stochastic case), an <u>instance of a scheduling problem</u> is given by a directed, acyclic graph G. The set of vertices of G is denoted by J(G). The vertices correspond to the jobs. We shall use the terms job and vertex interchangeably. The set of arcs of G is denoted by $A(G) \subseteq J(G) \times J(G)$. The arcs of G determine the precedence constraints among the jobs of G. We say that job J_i is a <u>predecessor</u> of job J_j if there is a directed path from J_i to J_j in G (and J_j is a <u>successor</u> of job J_i). We say that J_i is an <u>immediate predecessor</u> of job J_j if $(J_i, J_j) \varepsilon A(G)$ (J_j is an <u>immediate successor</u> of job J_i).

A job is called an <u>initial</u> job if it has no predecessors and is called <u>terminal</u> if it has no successors. Let I(G) denote the set of initial jobs of G.

Let U be a subset of J(G). We define the subgraph G/U of G as follows. $J(G/U) = J(G) - U$ and $A(G/U) = A(G) - \{(J_i, J_j) \mid (J_i, J_j) \varepsilon A(G)$ and either J_i or J_j or both are in $U\}$. The problem instance G/U is the restriction of the problem instance G to the jobs in J(G)-U.

A <u>policy</u> π is a mapping from scheduling problems into jobs such that

1. $\pi(G) \subseteq I(G)$;
2. if $|I(G)| > m$ (the number of processors) then $|\pi(G)| = m$ otherwise $\pi(G) = I(G)$.

A policy is used to make scheduling decisions at <u>decision points</u>. The decision points correspond to time zero and subsequent job completion times.

A <u>schedule</u> is an assignment of machines to jobs determined by a policy. Let G be a given scheduling problem and π a policy. Beginning at time zero, we assign machines to all the jobs in $\pi(G)$. Let t be the earliest time that one of the jobs in $\pi(G)$ completes. It is possible that more than one job completes at time t. Let $J_1, \ldots, J_k$ $(k \geq 1)$ be the jobs that complete at time t. Since $J_1, \ldots, J_k$ have completed their processing we remove them from G by setting $G \leftarrow G/\{J_1, \ldots, J_k\}$. We now (at time t) assign the machines to the jobs in $\pi(G)$. Scheduling continues until there are no jobs left in G.

We note that policies allow preemptions at task completion times and do not use any scheduling "history" in making decisions. That is, the machine assignments at a decision point do not depend on the current time or any previous scheduling decisions. Although we could have defined policies which take "history" into account, they would not be any "better" than the policies we have chosen for the criteria we study in this paper.

We define $c_j(\pi,G)$ to be the completion time of job J_j when policy π is applied to the scheduling problem G. Of course, $c_j(\pi,G)$ is a random variable in the stochastic case. We define the lateness of job J_j as $d_j-c_j(\pi,G)$ where d_j is the deadline associated with job J_j. We define $m\ell(\pi,G)$, the maximum lateness, as $m\ell(\pi,G)=\max_j\{d_j-c_j(\pi,G)\}$. The makespan of schedules obtained by applying π to G is defined as $ms(\pi,G)=\max_j\{c_j(\pi,G)\}$.

In order to define a highest-level-first policy we define a level function λ from J(G) into the positive integers. If J_i is a terminal job then $\lambda(J_i)=1$. If J_j is any other job then

$$\lambda(J_j) = \max\{\lambda(J_i)\,|\,(J_j,J_i)\varepsilon A(G)\} + 1.$$

The λ value of a job J_i is the number of jobs in a longest path from J_i to some terminal job.

We say that π is a highest-level-first ($h\ell f$) policy if it assigns machines to those initial jobs with the largest λ values.

2. DETERMINISTIC PROCESSING TIMES

2.1 In-tree Precedence Constraints and m Machines

G is an in-tree (in-forest) scheduling problem if for any job J_i there exists a unique path from J_i to a terminal job. A fundamental result for problems with in-tree precedence constraints was given by T. C. Hu [1]. He showed that $h\ell f$ policies minimize the makespan of the resulting schedules. We shall present some recent work of Brucker, Garey and Johnson [2] for a somewhat more general problem and from which the $h\ell f$ result can easily be derived.

Let G be a scheduling problem and for each job J_j in G let d_j be a positive number called the deadline of job J_j. Our goal is to find a scheduling policy which minimizes the maximum lateness. In [2] they give a deadline modification algorithm which determines, from the original set of deadlines, a new set of deadlines d'_j called modified deadlines. The algorithm for computing the modified deadlines follows:

1. If J_i is a terminal job of G set $d'_i \leftarrow d_i$.

2. Select a job J_i in G which has not been assigned it modified deadline and whose immediate successor J_j has been assigned its modified deadline. If no such job exists then halt.

3. Set $d'_i \leftarrow \min\{d_i, d'_j - 1\}$ and return to step 2.

We call π an <u>earliest-deadline-first</u> (edf) policy if π assigns machines to those uncompleted initial jobs in G with the smallest modified deadlines.

<u>Theorem 1</u> [2] Let G be an in-tree scheduling problem with deadlines. Then a schedule exists which meets all the original deadlines if and only if the edf policy π based on the modified deadlines determines a schedule which meets all the original deadlines.

The above theorem guarantees that π will construct a schedule with $m\ell(\pi,G) \leq 0$ if and only if such a schedule exists.

To approach the problem of minimizing $m\ell(\pi,G)$ we note that if some number h were added to all the original deadlines then the new modified deadlines would be given by d'_i+h where the d'_i s are the original modified deadlines. Therefore the edf policy π determined by the original modified deadlines is also an edf policy with respect to the new modified deadlines. Since π is independent of h, the schedule determined by π has $m\ell(\pi,G)=0$ when h is the minimum possible maximum lateness. Therefore the edf π minimizes $m\ell(\pi,G)$ over all policies.

If we were to set all the deadlines d_i equal to n and compute the modified deadlines then the edf policy π determines a schedule with the minimum possible $ms(\pi,G)$. This is true because, as above, the relative ordering of the modified deadlines would not change if all the original deadlines were set to n+h for any value of h.

The interesting point is that $\lambda(J_j) = n+1-d'_j$ for the modified deadlines computed from the original deadlines d_i=n for i=1,...,n. Therefore the edf policy π is also an $h\ell f$ policy.

<u>Theorem 2</u> [1] Let G be an in-tree scheduling problem and π an hℓf policy. Then $ms(\pi,G) \leq ms(\pi',G)$ for all policies π'.

2.2 Out-tree Precedence Constraints and m Machines

In this section we consider deterministic scheduling problems in which all the processing times are equal to one and the precedence constraints form an out-tree (out-forest). We say that G is an out-tree scheduling problem if for every job J_i there is a <u>unique</u> path from some initial job to J_i.

The level function λ defined earlier determines distances to terminal jobs. We now define a related function ρ which gives the distance from an initial job. Let G be an out-tree scheduling problem. If J_i is an initial job then $\rho(J_i) = 1$. If J_i has as immediate predecessor J_j, then $\rho(J_i) = \rho(J_j) + 1$.

We shall show that a hℓf policy π applied to an out-tree scheduling problem G gives a schedule with shortest possible makespan. Clearly if the number of initial jobs does not exceed m, the number of machines, all policies make the same decision. This, of course, is true until we reach a decision point at which the number of initial jobs exceeds m. Suppose therefore that $|I(G)| > m$. In order to continue with the analysis we define the following sets. Let Si be the set of all jobs J_j such that $\rho(J_j) = i$ for i=1,2,... . Let Ni = $|Si|$, the number of elements in Si for i=1,... .

By assumption, we have N1 > m. Now suppose that

$$(1) \qquad \sum_{i=1}^{p} Ni > pm$$

for p=1,...,k-1 and

$$(2) \qquad \sum_{i=1}^{k} Ni \leq km$$

for $k \geq 2$.

The above equations imply that we have enough jobs to keep all m machines busy for k-1 time units and during the kth interval we could keep $\sum_{i=1}^{k} Ni - (k-1)m$ machines busy.

We will show (Lemma 1) that hℓf policies assign machines in such a manner that all jobs in S1,...,Sk will be completed in the first k time units. Note that the jobs (if there are any) in Sk cannot be processed any earlier than the kth time unit. Let U←S1 ∪ S2 ∪ ... ∪Sk and set G←G/U. If G is not null, then by the above reasoning, the initial jobs in G could not be processed before time k+1 under any policy. Consequently, we are at a "regeneration" point where the above arguments may be repeated.

<u>Lemma 1</u> Let G be an out-tree scheduling problem and suppose that equations (1) and (2) hold. If π is a hℓf policy applied to G then all the jobs in S1 ∪ S2 ∪ ... ∪ Sk will be completed within the first k time units.

<u>Proof</u>: The proof is by induction on k.

Basis: (k=2) It is easy to see that the only way for equations (1) and (2) to hold is for the initial jobs which were <u>not</u> processed in the first time interval to be terminal jobs. Thus all the jobs in S1 ∪ S2 which were not processed in the first time interval are available for processing during the second interval.

Step: Assume k>2. Let A1 be the jobs in S1 which we processed in the first interval. Set B1 = S1-A1. Let Ai be the descendants of the jobs in A1 with ρ value equal to i for i=2,3,...,k. Similarly, let Bi = Si-Ai.

The m jobs in A1 are processed in the first time interval. Our goal is to show that equations (1) and (2) still hold for G/A1 with k' = k-1.

Set N'i = |A(i+1)| + |Bi| for i=1,...,k-1.

Equations (1) and (2) for G/A1 are

$$(3) \qquad \sum_{i=1}^{p} N'i > pm, \quad p=1,...,k'-1,$$

and

$$(4) \qquad \sum_{i=1}^{k'} N'i \leq k'm.$$

Our goal is to verify (3) and (4). This can be done by noticing that (1) and (2) can be written as follows.

$$(1') \qquad \sum_{i=1}^{p} N'i > pm - |B(p+1)|$$

for p=1,...,k'-1 and

$$(2') \qquad \sum_{i=1}^{k'} N'i \leq k'm - |Bk|.$$

Equation (2') implies equation (4). If $|B(p+1)| = 0$ then the corresponding equation in (3) is verified. If $|B(p+1)| > 0$ then we can argue that $|A1| \geq m,\ldots,|A(p+1)| \geq m$ and consequently the corresponding equation in (3) is verified. □

Theorem 3 [3] Let G be an out-tree scheduling problem and π an hℓf policy. Then π minimizes $ms(\pi,G)$ over all policies.

3. STOCHASTIC PROCESSING TIMES

3.1 In-tree Precedence Constraints and m=2

In this section we assume G is an in-tree scheduling problem and the processing times of the jobs in J(G) are independent and identically distributed random variables with distribution function F(t). Furthermore we take the number of machines to be 2. The results in this section were first obtained by Chandy and Reynolds [4]. They showed that a policy minimizes the expected value of the makespan if and only if it is hℓf. They also showed, by exhibiting a counterexample, that hℓf policies are not necessarily optimal when $m \geq 3$. The proof that hℓf is optimal when m=2 is based on a structural characterization of in-tree scheduling problems which they termed "flatness".

Let G be an in-tree schelduing problem. Let Mi be equal to the number of jobs with λ value equal to i for i=1,2,... . Then define $S(G,i) = \sum_{k \geq i} Mk$ for i=1,2,... .

We say that G is flatter than H if $S(G,i) \leq S(H,i)$ for $i \geq 1$. This is denoted by $G \underset{\sim}{\nabla} H$. We say that G is as flat as H if $S(G,i) = S(H,i)$ for $i \geq 1$. This is denoted by $G \sim H$. And finally, we say that G is strictly flatter than H (denoted by $G \nabla H$) if $G \underset{\sim}{\nabla} H$ and for some i we have $S(G,i) < S(H,i)$.

Our objective is to minimize the expected makespan. Let π be some policy. Then we can write

$E(\underset{\sim}{ms}(\pi,G)) = 0$ if G is null,

$E(\underset{\sim}{ms}(\pi,G)) = 1+E(\underset{\sim}{ms}(\pi,G/\{J_i\}))$ if G has one initial job J_i,

$E(\underset{\sim}{ms}(\pi,G)) = 1/2 + 1/2E(\underset{\sim}{ms}(\pi,G/\{J_i\})) + 1/2E(\underset{\sim}{ms}(\pi,G/\{J_j\}))$ if $\pi(G) = \{J_i,J_j\}$.

The main result which relates flatness to our scheduling criterion follows.

<u>Theorem 4</u> [4] Let G and H be in-tree scheduling problems, m=2 and let π be a policy which minimizes the expected makespan.

a) $H \sim H \Rightarrow E(\underset{\sim}{ms}(\pi,G)) = E(\underset{\sim}{ms}(\pi,H))$.

b) $G \underline{\nabla} H \Rightarrow E(\underset{\sim}{ms}(\pi,G)) \leq E(\underset{\sim}{ms}(\pi,H))$.

c) $G \nabla H \Rightarrow E(\underset{\sim}{ms}(\pi,G)) < E(\underset{\sim}{ms}(\pi,H))$.

Using Theorem 4 and some of the properties of flatness it is not difficult to prove

<u>Theorem 5</u> [4] Let G be an in-tree scheduling problem and m=2. Then π minimizes the expected value of the makespan if and only if π is an hℓf policy.

3.2 Out-tree Precedence Constraints

When G is a scheduling problem with out-tree precedence constraints there are no known simply-specified policies which are optimal for $m \geq 2$. It is known that hℓf is not necessarily optimal.

REFERENCES

1. T.C. HU (1961) Parallel sequencing and assembly line problems. *Oper. Res.* <u>9</u>,841-848.
2. P. BRUCKER, M.R. GAREY, D.S. JOHNSON (1977) Scheduling equal-length tasks under treelike precedence constraints to minimize maximum lateness. *Math. Oper. Res.* <u>2</u>,275-284.
3. G.I. DAVIDA, D.J. LINTON (1976) A new algorithm for the scheduling of tree structured tasks. *Proc. 1976 Conf. Information Sciences and Systems*, Johns Hopkins University, 543-548.
4. K.M. CHANDY, P.F. REYNOLDS (1975) Scheduling partially ordered tasks with probabilistic execution times. *Proc. 5th Symp. Operating Systems Principles*, 169-177; *Operating Systems Reviews* <u>9</u>.5.

ON THE EVALUATION OF NON-PREEMPTIVE STRATEGIES IN STOCHASTIC SCHEDULING

K.D. Glazebrook

University of Newcastle upon Tyne

A collection of stochastic jobs is to be processed by a single machine. The jobs must be processed in a manner which is consistent with a precedence relation but the machine is free to switch from one job to another at any time. Such switches are costly, however. A general model is proposed for this problem. Sufficient conditions are given which ensure that there is an optimal strategy given by a fixed permutation of the job set. These conditions are then used as a starting point for the important task of evaluating permutations as strategies in more general circumstances where no permutation is optimal.

1. INTRODUCTION

A single machine is available to process a collection J of jobs which evolve randomly in time as the machine processes them. Processing must be in accord with partial ordering R. Hence if $(i, j) \in R$, $i \in J$, $j \in J$, then job i must be completed before the processing of job j can begin. The machine is free to switch from one job to another at any time. Costs are incurred when such switching takes place and as processing occurs. The objective is to find a strategy for allocating the machine to the (uncompleted) jobs which minimises the total expected cost incurred.

Such stochastic scheduling problems have been studied by Bruno and Hofri (1), Gittins (2), Gittins and Glazebrook (3), Glazebrook (4), (5), (6), Meilijson and Weiss (12), and Sevcik (14), among others. In some special cases, it has proved possible to give optimal strategies explicitly. For example, some progress

M. A. H. Dempster et al. (eds.), Deterministic and Stochastic Scheduling, 375–384.

has been made by Glazebrook (4) and Meilijson and Weiss (12) in connection with problems where the precedence relation R has a digraph representation in the form of an out-tree. In general, though, the quest for explicit solutions has been a difficult one.

Here our aim is to give sufficient conditions which ensure that there is an optimal strategy for processing which is a fixed permutation of job set J, that is, a job once started is processed through to completion. Hence, when faced with a problem in which these conditions hold there is a reduction to the simpler problem of finding an optimal permutation. In many cases, there are algorithms for this latter problem (see Glazebrook and Gittins (10), Kadane and Simon (11) and Sidney (15)). These sufficient conditions are described in Section 2.

The question then arises of whether we can say anything about how well permutations perform as processing strategies when these sufficient conditions fail to hold. An indication is given in Section 3 of how the sufficient conditions can give rise to general methods for evaluating permutations.

2. PROBLEM FORMULATION

We formulate the stochastic scheduling problem described in Section 1 as a discrete-time controlled stochastic process with the following special features:

2.1 State space

The state of the process at integer time t is $x(t) = \{x_1(t), \ldots, x_K(t)\}$ where $|J| = K$ and where $x_j(t)$ is the state of job j at time t. The state space for job j is denoted Ω_j. We further denote by $\Gamma_j \subseteq \Omega_j$ the completion set of job j, namely that job j is completed when it enters Γ_j for the first time.

2.2 Action space

The set of decision times is the non-negative integers $\mathbb{N}$. Allocation of the machine to job j is called action a_j. The set of admissible actions at integer time t, A(t), is given by

$$A(t) = \{a_j;\ x_j(t) \notin \Gamma_j \text{ and } \nexists\ i \text{ such that } x_i(t) \notin \Gamma_i, (i,j) \in R\}.$$

Hence at time t the machine can process any job which has not been completed and which is not preceded by any uncompleted job.

2.3 Transitions of the process

If action a_j is taken at time t then $x_i(t+1) = x_i(t)$, $i \neq j$, and $x_j(t+1)$ is determined by a probabilistic law of motion $P_j\{x_j(t)\}$. Hence if at time t job j is allocated to the machine, then only the state of that job changes, and that in a Markov fashion.

2.4 Cost structure

The cost of processing job j at time t is written $C_j\{x_j(t), x_j(t+1)\}$. Switching costs are also incurred as follows: If action a_i is taken at time 0 we incur a set-up cost S(i). If job j is the last to be completed then a tear-down cost T(j) is incurred at its completion. If action a_i is taken at time t and action a_j is taken at time t+1 where $i \neq j$, a switching cost B(i, j) is incurred at t+1. The following consistency conditions apply:

$$S(i) \leq S(j) + B(j,i), \quad i \neq j \in J.$$

$$B(i,j) \leq B(i,k) + B(k,j), \quad i \neq k \neq j \neq i \in J.$$

$$T(i) \leq B(i,j) + T(j), \quad i \neq j \in J.$$

The interpretation of these conditions is plain. All costs are discounted at rate a, $0 < a < 1$, and are assumed bounded.

2.5 Optimal strategies

An optimal strategy is any rule for choosing admissible actions which leads to a minimal total expected cost.

This model generalises those contained in previous studies by Bruno and Hofri (1), Gittins (2), Glazebrook (5), Glazebrook and Gittins (10), Nash (13) and Sevcik (14).

As is explained in the introduction, our interest is in setting down conditions which are sufficient to ensure that there is an optimal strategy for allocating the machine to the jobs which is such that a job, once started, is processed through to completion. We give two conditions which must apply to each job in J. The first requires that (informally) as we process a job we do not make the time when rewards from other jobs will accrue more remote (i.e., the job is "shortening"). The second requires that (informally) as we process a job then the future prospects for that job (in terms of likely rewards) are progressively improved.

These conditions are now stated formally as follows: Assume that action a_j is taken from time 0 until job j's completion, which occurs at (random) time T_j. Define function $r_j : \Omega_j \to [0, 1]$ (expected remaining discounted processing time) by

$$r_j(x) = \begin{cases} E\{a^{T_j} \mid x_j(0) = x\}, & x \in \Omega_j \backslash \Gamma_j \\ 1, & x \in \Gamma_j. \end{cases}$$

We define job j to be <u>shortening</u> if under any strategy

$$P[r_j\{x_j(s)\} < r_j\{x_j(t)\}] = 0 \text{ for any } s > t.$$

Define function $\gamma_j : \Omega_j \to \mathbb{R}$ (Gittins index for job j) by

$$\gamma_j(x) = \begin{cases} E[\sum_{t=0}^{T_j - 1} a^t C_j\{x_j(t), x_j(t+1)\} \mid x_j(0) = x]\{1 - r_j(x)\}^{-1}, & x \in \Omega_j \backslash \Gamma_j, \\ M(1-a)^{-1}, & x \in \Gamma_j, \end{cases}$$

where M is a lower bound on the costs (see (2.4)). We now define job j to be <u>improving</u> if $\gamma_j\{x_j(0)\} \leq 0$ and if under any strategy

$$P[\gamma_j\{x_j(s)\} > \gamma_j\{x_j(t)\}] = 0 \text{ for any } s > t.$$

<u>Theorem 1</u>. If all the jobs in J are both shortening and improving there is an optimal allocation strategy for the stochastic scheduling problem which is given by a fixed permutation of J.

The proof of this result is to be found in Glazebrook (9). Under certain circumstances Theorem 1 may be extended in the following ways: replacement of switching costs by switching times, more elaborate switching cost structure, replacement of discounted costs by costs which are linear in time. It is also possible in some cases to establish a continuous time version of Theorem 1. The details of this appear in Glazebrook (7). Lastly, it becomes plain on close examination of the proof of Theorem 1 that the precedence relation can in fact be more general than a partial ordering.

We can relax the sufficient conditions described in relation to Theorem 1 considerably if we are prepared to specialise to the <u>simple jobs model</u>. Here we assume that the jobs in J have independent positive integer-valued random processing times $\{P_1, \ldots, P_K\}$ where $|J| = K$. Costs are incurred at switches and rewards are earned when jobs are completed. Hence, in terms of the general model described in (2.1) - (2.5) we take $x_j(t)$, the state of job j at time t, to be $\{y_j(t), z_j(t)\}$ where $y_j(t)$ is the number of times action a_j is taken prior to t and $z_j(t)$ is given by

$$z_j(t) = \begin{cases} 1, & \text{if job } j \text{ has been completed by time } t, \\ 0, & \text{otherwise.} \end{cases}$$

Plainly we have $\Gamma_j = \mathbb{N} \times \{1\}$. Switching costs are as in (2.4) and processing costs for job j are

$$C_j[(y,0),\ (y+1,0)] = 0, \quad y \in \mathbb{N},$$

$$C_j[(y,0),\ (y+1,1)] = -K(j), \ \{K(j) \geq 0\}, \ y \in \mathbb{N}.$$

Many studies to date (for example, Gittins (2), Gittins and Glazebrook (3), Glazebrook (5), Nash (13), and Sevcik (14)) have included versions of the simple jobs model.

Note that, in the simple jobs model, job j is improving if and only if it is shortening and so the two conditions of Theorem 1 become one. Further, it is not difficult to show that if random processing time P_j has a non-decreasing hazard rate then it is improving. Hence, Theorem 1 is seen to apply to simple jobs with the following processing time distributions: exponential, Weibull with index greater than one, gamma with parameter greater than one, truncated normal.

However, the simplification in the stochastic structure enables us to go considerably further. Define function $r_j : \mathbb{N} \to [0, 1]$ (expected remaining discounted processing time) by

$$r_j(x) = \begin{cases} E(a^{P_j - x} \mid P_j > x) & \text{if } P(P_j > x) > 0, \\ 1, & \text{otherwise.} \end{cases}$$

If we assume that $K(j) \geq \max\{T(j), \max_{k \neq j} B(j,k)\}$, $j \in J$, namely that completion rewards are always greater than the switching costs which succeed them, then we obtain the following result.

<u>Theorem 2</u>. If each job $j \in J$ in the simple jobs model satisfies

$$r_j(x) \geq r_j(0), \quad x \in \mathbb{N}, \tag{1}$$

there is an optimal allocation strategy given by a fixed permutation of J.

The proof of Theorem 2 is in Glazebrook (9). There are generalisations both to problems in continuous time and with linear costs.

3. EVALUATING NON-PREEMPTIVE STRATEGIES

It may be that we are faced with a stochastic scheduling problem in which the sufficient conditions described in Section 2, do not hold. It could be, for example, that we choose to model our jobs in such a way that an initial period during which processing brings an improvement in the job's future prospects is succeeded by a period of deterioration. Hence for example, our jobs are not improving in the sense required for Theorem 1. Evidently, then, we cannot conclude on the basis of Theorem 1 that there is a non-preemptive strategy which is optimal. However, what we can sometimes do, is to use Theorem 1 and results like it to say how well an optimal non-preemptive strategy performs in relation to an optimal strategy.

3.1 The general method

The general method may be described informally as follows: Given the stochastic scheduling problem P of interest, find another scheduling problem P' such that (i) the total expected cost of an optimal strategy for P' is less than that for P, (ii) there is an optimal strategy for P' determined by a fixed permutation π, say, and (iii) the permutation π may be viewed as a strategy for P. Denote by $V(P)\{V(P')\}$ the total expected cost of an optimal strategy for $P(P')$, by $V(P,\pi)\{V(P',\pi)\}$ the total expected cost incurred when adopting strategy π for $P(P')$ and by $\overline{V}(P)$ the total expected cost of an optimal non-preemptive strategy for P. It follows easily that

$$\overline{V}(P) - V(P) \leq \overline{V}(P) - V(P') = \overline{V}(P) - V(P',\pi)$$
$$\leq V(P,\pi) - V(P',\pi).$$

Hence $\overline{V}(P) - V(P',\pi)$ and $V(P,\pi) - V(P',\pi)$, which can both in principle be computed, are upper bounds on $\overline{V}(P) - V(P)$, the quantity of interest. It is sometimes more convenient to use the second of these bounds, although it is weaker.

Plainly the value of the theoretical results in Section 2 for the above procedure is to suggest problems P' which have an optimal strategy given by a permutation, as required by (ii) above. (A further theoretical question of interest is how to obtain P' satisfying (i) also - see Glazebrook (8)). We now proceed to exemplify the method by an example.

3.2 The simple jobs model with initially improving jobs

We consider the simple jobs model with zero switching costs. In order to simplify the presentation of the results we shall now consider continuous time problems. These can have either discounted costs or linear costs. Job j, then, has a non-negative processing time P_j whose distribution is determined by a smooth hazard function ρ_j. In the discounted costs case, if job j is completed at time t the reward (negative cost) is $K(j)\exp(-\alpha t)$ $(\alpha > 0)$ and in the linear costs case the reward is $-K(j)t$. Theorem 2C will denote the continuous time version of Theorem 2 which holds in this case (a being replaced by $\exp(-\alpha)$ in the original statement). Theorem 3C refers to the continuous time version of the simple jobs model with linear completion costs and no switching costs, which we now quote.

Define function $s_j: \mathbb{R}^+ \to \mathbb{R}^+$ (expected remaining processing time) by

$$s_j(x) = \begin{cases} E(P_j - x \mid P_j > x) \text{ if } P(P_j > x) > 0 \\ 0, \text{ otherwise.} \end{cases}$$

<u>Theorem 3C</u>. If each job $j \in J$ satisfies

$$s_j(x) \leq s_j(0), \quad x \in \mathbb{R}^+, \tag{2}$$

there is an optimal allocation strategy given by a fixed permutation of J.

Our aim is to evaluate permutations as strategies for the simple jobs model with no switching costs in the case where each job is <u>initially improving</u> in the sense that $\rho_j'(0) \geq 0$. This seems an appropriate condition to enforce in some practical contexts but is not a sufficiently strong one to ensure that appropriate versions of conditions (1) and (2) hold. Hence we cannot conclude that in problems with initially improving jobs there is an optimal strategy which is non-preemptive, indeed it is not difficult to give an example where that is not the case. We develop a method for evaluating permutations as strategies based on the ideas described in (3.1).

We proceed as follows: From our original scheduling problem

P, we construct a new problem P′ identical in all respects to P save only that the original processing times $(P_1,\ldots,P_K)$ are replaced by $(P_1^*,\ldots,P_K^*)$. New processing time P_j^* is processing time P_j truncated at time τ_j, which is chosen such that

$$E\{\exp(-\alpha P_j^*)\} - E\{\exp(-\alpha P_j)\}$$
$$= \sup_{x>0}\left(\exp\{-\alpha x - \int_0^x \rho_j(t)dt\}\left[1-\exp\{-\alpha x-\int_0^x \rho_j(t)dt\}\right]^{-1}\{r_j(0)- r_j(x)\}\right) \quad (3)$$

in the discounted costs case. In the linear costs case τ_j is chosen such that

$$E(P_j) - E(P_j^*)$$
$$= \sup_{x>0}\left(P(P_j > x)\left[P(P_j \le x)\right]^{-1}\{s_j(x) - s_j(0)\}\right). \quad (4)$$

Note that the expressions on the right of (3) and (4) yield natural measures of the extent to which job j fails to satisfy conditions (1) and (2) respectively. Note also that the condition $\rho_j'(0) \ge 0$ is sufficient to ensure the existence of $\tau_j > 0$ defined according to (3) or (4).

τ_j is chosen according to (3) or (4) in order to ensure that new processing time P_j^*, with associated functions r_j^* and s_j^* satisfies whichever is appropriate of conditions (1) or (2) (in both cases τ_j is the largest value with this property). Hence, whether problem P has linear or discounted costs, it will follow from Theorems 2C and 3C that there is an optimal strategy for P′ given by a fixed permutation. It is also trivial to demonstrate that the total expected cost of an optimal strategy for P′ is less than that for P under both kinds of costs. Hence we have achieved objectives (i) and (ii) set out in paragraph (3.1). Theorems 4 and 5 follow easily and refer to the cases with discounted costs and linear costs respectively. Full proofs may be found in Glazebrook (8).

<u>Theorem 4</u>. (Discounted costs). If all the jobs in J are initially improving

$$\{V(P) - \overline{V}(P)\}\{V(P)\}^{-1} \le \{\prod_{j\in J} (1+R_j)\}-1, \text{ where}$$

$$R_j = \sup_{x>0}\left(\exp\{-\alpha x-\int_0^x \rho_j(t)dt\}\left[1-\exp\{-\alpha x-\int_0^x \rho_j(t)dt\}\right]^{-1} \times\right.$$
$$\left.\times \{r_j(0) - r_j(x)\}\{r_j(0)\}^{-1}\right)$$

Theorem 5. (Linear costs) If all the jobs in J are initially improving

$$\{\overline{V}(P) - V(P)\}\{V(P)\}^{-1} \le \max_{j\in J} (S_j\{s_j(0) - S_j\}^{-1}), \text{ where}$$

$$S_j = \sup_{x>0} (P(P_j>x)[P(P_j \le x)]^{-1}\{s_j(x) - s_j(0)\}.$$

Note that, for any specific problem, it may be possible to improve the bounds given in Theorems 4 and 5 by exploiting features specific to that problem.

3.3 Lognormal jobs

Processing time P_j is said to have a lognormal distribution with parameters μ_j and σ_j if log P_j is normal with mean μ_j and standard deviation σ_j. Lognormal distributions are interesting in that they furnish us with probably the best-known examples of distributions having non-monotone hazard rates. However the hazard rates are always increasing at 0 and so all lognormal jobs are initially improving. Glazebrook (8) has reported some calculations based on Theorem 5 which demonstrate that in problems involving lognormal jobs with linear costs very little is lost by restricting to non-preemptive strategies provided that each job j has a small parameter σ_j. For example if no job j is such that $\sigma_j>0.5$ then the percentage loss is never greater than 2×10^{-5}. For parameter values of 0.8 and 1, the losses are no greater than 1.54% and 12.08% respectively. The upper bound given in Theorem 5 increases rapidly above the parameter value 1.

The position is similar in problems with lognormal jobs where the costs are discounted.

REFERENCES

1. Bruno, J., and Hofri, M.: "On scheduling chains of jobs on one processor with limited preemption" 1975, SIAM J. Comput. 4, pp. 478-490.
2. Gittins, J.C.: "Bandit processes and dynamic allocation indices", 1979, J. Roy. Statist. Soc. B41, pp. 148-177.
3. Gittins, J.C. and Glazebrook, K.D.: "On Bayesian models in stochastic scheduling", 1977, J. Appl. Prob. 14, pp. 556-565.
4. Glazebrook, K.D.: "Stochastic scheduling with order constraints", 1976, Int.J. Systems Sci. 7, pp. 657-666.
5. Glazebrook, K.D.: "On stochastic scheduling with precedence relations and switching costs", 1980, J. Appl. Prob. 17, pp. 1016-1024.

6. Glazebrook, K.D.: "On non-preemptive strategies in stochastic scheduling", 1981, Nav.Res. Logist. Qu. (to appear).
7. Glazebrook, K.D.: "On non-preemptive strategies for stochastic scheduling problems in continuous time", 1981, Int. J. Systems Sci. (to appear).
8. Glazebrook, K.D.: "Methods for the evaluation of permutations as strategies in stochastic scheduling problems", (submitted).
9. Glazebrook, K.D.: "On the evaluation of fixed permutations as strategies in stochastic scheduling", (submitted).
10. Glazebrook, K.D. and Gittins, J.C.: "On single-machine scheduling with precedence relations and linear or discounted costs", 1981, Oper. Res. 29, pp. 161-173.
11. Kadane, J.B. and Simon, H.A.: "Optimal strategies for a class of constrained sequential problems", 1977, Ann. Statist. 5, pp. 237-255.
12. Meilijson, I. and Weiss, G.: "Multiple feedback at a single server station", 1977, Stoch Proc. Appl. 5, pp. 195-205.
13. Nash, P.: "Optimal allocation of resources between research projects", 1973, Ph.D. thesis, Cambridge University.
14. Sevcik, K.C.; "The use of service-time distributions in scheduling", 1972, Tech. Rept. CSRG-14, University of Toronto.
15. Sidney, J.B.: "Decomposition algorithms for single machine sequencing with precedence relations and deferral costs", 1975. Oper. Res. 23, pp. 283-298.

SEQUENTIAL OPEN-LOOP SCHEDULING STRATEGIES

P. Nash, R.R. Weber

Cambridge University

ABSTRACT

For certain scheduling problems with pre-emptive processing, a dynamic programming formulation reduces the problem to a sequence of deterministic optimal control problems. Simple necessary and sufficient optimality conditions for these deterministic problems are obtainable from the standard results of optimal control theory, and sometimes lead to analytic solutions. Where this does not happen, then as with many dynamic programming formulations, computational solution is possible in principle, but infeasible in practice. After a survey of this approach to scheduling problems, this paper discusses a simplification of the method which leads to computationally tractable problems which can be expected to yield good, though sub-optimal, scheduling strategies. This new approach is based on the notion of sequential open-loop control, sometimes used in control engineering to solve stochastic control problems by deterministic means, and is not based on dynamic programming.

1. INTRODUCTION

The basic tools of most of optimization theory are optimality conditions obtained by variational methods. A fundamental difficulty with scheduling problems is their combinatorial nature, which usually makes their solution by variational techniques impossible. Thus for most scheduling problems, first- and second-order optimality conditions analogous to those of mathematical programming are unavailable. For one class of problems, namely those in which scheduling is completely pre-emptive, this is not the case. In such problems, processor

M. A. H. Dempster et al. (eds.), Deterministic and Stochastic Scheduling, 385–397.

effort can be regarded as infinitely divisible at each point in time, and the allocation of effort as instantaneously variable. Under these conditions, variational methods can be applied. As a bonus, we find that when we formulate such problems appropriately, they can be approached by completely deterministic methods.

In this paper, we outline a method whereby scheduling problems of this type can be reduced to deterministic optimal control problems. The application of the maximum principle to these problems leads to necessary conditions for optimality of a schedule, while the Hamilton-Jacobi-Bellman equation for the problem gives a corresponding sufficient condition. For a number of cases, these conditions enable one to derive the optimal scheduling strategy. This is most easily done for static single-processor problems, and becomes increasingly difficult as the structure of the problem becomes more complex. Extending results from the static to the dynamic case requires more complicated notation and proofs, and when we examine the parallel-processor case, the arguments needed to produce analytic results become very delicate indeed. The formulation is based on a particular dynamic programming approach, and when analytic results are not obtainable, solution by direct computation is not a practical proposition. Without going into formal proofs in detail, we indicate the features of the problem which make its solution difficult, and these lead us to propose the investigation of what control engineers call sequential open-loop strategies.

To apply an sequential open-loop strategy to a scheduling problem, we compute an allocation of processor effort for all future times, by optimizing an objective functional on the assumption that the chosen allocation will be followed - even to the extent of leaving resources idle - irrespective of the realizations of the job completions. This allocation is followed for the duration of some review period (possibly random), then a new allocation is computed in the same way, after updating all the probability distributions. This new allocation is put into effect during the next review period, and so on.

Such a strategy will be sub-optimal, but it will approximate to optimality in some cases. In particular cases, indeed, such a strategy is optimal if the review times are suitably chosen. Where we can have some hope that such a strategy will perform well, it has considerable advantages: it can be obtained analytically at least as often as the optimal closed-loop strategy, while being relatively easy to compute even if not so obtainable.

The next section describes the reduction of the scheduling problem to a problem in deterministic optimal control, and reviews some of the results that can be obtained from this formulation. The exposition will be brief, as this material has appeared elsewhere

[5], [6], [8]. In the following sections, we examine sequential open-loop schedules for a number of examples and compare them with known optimal strategies.

2. AN OPTIMAL CONTROL FORMULATION

For simplicity, we consider first the single-server problem with no arrivals. At time $t=0$ there are n jobs waiting to be processed. Processing effort is available at a constant rate, which may be taken to be unity. The amount of processing needed to complete job i is a random variable whose distribution is known, and has distribution F_i with density f_i. We define

$$P_i=1-F_i,\ \rho_i=f_i/P_i,\ i=1,2,\ldots,n,$$

the survivor functions and completion rate functions respectively. On completion of each job, a reward is received which depends on the job and on the time of its completion. The reward received if job i is completed at time t is $r_i(t)$. (We assume that the functions F_i, f_i, ρ_i and r_i are all well behaved.) We seek to allocate processing effort between jobs over time so as to maximize the expected total reward.

The state of the set of jobs at time t is described by the pair $(\Gamma(t),X(t))$, where $\Gamma(t)$ is a list, possibly empty, of the indices of the jobs that have been completed by time t, and $X(t)=(X_1(t),X_2(t),\ldots,X_n(t))$ is a vector whose i'th component is the amount of processing received by job i during $(0,t)$. A scheduling strategy U is a rule which determines, for any state (Γ,X) and time t an allocation $U(\Gamma,X,t)$ of the total processing effort. We use $\Gamma(t|U)$ and $X(t|U)$ to denote the components of the process $(\Gamma(t),X(t))$ when a fixed strategy U is used.

A crucial feature of the controlled semi-Markov process $(\Gamma(t),X(t))$ is that for any fixed strategy U, the evolution of $X(t|U)$ is completely deterministic between jumps in $\Gamma(t|U)$. Any strategy U can therefore be implemented in the following way. Define, for each (Γ,X) and t,

$$u^\Gamma(X,t;s)=U(\Gamma,x^\Gamma(X,t;s),s),\quad t\leqslant s<\infty,$$

where $x^\Gamma(X,t;s)$ is determined by

$$\dot{x}^\Gamma(X,t;s)=u^\Gamma(X,t;s),\ t\leqslant s<\infty,\ x^\Gamma(X,t;t)=X.$$

Let $C_0,C_1,\ldots,C_n$ denote the initial time and the times of the first, second, etc. job completions. Let $\Gamma_0,\Gamma_1,\ldots,\Gamma_n$ denote the lists of completed jobs initially and at times just after the first, second, etc. job completions. Then U is implemented by using the allocation

$$u^{\Gamma 0}(X(C_0),C_0;s)$$

at all times s up to the first job completion, then using the allocation

$$u^{\Gamma 1}(X(C_1),C_1;s)$$

for all times between the first and second completions, then switching to

$$u^{\Gamma 2}(X(t_2)C_2;s),$$

and so on.

This idea allows us to write the following recursive expression for expected total reward obtained using strategy U and starting at time t in state (Γ,X):

$$V^{\Gamma}(X,t|U)=\int_t^{\infty}\Big(\sum_{i\in\bar{\Gamma}}\Big\{\Big(\prod_{\substack{k\neq i\\ k\in\bar{\Gamma}}}P_k(x_k^{\Gamma}(X,t;s))\Big)r_i(s)$$

$$+V^{\Gamma,i}(x^{\Gamma}(X,t;s)),s|U)\Big\}f_i(x_i^{\Gamma}(X,t;s))u_i^{\Gamma}(X,t;s)\Big)ds,$$

with the obvious terminal condition that the expected value in states with no uncompleted job is identically zero. This makes it clear that for any initial state, the scheduling problem can be viewed as one of choosing an optimal sequence

$$\{u^{\Gamma i}(X(C_i),C_i;s)\colon i=0,1,2,\ldots,n-1\}$$

of deterministic controllers. If we apply the dynamic programming optimality principle to this sequence of choices, we see that U is optimal if and only if each such controller $u^{\Gamma}(X,t;s)$ solves a deterministic optimal control problem

$$P^{\Gamma}(X,t)\text{: Maximize } \int_t^{\infty}\Big(\sum_{i\in\bar{\Gamma}}h_i^{\Gamma}(x(s),s)u_i(s)\Big)ds$$

$$\text{subject to } \dot{x}(s)=u(s),\ t\leqslant s<\infty,\ x(t)=X,$$

$$\sum_i u_i(s)=1,\ t\leqslant s<\infty,$$

where we define for each Γ,x,s

$$h_i^{\Gamma}(x,s)=\Big\{\Big(\prod_{\substack{k\neq i\\ k\in\bar{\Gamma}}}P_k(x_k)\Big)r_i(s)+V^{\Gamma,i}(x,s|U)\Big\}f_i(x_i).$$

Not only is this problem deterministic, but it has a

particularly simple form, especially in its dynamics. We quote two theorems which show that this simplicity leads to equally simple necessary and sufficient conditions for optimality. Let P1 be the problem

$$P1(X,t): \text{ Maximize } \int_t^\infty \Big(\sum_i h_i(x(s),s)u_i(s)\Big)ds$$

$$\text{subject to } \dot{x}(s)=u(s),\ t\leqslant s<\infty,\ x(t)=X,$$

$$\sum_i u_i(s)=1,\ t\leqslant s<\infty,\ u(s)\,\varepsilon\,\Omega,\ 0\leqslant s<\infty.$$

Theorem 1. A necessary condition for the optimality of u in P1 is that , for each time t,

$$\sum_i u_i(t)\int_t^\infty \Big(\sum_j H_{ij}(x(s),s)u_j(s)\Big)ds = \max_{\substack{\omega\varepsilon\Omega \\ \Sigma\omega_i=1}} \Big\{\sum_i \omega_i \int_t^\infty \Big(\sum_j H_{ij}(x(s),s)u_j(s)\Big)ds=0,$$

where

$$H_{ij}(x,s)=\frac{\partial h_j(x,s)}{\partial x_i}-\frac{\partial h_i(x,s)}{\partial x_j}+\frac{\partial}{\partial s}(h_j(x,s)-h_i(x,s)).$$

This condition is just a statement of the maximum principle for P1, after integrating the co-state equations to give the co-state variables in terms of the states and controls. The details are in [5].

Theorem 1 is a necessary condition, and relates to open-loop controllers, that is controllers specified as functions of time for particular sets of initial data. We can ask for a closed-loop or feedback controller, which assigns to each state a control action. The open-loop controllers are then just the implementations of the closed-loop controller on trajectories with particular initial data. By using dynamic programming on P1, we obtain a necessary and sufficient condition for optimality of a feedback controller. Let $V(x,t|u)$ denote the value of the objective functional in P1 starting from state x and time t, using control u.

Theorem 2. A necessary and sufficient condition for the feedback controller $u(x,t)$ to be optimal for P1 is

$$\sum_i u_i(x,t)\Big\{\frac{\partial V(x,t|u)}{\partial x_i}+h_i(x,t)\Big\} \geq \sum_i \omega_i\Big\{\frac{\partial V(x,t|u)}{\partial x_i}+h_i(x,t)\Big\}$$

for all $\omega\varepsilon\Omega$ such that $\Sigma\omega_i=1$, for all x and t.

Again, the proof of this result is a straightforward application of the techniques of optimal control theory. The details can be found in [8].

The deterministic problem $P^{\Gamma}(X,t)$ is of the same form as P1, and theorems 1 and 2 may therefore be applied to it to obtain necessary and sufficient optimality conditions. To solve the scheduling problem with the usual initial data - no jobs completed and no processing as yet carried out, starting at time 0 - we need to solve P(0,0). A snag is that the conditions provided by the theorems will involve the derivatives of the optimal value functions V^i for the (n-1)-job scheduling problems arising at the first completion. It is readily shown [5] that these derivatives can be written in terms of the co-states for the (n-1)-job problems. When the optimal scheduling strategy has a structure which is independent of the number of jobs (when it is given by a priority index, for example), it is usually possible to prove optimality by using this as the basis of an inductive argument. When this is not the case, the dependence of the objective in P(0,0) on the functions $\{V^i\}$ means that any computational procedure based on this approach would have to be recursive, and would involve the solution of all 2,3,...,(n-1) job sub-problems over a region of the state space of the n job problem. While possible in principle, there is little hope that such a procedure could be practical.

This is a pity, because the formulation permits of great generality. For example, we can deal with parallel processors just by changing the constraints to be satisfied by the deterministic controllers. For the multi-server case, we just add the constraints

$$u_i(s)\leq 1/m, \quad i=1,2,\ldots,n.$$

Some results for multi-server problems obtained in this way are described in [8]. Precedence relations between jobs can be modelled by allowing the control constraint set Ω to depend on Γ, so that an optimal controller u^{Γ} for P^{Γ} has to satisfy

$$u^{\Gamma}(t)\in\Omega^{\Gamma}, \quad 0\leq t<\infty.$$

This does not affect the form of the deterministic control problems, so that theorems 1 and 2 may still be applied. Queueing problems are included simply by noting that arrivals constitute just another type of random jump, so that the set of times at which deterministic controllers have to be chosen now includes arrival instants as well as completion instants. We thus have to solve a set of problems $P^{\Gamma;\Delta}$, where Δ is the list of jobs which have arrived or been released. Similarly, reneging of customers in a queue is included by adding times at which customers renege to the set of decision times. In fact, almost any effect of this sort, which just adds more jumps to the basic process, can be incorporated in this way, and the objective functional and dynamics in the associated deterministic control problem are always formally the same as those of P1: only the specific definition of the functions h_i changes. Various results involving the determination of optimal strategies can be extended by these means; some of these extensions are described in [6].

In the remainder of this paper, we investigate a possible method of using this formulation to generate scheduling strategies in problems where the full deterministic control problem derived from the optimality principle as above cannot be solved analytically. We note that most of the difficulties arise because of the dependence of the optimality conditions on the lower-order sub-problems, and consider a formulation in which these dependencies do not occur.

3. OPEN-LOOP FORMULATIONS

The basic observation we make is that if we pretend that the evolution of the system will be unobserved, or that at any rate we are unable to act on any observations, then we are faced with the problem of finding an open-loop controller, which is simply an allocation of processor effort for all future times, which will be followed exactly, even if it implies allocating effort to already completed jobs. This is equivalent to controlling the probability dynamics of the process, which is the approach used in [1] and [7], although neither of these papers makes explicit reference to the fact that the control is open-loop.

The approximation involved in open-loop control is obviously a gross one, and the allocations produced would become more and more ridiculous as jobs were completed. A compromise between a full closed-loop solution and the completely open-loop one that has been used in a number of control applications is open-loop feedback or sequential open-loop control [2]. To implement a sequential open loop controller, we solve an open-loop problem at each of a succession of review times $\tau_0, \tau_1, \ldots$, and employ the resulting controller between succesive review times. The review times may be predetermined, or state-dependent(and hence random). Each time we solve for a new open-loop controller, we incorporate up-to-date information about the system state.

For this to be a useful approach to scheduling and allocation problems, a number of criteria must be met. First, we must have some expectation that the scheduling strategies that result are near-optimal, and better than those that might result from the application of one or other of the well-known scheduling heuristics. Secondly, where the resulting open-loop problems are not soluble analytically, they must be computationally tractable. Thirdly, their applicability should be reasonably general.

3.1 Sequential open-loop scheduling

We begin by presenting an open-loop analogue of problem $P^{\Gamma}(X,t)$.

$$OLP^{\Gamma}(X,t):\ \text{Maximize} \int_t^{\infty} \Big(\sum_{i\varepsilon\Gamma} r_i(t')f_i(x(s))u_i(s)\Big)ds'$$

$$\text{subject to}\quad \dot{x}(s)=u(s),\ t\leqslant s<\infty,\ x(t)=X,$$

$$\sum_i u_i(s)=1,\quad t\leqslant s<\infty.$$

This is formally the same as P1, so that theorems 1 and 2 apply. Note also that this is a different problem from that of maximizing the reward for the first completion, which would be formulated

$$MFC^{\Gamma}(X,t):\ \text{Maximize} \int_t^{\infty} \sum_i \{ \prod_{j\neq i} P_j(x_j(s))\} r_i(s)f_i(x(s))u_i(s)ds$$

$$\text{subject to}\quad \dot{x}(s)=u(s),\ x(t)=X,$$

$$\sum_i u_i(s)=1,\quad t\leqslant s<\infty.$$

We could, of course, as well use the latter objective functional as that of OLP. We are just seeking to generate a sequence of open-loop controllers by solving a surrogate optimization problem, and the precise form of that problem is obviously open to choice. The main reason for concentrating on OLP is its extreme simplicity.

Because OLP no longer contains the recursion implicit in P, the proof of analytic results about OLP, where these exist, is in general much simpler than for P. An example is the proof of theorem 4 of this paper. From the computational point of view, OLP is a particularly simple member of a class of standard problems in optimal control. Well-developed algorithms are available for the solution of such problems. We will not go further into this aspect here, but in the rest of this paper we compare the performance of sequential strategies derived from OLP with known optimal solutions.

Consider, for example, the case where

$$r_i(t)=w_i e^{-\alpha t},\ i=1,2,\ldots,n,$$

where the w_i's are positive constants. The solution of P in this case given by a dynamic priority index ν which is a particular form of the Gittins index [4] for multi-armed bandits:

$$\nu_i(x)=\sup_{x'>x} \left\{ \frac{\int_x^{x'} w_i f_i(\zeta)e^{-\alpha\zeta}d\zeta}{\int_x^{x'} P_i(\zeta)e^{-\alpha\zeta}d\zeta} \right\}.$$

For this class of reward functions, OLP is a deterministic, continuous time multi-armed bandit. Its solution is given by a dynamic priority index β, where

$$\beta_i(x)=\sup_{x'>x}\{\frac{\int_x^{x'} w_i f_i(\zeta)e^{-\alpha\zeta}d\zeta}{\int_x^{x'} e^{-\alpha\zeta}d\zeta}\}.$$

Thus the sequential open-loop controller for P derived from OLP works as follows. For each $\tau\varepsilon(\tau_r,\tau_{r+1})$, compute $\beta_i(X_i(\tau))$ for each job in $\bar{\Gamma}(\tau)$, on the basis of the current data. Over the interval (τ_r,τ_{r+1}), schedule using the dynamic priority index β. This heuristic has the property that, as the intervals $\{(\tau_r,\tau_{r+1}): r=1,2,\ldots\}$ tend to zero in length, it tends to a closed-loop controller for P which is based on the priority index γ, where

$$\gamma_i(x)=\beta_i(x)/P_i(x).$$

Theorem 3.

(i) If $\rho_i(x)$ is a monotonic, non-increasing function of x, then $\gamma_i(x)=\nu_i(x)$ for all x.

(ii) If the functions f_i are all identical, with increasing completion rate ρ, and the weights w_i all equal, then ν and γ determine identical scheduling strategies.

Proof.

(i) If ρ_i is a decreasing function, so is f_i. Thus

$$\gamma_i(x)=w_i\rho_i(x)=\nu_i(x).$$

(ii) When ρ is increasing in x, then so is ν. It therefore suffices to show that this is also true of γ. This is obvious for all values of x for which $\alpha(x)=\rho(x)$. Otherwise, suppose that in the definition of $\beta_i(x)$, the supremum on the RHS is achieved by some value $x'=\zeta(x)(>x)$. Then

$$\gamma(x)>\rho(x).$$

Hence, for some $\delta>0$, and all $\xi\varepsilon(x,x+\delta)$,

$$\gamma(\xi)=\sup_{x'>\xi}\{\frac{\int_\xi^{x'} w_i f_i(\zeta)e^{-\alpha\zeta}d\zeta}{P_i(\xi)\int_\xi^{x'} e^{-\alpha\zeta}d\zeta}\}$$

$$\geqslant\frac{\int_\xi^{\zeta(x)} w_i f_i(\zeta)e^{-\alpha\zeta}d\zeta}{P_i(\xi)\int_x^{\zeta(x)} e^{-\alpha\zeta}d\zeta}$$

$$> \frac{\int_x^{\zeta(x)} w_i f_i(\zeta) e^{-\alpha\zeta} d\zeta}{P_i(x) \int_x^{\zeta(x)} e^{-\alpha\zeta} d\zeta} = \gamma(x).$$

This theorem shows that the sequential open-loop strategy is the same as the optimal strategy in the limit of small review period, for these particular classes of completion time distribution, and a single processor with this reward structure. We call a sequential open-loop strategy with this property asymptotically optimal.

The theorem extends to the problem of minimizing the weighted flowtime, by taking the limit of the discounted case as $\alpha \to 0$. Part (i) of the theorem extends to a more general class of reward functions. Suppose that

$$r_i = w_i r + d_i, \quad i = 1, 2, \ldots, n,$$

where w_i is a positive constant, d_i is a constant and r is a decreasing function of time. Then the optimal controller for this version of OLP is still given by the dynamic priority index whose value is $w_i f_i$. This result can be proved by the straightforward application of theorem 1 to OLP. In the limit of small review periods, this becomes the strategy determined by the priority index $w_i \rho_i$. This can be shown [5] to be optimal for P, again via theorem 1.

We now consider a class of single-processor problems for which a full optimal strategy is hard to find. Suppose that the jobs represent customers in a queue, and that customers will become dissatisfied and leave (renege) at random times if their service is not already complete. This is equivalent to giving each job a randomly distributed due date, after which no reward can accrue from the completion of that job. Some results concerning the optimal strategy for a variant of this problem when the jobs have a common deadline and the processing time distributions are exponential are in [3]. Let G_i denote the distribution of the time that the i'th customer is prepared to wait, with density g_i and survivor function Q_i. Then the open-loop problem is

$$\mathrm{OLR}^{\Gamma}(X,t): \text{ Maximize } \int_t^{\infty} \Big(\sum_{i \in \Gamma} r_i(s) Q_i(s) f_i(x(s)) u_i(s) \Big) ds$$

$$\text{Subject to } \dot{x}(s) = u(s), \quad t \leq s < \infty, \quad x(t) = X,$$

$$\sum_i u_i(s) = 1, \quad t \leq s < \infty.$$

To maximize the weighted number of satisfied customers, we take $r_i \equiv w_i$ for all i. Then OLR and OLP are the same, with Q_i replacing r_i. As before, the optimal control for OLR when the f_i are decreasing is myopic, provided the derivatives of the Q_i are proportional.

Suppose the customers are all equally impatient, the distributions $\{G_i\}$ being identical and negative exponential. Then the results given by theorem 3 for the asymptotic form of the sequential open-loop strategy hold. This provides an example where the sequential open-loop strategy need not be asymptotically optimal. If n=2, and

$$f_i(x)=\lambda_i e^{-\lambda_i x}, \quad i=1,2,$$

then it is optimal to process job 1 first if

$$w_1 \lambda_1 \left(1+\frac{\alpha}{\alpha+\lambda_1}\right) \geqslant w_2 \lambda_2 \left(1+\frac{\alpha}{\alpha+\lambda_2}\right) .$$

The asymptotic form of the sequential open-loop strategy is to process the jobs in $w_i \lambda_i$ order. These two indices are equivalent when $w_1=w_2$, and asymptotically equivalent for large λ_1, λ_2 for any weights, but one may readily find values of the parameters for which this is not so.

We remark that for the variant of the problem where there is a single deadline, the sequential open-loop strategy is asymptotically optimal if the conditions of either theorem 3(i) or (ii) hold for the completion time distributions. This follows from the formal equivalence of the full problem with problem P, and of the open loop problem with OLP, with r_i replaced in each case by Q, the survivor function of the deadline distribution.

As a final example, we consider a multi-server problem. Suppose we are interested in minimizing the flowtime for the set of n jobs on m identical machines. Then we add to OLP the constraints

$$u_i(s) \leqslant 1/m, \quad i=1,2,\dots,n, \quad t \leqslant s < \infty.$$

Suppose that all the jobs have identical completion time distributions, but have possibly received different amounts of processing. Using theorem 2, it can be shown [8] that if the common completion time distribution has monotone completion rate, then the strategy of processing at each time the jobs with currently greatest completion rate minimizes the expected flowtime. If the common completion time density is log-convex or log-concave, then this strategy minimizes the flowtime in distribution. These results are extensions of similar results for the single server problem P, but their proof is much more difficult. In contrast, the corresponding open-loop results extend easily to the multi server case when f is a decreasing function.

Theorem 4. Suppose the common completion rate distribution of the n jobs has monotone completion rate. The the sequential open-loop strategy asymptotically minimizes the expected flowtime.

Proof.

What we actually prove is that f is a priority index for the problem under these conditions; the result proper then follows as in theorem 3. Observe that, under our assumptions, two jobs which have received the same amount of processing are identical. Thus we can solve OLP with a controller which preserves the initial ordering of the values of f_i, $i=1,2,\ldots,n$, which we may as well take to be the same as the lexicographic order. A simple calculation shows that if $i<k$, then

$$\int_t^\infty \Big(\sum_j H_{ij}(x(s),s)u_j(s)\Big)ds \leqslant \int_t^\infty \Big(\sum_j H_{kj}(x(s),s)u_j(s)\Big)ds$$

with equality only if i and k are identical at time t and receive identical treatment thereafter. The result follows by the maximizing property of the optimal u.

4. CONCLUSION

The use of optimal control theory is a powerful analytical tool in the solution of scheduling problems when the processing can be pre-emptive. Its use in computing optimal scheduling strategies where these are not obtainable analytically is not likely to be practicable. If we seek sequential open-loop strategies, by finding solutions to intuitively plausible surrogate problems, we are faced with control problems which have analytical solutions over a somewhat wider range of conditions than the closed-loop counterpart. Where these problems do not yield an analytic solution, numerical solution should not be difficult in practice. The examples considered here give us some grounds for hope that the resulting strategies will be effective. At the same time, we have shown that the optimal strategies for a range of problems can be characterized as the limit of sequential open-loop controllers.

Before the use in practice of open-loop sequential scheduling can be recommended, a number of further points have to be examined. Chiefly, their performance on problems with more difficult structures than those considered here needs to be tested, probably by computer. Algorithms for solving deterministic optimal control problems are numerous, and it would be worth knowing which perform well on problems like OLP.

REFERENCES

[1] Alam, M.: 1969, An application of modern control theory to a time-dependent queueing system for optimal operation, Int. J. Sys. Sci. 10, pp. 693-700.
[2] Aoki, M.: 1967, "Optimization of Stochastic Systems", Academic Press, New York.
[3] Derman, C., Lieberman, G., and Ross, S.M.: 1978, A renewal decision problem, Management. Sci. 24, pp. 554-561.
[4] Gittins, J.C.: 1976, Bandit processes and dynamic allocation indices, J.R. Stat. Soc. B 41, pp. 148-177.
[5] Nash,P. and Gittins, J.C.: 1977, A Hamiltonian approach to stochastic resource allocation, Adv. App. Prob. 9, pp. 55-68.
[6] Nash, P.: 1969, Controlled jump process models for stochastic scheduling, Int. J. Control 29, pp. 1011-1025.
[7] Scott, C.H. and Jefferson, T.R.: 1976, Optimal regulation of service rate for a queue with finite waiting room, J.O.T.A. 20, pp. 245-250.
[8] Weber,R.R.: 1982, Scheduling jobs with unknown processing requirements on parallel machines to minimize makespan and flowtime, J. App. Prob. 19, (to appear).

The basic optimal control theory underlying these papers can be found in

Lee, E.B. and Markus, L: 1968, "Foundations of Optimal Control Theory". Wiley, New York.

The theory of maximum principles for discrete state, continuous-time Markov processes is discussed in, for example,

Rishel, R.: 1974, in "Lecture notes in Economics and Mathematical systems", Vol 107. Springer, Berlin.

Sworder, D.D.: 1969, I.E.E.E. Trans. Autom. Control 14, 9.

The following paper considers the reduction to discrete semi-Markov decision problems of controlled jump process models in general.

Yushkevitch, A.A.: 1981, On reducing a jump-controllable Markov model to a model with discrete time, SIAM J. Prob. Theory and Appl. 25.

ON THE DELAY FUNCTIONS ACHIEVABLE BY NON-PREEMPTIVE SCHEDULING STRATEGIES IN M/G/1 QUEUES

I. Mitrani

University of Newcastle upon Tyne

ABSTRACT

For a queueing system in equilibrium, the delay function, W(x), is defined as the expected time spent in the queue by a job of length x. In an M/G/1 queue where the arrival rate and the distribution of job lengths are given, the delay function depends on the scheduling strategy employed. This note addresses the problem of characterising the set of delay functions that are achievable by non-preemptive scheduling strategies. An integral equality constraint and a set of integral inequalities which the functions W(x) must satisfy are given, as well as the extreme elements of the set of achievable functions.

1. BACKGROUND TO THE PROBLEM

Consider a single server queueing system where each customer (job, from now on), announces on arrival the exact amount of service that he will require (job length, from now on). The scheduling strategy is allowed to make use of this information when allocating the server. An object of interest in such a system is the delay function W(x), defined for $x \geq 0$ as the expected waiting time (that is, time spent in the queue, not in service) for a job of length x, assuming that the system is in equilibrium.

Given the arrival process and the distribution of job lengths, the delay function depends only on the job scheduling strategy. A pre-specified function W(x), $x \geq 0$, is said to be achievable in that queueing system if there exists a scheduling

M. A. H. Dempster et al. (eds.), Deterministic and Stochastic Scheduling, 399–404.

strategy whose delay function is W(x). A problem of characterisation then arises: what is the set of achievable delay functions? That problem seems to have been posed first by Kleinrock, Muntz and Hsu [3], who obtained upper and lower bounds for the delay functions achievable in M/G/1 systems by a special family of scheduling strategies (of the Processor-Sharing type). There are, of course, many functions which are between those bounds and yet are not achievable.

The general characterisation problem is still open. The purpose of this short note is to suggest a possible solution for the case of an M/G/1 system where the scheduling strategies are non-preemptive, i.e. services, once started, are always allowed to complete without interruption. A set of conditions will be given, together with a plausible argument indicating that they are necessary and sufficient for achievability. The set of achievable delay functions, which is convex, can also be characterised by specifying its extreme elements. These are (again by an intuitive argument) the delay functions corresponding to the fixed priority scheduling strategies.

The proposed development is based on the following result [1, 2], valid for M/G/1 systems with K job classes, where the scheduling decisions may be based on class affiliation (and hence on expected job lengths) but not on exact job lengths.

Denote by λ_i, $1/\mu_i$ and M_{2i} the arrival rate, average job length and second moment of the job length for class i jobs (i=1, 2,...,K). Let W_i be the expected steady-state waiting time for class i jobs and $\underline{W}$ be the vector $(W_1, W_2, \ldots, W_K)$. A pre-specified vector $\underline{W}$ is achievable in that system if, and only if,

$$\sum_{i=1}^{K} \rho_i W_i = w_o \rho/(1-\rho) \qquad (1)$$

and

$$\sum_{i\in g} \rho_i W_i \geq w_o \left(\sum_{i\in g} \rho_i\right) \Big/ \left(1 - \sum_{i\in g} \rho_i\right) \qquad (2)$$

for every proper and non-empty subset $g \subset \{1,2,\ldots,K\}$, where

$$\rho_i = \lambda_i/\mu_i, \quad \rho = \sum_{i=1} \rho_i \text{ and } w_o = \tfrac{1}{2}\sum_{i=1} \lambda_i M_{2i} .$$

Moreover, if $\underline{W}_1, \underline{W}_2, \ldots, \underline{W}_{K!}$ are the waiting time vectors of the K! non-preemptive priority scheduling strategies that can be operated with the K job classes, and if H is the set of vectors $\underline{W}$ which can be expressed as convex combinations of the type

$$\underline{W} = \sum_{j=1}^{K!} \alpha_j \underline{W}_j \; ; \; \alpha_j \geq 0 \; ; \; \sum_{j=1}^{K!} \alpha_j = 1 \; , \qquad (3)$$

then H coincides with the set of all achievable waiting time vectors.

The idea now is to apply this result to the characterisation of achievable delay functions in an M/G/1 system. The restriction that the scheduling decisions must not be based on information about exact job lengths can be circumvented by introducing 'artificial' job classes.

2. CONJECTURED SOLUTION

Jobs arrive into the system in a Poisson stream at rate λ. The lengths of consecutive jobs are independent and identically distributed random variables, with distribution function $F(x)$. The first two moments of $F(x)$ are denoted by $1/\mu$ and M_2, respectively. Steady-state is assumed, for which the necessary and sufficient condition is $\rho = \lambda/\mu < 1$. The restrictions on the scheduling strategies are a) they must be non-preemptive, b) they must be work-conserving (i.e. no job departs before being served and the server is never idle when there are jobs in the system) and c) the scheduling decisions during one busy period must not depend on what happened during previous busy periods (this last restriction can be relaxed).

Consider first the case when $F(x)$ is a step function with a finite number of jumps, at points $x_1, x_2, \ldots, x_K$ $(K>1)$. One can then classify each job according to its length: there are K job classes, class i consisting of the jobs of length x_i $(i=1,2,\ldots,K)$. Scheduling decisions, instead of depending on job lengths, can be said to depend on class indices, which allows the result of the previous section to be applied directly. The delay functions $W(x)$ can in this case be treated as vectors $(W(x_1), W(x_2), \ldots, W(x_K))$. Bearing in mind that jobs of length x_i arrive at rate $\lambda dF(x_i) = \lambda[F(x_i)-F(x_i-)]$ $(i=1,2,\ldots,K)$, we can say that a given delay function is achievable if, and only if,

$$\sum_{i=1}^{K} \lambda x_i W(x_i) dF(x_i) = w_o \rho/(1-\rho) \qquad (4)$$

and

$$\sum_{i \in g} \lambda x_i W(x_i) dF(x_i) \geq w_o \sum_{i \in g} \lambda x_i dF(x_i) \Big/ \left[1 - \sum_{i \in g} \lambda x_i dF(x_i)\right] \qquad (5)$$

for every proper and non-empty subset $g \subset \{1,2,\ldots,K\}$, where

$$\rho=\lambda/\mu = \sum_{i=1}^{K} \lambda x_i dF(x_i) \text{ and } w_o = \tfrac{1}{2}\sum_{i=1}^{K} \lambda x_i^2\, dF(x_i) = \tfrac{1}{2}\lambda M_2.$$

The vertices of the set of achievable delay functions are the K! functions corresponding to the K! non-preemptive strategies which assign fixed (different) priorities to jobs of different length and schedule accordingly. For example, the delay function corresponding to the ordering 'top priority to x_1-jobs, second top to x_2-jobs, etc.', is given by

$$W(x_i) = w_o/\{[1-\sum_{j=1}^{i-1} \lambda x_j dF(x_j)]\,[1-\sum_{j=1}^{i} \lambda x_j dF(x_j)]\}\ ,\qquad i=1,2,\ldots,K \quad (6)$$

(when i=1, the first sum in the denominator is zero by definition). Every achievable delay function can be expressed as a convex combination of those K! vertices (in fact, it can be expressed as a convex combination of not more than K of the vertices).

It is fairly obvious that this result can be generalised to the case when F(x) is a step function with denumerably many jumps (i.e., any discrete distribution). The proofs in [1,2] should carry over, replacing where necessary finite sums by infinite series. We would go one step further and conjecture that a similar result will hold for arbitrary distributions. The characterisation statement in the general case will be:

A delay function W(x) is achievable in an M/G/1 system if, and only if,

$$\int_0^\infty \lambda x W(x) dF(x) = w_o \rho/(1-\rho) \qquad (7)$$

and

$$\int_g \lambda x W(x) dF(x) \geq w_o [\int_g \lambda x dF(x)] / [1-\int_g \lambda x dF(x)] \quad (8)$$

for every non-empty subset $g \subset [0,\infty)$ which is measurable with respect to F(x). When g consists of a single point, x, (8) should be interpreted as

$$W(x) \geq w_o/[1 - \lambda x dF(x)]\ . \qquad (9)$$

If, in (8), g is replaced by its complement, $[0,\infty)-g$, denoted again by g, then, taking (7) into account, the inequalities become

$$\int_g \lambda x W(x) dF(x) \leq w_o [\int_g \lambda x dF(x)]/\{(1-\rho)[1-\rho+\int_g \lambda x dF(x)]\}. \quad (10)$$

When g consists of a single point, x, (10) should be interpreted as

$$W(x) \leq w_o/[(1-\rho)(1-\rho+\lambda x dF(x))]. \quad (11)$$

In particular, if $F(\cdot)$ is continuous at point x, then (9) and (11) become

$$w_o \leq W(x) \leq w_o/(1-\rho)^2. \quad (12)$$

These last inequalities simply state that the expected delay for jobs of length x cannot fall below the value it would have if those jobs were given top priority and cannot rise above the value it would have if they were given bottom priority.

One consequence of conditions (7) and (8) is that if W(x) is achievable, then any function which coincides with W(x) everywhere except on a set of measure zero with respect to F(x), and on that set satisfies (12), is also achievable. Intuitively, the scheduling of jobs whose contribution to the total load is zero can be changed without affecting the expected delays of other jobs.

The set H of achievable delay functions is convex. This can be demonstrated as in [1,2]: if $W_1(x)$ and $W_2(x)$ are achievable, by scheduling strategies S_1 and S_2 respectively, then for every $\alpha \in (0,1)$ the function $W(x)=\alpha W_1(x)+(1-\alpha)W_2(x)$ is achievable, for example by the scheduling strategy which at the beginning of every busy period decides to operate S_1 with probability α and S_2 with probability $(1-\alpha)$. The extreme elements of H are the delay functions corresponding to the strategies which schedule jobs strictly according to their length. A priority ordering is defined by a one-to-one mapping, φ, of $[0,\infty)$ onto itself: jobs of length x have priority over jobs of length y if $\varphi(x)<\varphi(y)$. The delay function corresponding to the priority ordering φ is given by

$$W_\varphi(x)=w_o/\{[1-\int_0^{z-} \lambda \varphi^{-1}(u) dF(\varphi^{-1}(u))]\,[1-\int_0^{z} \lambda \varphi^{-1}(u) dF(\varphi^{-1}(u))]\}, \quad (13)$$

where $z=\varphi(x)$; the two integrals in the denominator are equal if $F(\cdot)$ is continuous at point x (this is a generalisation of the Shortest-Processing-Time-first expression, [2]). Of course, the mapping φ should be such that the right-hand side of (13) is defined.

Every achievable delay function W(x) can be expressed as a convex combination of a finite number of extreme elements:

$$W(x) = \sum_{j=1}^{M} \alpha_j W_{\varphi_j}(x) \; ; \; \sum_{j=1}^{M} \alpha_j = 1 \; . \tag{14}$$

3. CONCLUDING REMARKS

Two points should, perhaps, be emphasised. First, the generalisation to arbitrary distribution functions, if true, is not easy to prove. The main reason for this is the inability to determine (as in the discrete case) the extreme elements of the set H of achievable delay functions as intersections of planes. There are also other issues which have to be considered, such as the closeness of H with respect to a suitably defined topology. Second, the practical utility of the characterisation in the general case is limited. Conditions (8) may be difficult to check. In practice, job lengths are approximated by rational numbers, or even by integers, in which case a finite discrete approximation to F(x) is perfectly adequate. On the other hand, a characterisation in terms of the extreme elements of H opens the way to solving various synthesis problems. If a target delay function W(x) can be expressed (or approximated) as a convex combination of functions $W_{\varphi}(x)$, then one can construct a scheduling strategy to achieve W(x).

REFERENCES

[1] Coffman, E.G. and Mitrani, I., A Characterisation of Waiting Time Performance Realisable by Single Server Queues, Opns. Res., 28(3), pp. 810-821, 1980.
[2] Gelenbe, E. and Mitrani, I., Analysis and Synthesis of Computer Systems, Academic Press, 1980.
[3] Kleinrock, L., Muntz, R.R. and Hsu, J., Tight Bounds on Average Response Time for Processor-Sharing Models of Time-Shared Computer Systems, Inf. Processing, 71, TA-2, pp. 50-58, 1971.

MODELLING FOR MULTIPROCESSOR PROJECTS

Ulrich Herzog

Universität Erlangen-Nürnberg

Important performance problems for multiprocessor computer systems have been discussed, modelled and investigated since many years. These fundamental results are - although derived without experience with real systems - still of great importance.

Nowadays, however, there are several multiprocessor-projects operational. Experiences with such experimental systems give us a deeper insight and many impulses for performance modelling.

This contribution discusses two multiprocessor projects at the University of Erlangen-Nuremberg and related performance problems.

1. EGPA, THE ERLANGEN GENERAL PURPOSE ARRAY

Hierarchical structures may often be found in nature, and they have proved to be highly efficient in civilisation, too. Hierarchical structures allow to form complex systems which are transparent and rather flexible without excessive communication and coordination overhead. Typical examples within the scope of computer architecture are the multiprocessor system at the SUNY, X-Tree, and EGPA.

1.1. The EGPA-Architecture

EGPA, the Erlangen General Purpose Array (1), consists of several layers of processors building a (complete or truncated) pyramid structure: The lowest (A-) level processors execute the

M. A. H. Dempster et al. (eds.), Deterministic and Stochastic Scheduling, 405–410.

user processes. The upper (B-, C-) level processors mainly perform operating system functions, i.e. they are responsible for task scheduling, coordination of subprocesses, initiation of I/O-activities, etc.

Fast communication between processors as well as efficient code and data sharing is possible by means of a hardware interrupt system and multiport memories: Each processor may access its local memory, the memories of its four neighbours at the same pyramid-level and of its subordinates.

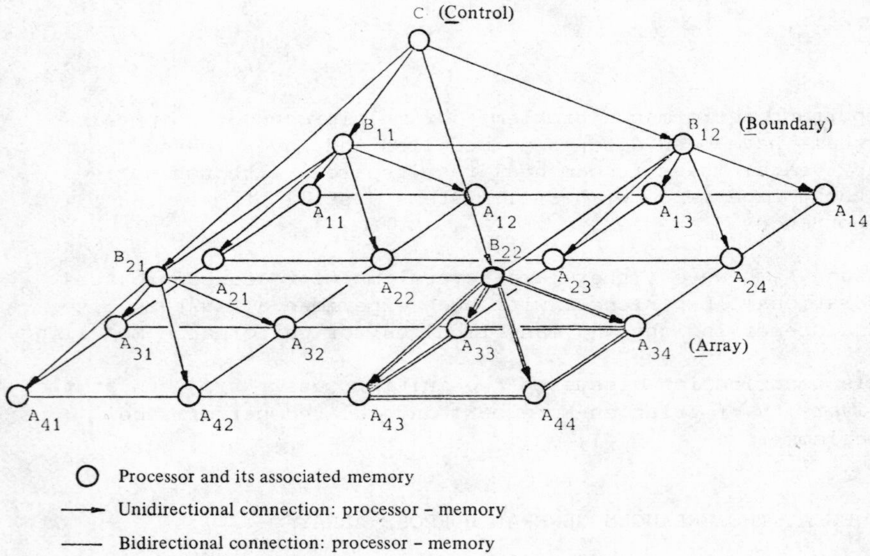

Figure 1: Interconnection structure of an EGPA system with 21 processors and (at the bottom right hand side) an elementary pyramid implemented with 5 computers AEG 80-60.

The EGPA-project is supported by the federal ministry of research and technology (BMFT) since January 1978, a pilot-pyramid with 5 AEG 80-60 control processors is operational. Various application programs have been implemented. A powerful hardware-monitor allows performance measurements for different components at the same time.

1.2. Special features of multiprocessor modelling

The flow of information in a multiprocessor computer system depends on the hardware components, the interconnection scheme and the operating system. It is, however, heavily influenced also by the internal structure of the algorithms implemented in the application programs. Therefore, many questions arise, such as:
- how to decompose a task into cooperating subtasks?
- how to distribute these subtasks among the processors?
- how to coordinate and synchronize their interactions?
- how to control data transfers and I/O activities?

Most of these problems are known since long ago. Usually however, system overhead due to interprocessor communication for signalling and data transfers is being neglected. Experiences with our pilot pyramid show these subjects to be of major importance for multiprocessor scheduling.

In several publications we recently surveyed these problems, characterized the internal structure of programs by means of directed graphs, modelled the traffic flow by a new class of queuing systems and analyzed these models under Markovian and non-Markovian assumptions, as well (2-6).

Many results are available for local and global performance characteristics such as
- Moments and distribution function (d.f.) of the synchronization time
- Moments and d.f. of the cycle time
- Utilization of processors
- System throughput
- Number of processors working simultaneously
- Moments and d.f. of waiting and response times
- Optimal number of parallel subprocesses.

We also considered and are still investigating the impact of memory interference, data dependencies, individual initiation of subprocesses, priorities and multilevel-hierarchies.

2. DIRMU, A DISTRIBUTED RECONFIGURABLE MULTIPROCESSOR KIT

Computer systems designed for a specific application are extremely powerful, however, often expensive and inflexible when the type of problems changes. With the advent of large scale integration there seems to be a possibility to build multiprocessor networks without the above disadvantages.

The objective of the DIRMU (7) is to implement a mudule-computer kit for dedicated and user configurable multimicrocomputer systems.

2.1. The Basic Concept

There are four major points united in the DIRMU-architecture and its implementation:

- cost savings via standard module computers and standardized system software
- high performance by specialisation (data flow concept)
- fault tolerance via self-testing, neighbourhood-checks and reconfiguration
- flexibility by specialized application software and plug-in module connections.

Figure 2 shows as an example a high-level-language training system. I/O-processors are responsible for user interactions. Compile-processors compile the HLL such as FORTRAN, ALGOL, COBOL; finally, the object code is processed by object-processors.

This project is supported by the German Research Society since 1981. Several experimental implementations and prototypes for basic elements (8086-based processors, multiport-memories, ... os kernel,... are available. Selected applications are going to be implemented next.

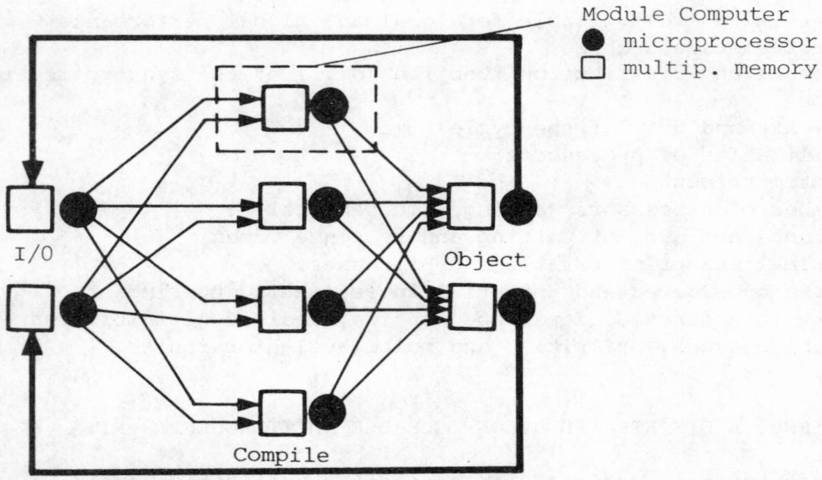

Figure 2: Example of a multimicroprocessor network.

2.2. Performance Modelling

When modelling and analysing networks with a given structure similar questions arise as discussed above. And communication and coordination overhead are the main problems, again.

Usually, however, the structure may be not as obvious as in our example, Fig. 2. Therefore, two major problems have to be solved when designing such specialized computer networks:

. Determine the structure of the network and
. Allocate subtasks and data

in order to guarantee performance requirements, fault tolerance and reasonable cost.

The manifold of questions to be attacked and solved simultaneously may be summarized by the following implementation methodology:

1) EXPLORE THE PROBLEM STRUCTURE AND DECOMPOSE IT
   - formal description of functions, code and data dependencies

2) DETERMINE THE COMPUTER SYSTEM STRUCTURE
   - number of module computers
   - load of each module (function allocation)
   - interconnection scheme

   considering
   - performance requirements
   - reliability aspects
   - problem structure

3) CONFIGURE THE SYSTEM
   - connect the computer modules
   - provide local operating systems with coordination parameters
   - load application code and data

4) RUN THE SYSTEM
   - run automatic self tests
   - run applications
   - monitor performance.

Again, there are many old (e.g.8) and more recent results (e.g.9) of importance available. But, as above, they have to be modified, extended or combined with each other. This is one of our major tasks at the moment (10-13). And there are still many important and challenging problems to be solved in the future.

REFERENCES

(1) HÄNDLER, W.; HOFMANN, F.; SCHNEIDER, H.J.: A General Purpose Array with a Broad Spectrum of Applications. In: Computer Architecture (Händler ed.), Informatik-Fachberichte 4, Springer-Verlag (1975), pp. 311-335.

(2) HERZOG, U.; HOFMANN, W.: Synchronization Problems in Hierarchically Organized Multiprocessor Computer Systems. Proceedings of the 4th International Symposium on Modeling and Performance Evaluation of Computer Systems, Feb. 6-8, 1979, Techn. University, Vienna; North-Holland, 1979, pp. 29-48.

(3) HERZOG, U.; HOFMANN, W.; KLEINÖDER, W.: Performance Modeling and Evaluation for Hierarchically Organized Multiprocessor Computer Systems. Conf. on Parallel Processing, Bellaire, Michigan, August 21-24, 1979.

(5) HERZOG, U.: Performance Characteristics for Hierarchically Organized Multiprocessor Computer Systems with Generally Distributed Processing Times. Proc. Ninth Intern. Teletraffic Congress, October 1979, Torremolinos, Spain.

(6) FROMM, H.J.: Zur Modellierung der Speicherinterferenz bei Hierarchisch organisierten Multiprozessorsystemen. Arbeitsberichte des IMMD, Universität Erlangen-Nürnberg, Bd. 13, No. 13, Erlangen, April 1980.

(7) HÄNDLER, W.; ROHRER, H.: Thoughts on a Computer Construction Kit. Elektr. Rechenanlagen, 1980, H. 1, pp. 3-13.

(8) CONWAY, R.W.; MAXWELL, W.L.; MILLER, L.W.: Theory of Scheduling. Reading, Mass.: Addison-Wesley, 1967.

(9) CHU, W.W. et al.: Task Allocation in Distributed Data Processing. Computer, Nov. 1980, pp. 57-68.

(10) HERZOG, U.; WOO, L.S.: Design and Performance Evaluation of distributed Multiprocessor Systems. - A Framework. Internal report, Yorktown Heights, October 1980.

(11) FROMM, H.J.: Programmstrukturen, Datenstrukturen, Optimale Ablaufsteuerung von Multiprozessor Systemen. Internal report, January 1981.

(12) FROMM, H.J.: Multiprocessor Scheduling. Internal report, February 1981.

(13) KLEINÖDER, W.: Programme für hierarchische Mehrrechnersysteme. Internal report, March 1981.

ADDRESSES OF AUTHORS

J. BRUNO
Department of Computer Science
University of California
Santa Barbara, CA 93106
U.S.A.

E.G. COFFMAN, JR.
Bell Laboratories
600 Mountain Avenue
Murray Hill, NJ 07974
U.S.A.

M.A.H. DEMPSTER
Department of Mathematics
Dalhousie University
Halifax, Nova Scotia
CANADA

M.L. FISHER
Department of Decision Sciences
The Wharton School
University of Pennsylvania
Philadelphia, PA 19104
U.S.A.

G.N. FREDERICKSON
Computer Science Department
Pennsylvania State University
University Park, PA 16802
U.S.A.

W. GAUL
Institut für Entscheidungstheorie
und Unternehmensforschung
Universität Karlsruhe (TH)
Kollegium am Schloss, Bau III
Postfach 6380
7500 Karlsruhe 1
F.R.G.

E. GELENBE
Laboratoire de Recherche en
Informatique
Université de Paris Sud
Bâtiment 490
Centre d'Orsay
91405 Orsay Cédex
FRANCE

J.C. GITTINS
Mathematical Institute
University of Oxford
24-29 St. Giles
Oxford OX1 3LB
U.K.

K.D. GLAZEBROOK
Department of Statistics
University of Newcastle upon Tyne
Newcastle upon Tyne NE1 7RU
U.K.

U. HERZOG
Institut für Mathematische Maschinen und Datenverarbeitung VII
Universität Erlangen-Nürnberg
Martensstrasse 3
8520 Erlangen
F.R.G.

E.L. LAWLER
Computer Science Division
591 Evans Hall
University of California
Berkeley, CA 94720
U.S.A.

J.K. LENSTRA
Mathematisch Centrum
P.O. Box 4079
1009 AB Amsterdam
THE NETHERLANDS

G.S. LUEKER
Department of Information and Computer Science
University of California
Irvine, CA 92717
U.S.A.

C.U. MARTEL
Department of Electrical and Computer Engineering
University of California
Davis, CA 95616
U.S.A.

I. MITRANI
Computing Laboratory
University of Newcastle upon Tyne
Newcastle upon Tyne NE1 7RU
U.K.

P. NASH
Engineering Department
University of Cambridge
Mill Lane
Cambridge CB2 1RX
U.K.

M.L. PINEDO
School of Industrial and Systems Engineering
Georgia Institute of Technology
Atlanta, GA 30332
U.S.A.

A.H.G. RINNOOY KAN
Econometric Institute H2-11
Erasmus University
P.O. Box 1738
3000 DR Rotterdam
THE NETHERLANDS

S.M. ROSS
Department of Industrial Engineering and Operations Research
Etcheverry Hall
University of California
Berkeley, CA 94720
U.S.A.

L.E. SCHRAGE
Graduate School of Business
University of Chicago
1101 East 58th Street
Chicago, IL 60637
U.S.A.

K.C. SEVCIK
Computer Systems Research Group
University of Toronto
121 St. Joseph Street
Toronto, Ontario M5S 1A1
CANADA

B. SIMONS
IBM Research K52/282
6500 Cottle Road
San Jose, CA 95193
U.S.A.

R.R. WEBER
Queen's College
Cambridge CB3 9ET
U.K.

G. WEISS
Department of Statistics
Tel-Aviv University
Ramat-Aviv
Tel-Aviv
ISRAEL 69978

P. WHITTLE
Statistical Laboratory
University of Cambridge
16 Mill Lane
Cambridge CB2 1SB
U.K.

NAME INDEX